Chaos

Chaos and strange attractors

A suite of programs by M. A. Muhamad and A. V. Holden that illustrate chaotic behaviour in maps and in differential systems is available to accompany this book: these programs are available only on 40– or 80–track 5.25″ discs for the Acorn BBC-B microcomputer.

Maps of the real line, the plane and the complex plane, with their associated bifurcation diagrams, strange attractors and Julia sets, are presented. Trajectories, strange attractors and reconstructed strange attractors are presented for nonlinear differential systems and forced systems described in Chapter 2: by changing the parameters of the equations the routes into chaos may be followed.

Further information from: Chaos and Strange Attractors, Manchester University Press (Software), Oxford Road, Manchester M13 9PL, UK.

Chaos

Edited by Arun V. Holden

Department of Physiology, The University, Leeds LS2 9NQ, UK

Princeton University Press
Princeton, New Jersey

Published by
Princeton University Press,
41 William Street, Princeton,
New Jersey 08540

Library of Congress cataloging in publication data
Chaos.
   (Nonlinear science)
   Includes index.
   1. Chaotic behavior in systems.   I. Holden, Arun V.,
1947–
QA402.C43   1986   003   86-004952
ISBN 0-691-08423-8 *cased*
ISBN 0-691-08424-6 *paperback*

Printed and bound by Princeton University Press, Princeton, New Jersey

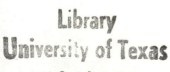

Contents

Acknowledgements

For permission to reproduce text figures from a number of publications we should like to thank:

Academic Press (*Journal of Theoretical Biology*) for Figs 8.8 and 9.9; the American Association for the Advancement of Science (*Science*) for Fig. 11.5; the American Physical Society (*Physics Reviews*) for Figs 11.1, 11.2, 11.3, 11.9, 11.10 and 14.8; Cambridge University Press for Figs 9.5 and 9.6; the Ecological Society of America (*Ecology*) for Figs 8.1, 8.2 and 8.3; Elsevier Science Publishing Company (*Mathematical Biosciences*) for Fig. 9.3; Federation of European Biochemical Societies (*FEBS Letters*) for Fig. 9.7; the Institute of Electrical Engineers, London (*Institute of Electrical Engineers Proceedings*) for Figs 5.2 and 10.2; New York Academy of Sciences (*Annals of the New York Academy of Sciences*) for Fig. 9.8; North-Holland Publishing Company (*Physica-D*) for Figs 13.5, 14.2 and 14.3, and (*Physics Reports*) for Figs 10.7, 10.8 and 10.10; Peter Peregrinus for Figs 9.1 and 9.4; Society for Experimental Biology (*Journal of Experimental Biology*) for Fig. 9.2; Springer Verlag for Fig. 6.8.

Part I

Prologue

1
What is the use of chaos?

M. Conrad

Departments of Computer Science and Biological Sciences,
Wayne State University, Detroit, Michigan 48202, USA

1.1 A functional question

Rössler's rotating taffy puller provides a beautiful image for appreciating
the origin of chaos in one of its simplest forms [16]. Stretching plus folding
lead to mixing by distancing neighbouring points and bringing distant points
into close proximity. The addition of rotation causes the point to follow a
highly irregular path, which Rössler aptly calls a 'disciplined tangle'. The
tangle will be different for each different choice of initial conditions;
nevertheless the overall impression given by any two different tangles is
basically the same.

Deterministic dynamical systems of three or more dimensions can exhibit
behaviours of the type generated by the rotating taffy machine. Despite
their determinism, the behaviours generated look extremely random. This
is what it means to say that such systems are effective mixing devices. The
discovery of chaos suggests that the question of whether a given random
appearing behaviour is at base probabilistic or deterministic may be
undecidable.

It is by now probably fair to say that many plausible dynamical models
for complex biological systems become chaotic for some choice of the
parameters. Chaotic solutions have been found for equations similar to
chemical kinetic equations [15], equations governing neurone dynamics [1,
9, 10], and population dynamics equations [13]. The question I have been
asked to address is what the function, if any, of such chaotic dynamics might
be. Such questions have a teleological ring. Nevertheless it is useful and
justified to look at living systems from the functional point of view. This is
due to the enormous asymmetry between existence and nonexistence. Some
biological systems are so organised that they remain in the game of life.
Others go out of existence. The asymmetry is simply that it is the existing

systems which are of interest to us. It is legitimate to ask what it is about the organisation of these systems that allows them to persist. As soon as we do so, we are adopting a functional point of view.

Complex biological systems, such as neurones and the immune system, are the end result of the long process of evolution by variation and natural selection. Physiological mechanisms which control the population dynamics are also subject to variation and selection. The equations suitable for describing, say, the neurone need not be the same for all organisms, and probably are not. In biology the equations are as much the product of evolution as traits such as eye colour. If biological dynamics could be recorded in the fossil record, it would undoubtedly show evolutionary changes as dramatic as those exhibited by bones and other biological structures. The dynamics and parameter values could be selected to exhibit chaos; or they could be selected to preclude chaos.

However justified and even necessary the functional question is from the biological point of view, it is replete with dangers. What is the function of Rössler's rotating taffy puller? To make taffy, to advertise taffy, to provide employment, to earn a profit, to inspire Otto Rössler? Or, in an emergency, to serve as a lever or as a weapon? All biological entities and machines are multifunctional. How they have contributed to staying in the game of life cannot be specified completely, and how they might contribute in the future is an open question. It is dangerous to suppose that natural selection wants this or that. What is selected changes in the course of evolution and not all the phenomena of life are controlled by selection.

This caveat applies to the functional interpretation of chaos. However, in one respect, chaos is simpler to analyse functionally than most biological structures or processes. This results from the fact that it is so difficult to distinguish deterministic chaos from highly random behaviour. In so far as chaos contributes to the variability of biological matter, any analysis of the functional significance of variability *a fortiori* applies to the phenomenon of chaos. Fortunately, a systematic theory of biological variability is available. This is adaptability theory.

1.2 Overview of adaptability theory

A thorough review of adaptability theory can be found in my book on this subject [7], and more limited reviews in a number of earlier papers [3,4,5]. It would be duplicative to re-present the theory here. It should be sufficient to state the central idea verbally and to indicate the connection to dynamical systems theory.

By adaptability I mean the ability of a system to continue to function in the face of an uncertain or unknown environment. The system of interest is usually a living system, say an organism, a population, or even a whole community. The environment is everything that influences this system. It may include other biological systems as well as the physical environment, and may in part be influenced by the activities of the system of interest. For simplicity I will call the system of interest the biological system and its surroundings the environment.

In adaptability theory the biological system is treated as a system with a set of states and a transition scheme, ω, governing the state-to-state transitions. The transition scheme is unknown, but notationally consists of a set of probabilities for the state of the biological system at time $t + 1$ given the state of both the biological system and environment at time t. For simplicity it is assumed that the state set is discrete and finite. The environment is also treated as a system with a set of states with a probabilistic, generally unknown, transition scheme, denoted by ω^*. In general, transition schemes have a deterministic aspect (for example, connected with the life cycle of the organism or the cycle of the year) and an indeterminate aspect (for example, connected with mutation or unpredictable weather patterns).

The fundamental quantities in adaptability theory are measures of behavioural uncertainty such as the following.

(1) $H(\omega^*)$ = behavioural uncertainty of the environment.
(2) $H(\hat{\omega})$ = potential behavioural uncertainty of the biological system ($\hat{\omega}$ is the transition scheme of the biological system in the most uncertain tolerable environment).
(3) $H(\hat{\omega}|\hat{\omega}^*)$ = potential behavioural uncertainty of the biological system given the behaviour of the environment. This increases as the ability to anticipate the environment increases or as the uncertainty which the biological system internally generates increases.
(4) $H(\hat{\omega}^*|\hat{\omega})$ = potential behavioural uncertainty of the environment given the behaviour of the biological system. This increases as the indifference to the environment increases, for example as the organism lives in a smaller region of space.

The main question of adaptability theory is: what is the relation between the statistical properties of the biological system and the statistical properties of the environment? Ignoring for the time being the all-important question of detailed statistical structure, it is possible to summarise the situation by the following simple formula

$$(1.1) \quad H(\hat{\omega}) - H(\hat{\omega}|\omega^*) + H(\hat{\omega}^*|\hat{\omega}) \to H(\omega^*)$$

The left-hand side represents the adaptability of the biological system. The right-hand side represents the actual uncertainty of the environment. The arrow represents a plausible evolutionary tendency of adaptability. All forms of adaptability are costly. Adaptabilities which are never used tend to disappear in the course of evolution, so the magnitude of the left-hand side tends to drop in the direction of the actual uncertainty of the environment.

The magnitude of the terms in eqn (1.1) could be individually high, yet the adaptability low. In this case there would be a great deal of biological variability, but not much of it would appear as adaptability. These reserves of variability can be converted to adaptability in the event of a crisis. Some of the variability which contributes to the magnitudes of the entropies serves to increase reliability or to enhance the evolutionary transformability of the system.

The transition-scheme description is connected to descriptions in terms of biological variables by utilising the fact of hierarchy. Ecosystems consist of compartments such as communities, populations, organisms, cells and genes. These are associated with variables such as locations and numbers of organisms, physiological states of cells, and base sequence in DNA. Let the symbol ω_{ij} represent the transition scheme of compartment i at level j in terms of its subcompartments at the next lower level. The uncertainty of the whole biological system being considered can be expressed in terms of a sum of effective entropies of each compartment:

$$(1.2) \quad H(\hat{\omega}) = \sum_{i,j} H_e\,(\hat{\omega}_{ij})$$

Each effective entropy is a sum of conditional and unconditional entropies

$$(1.3) \quad H_e\,(\hat{\omega}_{ij}) = f\,H(\hat{\omega}_{ij}) + \text{conditional terms}$$

where f is a normalising coefficient. The unconditional part is the behavioural uncertainty of the compartment considered in isolation, and the conditional parts express the correlation between this uncertainty and the modifiabilities of other compartments. The uncertainty taken in isolation will be called the modifiability and the conditional terms will be the independence terms.

The adaptability is not the sum of the modifiabilities. If a system is more decentralised (that is, if the parts are more independent), the adaptability is greater for given observable modifiabilities of the parts. If constraints are added to the system which decrease the conditional entropies, the adaptability must decrease or be compensated by enhanced anticipation, increased indifference, or development of new subsystems with high behavioural uncertainty (such as the central nervous system or the immune system). If no such compensation occurs, the niche must narrow or the community must absorb disturbances at the level of population fluctuations. This is an acceptable mode of adaptability for micro-organisms since these are fast growing.

Adaptability cannot always decrease. This would be incompatible with the tenure of life on Earth. Many factors control the rise and fall of adaptability in the course of evolution. When an evolutionary system loses too much adaptability, or when the uncertainty of its environment increases, it is likely to go into a crisis. The crisis instigates a series of changes which result in the renewal of the adaptability structure.

1.3 **Adaptability, dynamics and the place of chaos**

Many dynamical models in biology are of a continuous dynamical nature. The transition schemes of adaptability theory are discrete and probabilistic. It is possible to cross-correlate the two approaches by using the idea of a tolerance, that is of a relation on the states which is symmetric, reflexive, but not transitive [8]. State A is similar to state B, which is similar to state C. But A is not necessarily similar to C. In this way an aspect of continuity can be conferred on discrete transition schemes. Analogues of concepts such as neutral stability, asymptotic stability and structural stability can be defined. The question can then be asked: what is the relation between the components of adaptability theory and these different dynamical concepts of stability?

   Without going into details, the general situation can be pictured thus. The modifiability terms correlate to instabilities of dynamical models. This is because the alternative structures and modes of behaviour which contribute to modifiability can correspond to alternative weakly or strongly stable states. In this case disturbances are absorbed by instabilities. The modifiability terms can also correlate to asymptotic orbital stability. This occurs when the disturbance is dissipated by direct return to a strongly stable state. In this case the modifiability consists in a variation around the strongly stable state which in time is dissipated into the heat bath. The independence terms correlate either to structural stability or to weakening of the interaction of two systems. The correlation here is somewhat subtle and depends on whether one is near to the bifurcation points or far from them. If a system is structurally stable, it can undergo a variation in response to disturbance, which, however, does not qualitatively alter its structure or behaviour. In this case the behaviour of the system is less dependent on parametric compartments (see [7] for full details).

   The problem of adaptability theory is to ascertain how the different forms of adaptability will be allocated to different parts of a system given the constraints which are present (such as morphological constraints). If one compartment is to be maintained in a very certain state, this must be at the expense of some other compartment being in an uncertain state. The uncertainty serves to absorb the disturbance. Corresponding to this economics of adaptability components there is an economics of stability and instability in biological systems. This is due to the fact that adaptability components correlate to forms of stability and instability in dynamical descriptions. If the dynamics of one level of organisation, say the population level, is to be very stable, it is necessary for disturbances to be absorbed by instabilities at some other level. For example, neurobehavioural instabilities or genetic variability may protect the population dynamics from disturbance and therefore allow it to appear highly stable. Alternatively, the stability of the state of internal fluids in the vertebrate may be obtained at the expense of intricate dynamical instabilities in the nervous and immune systems.

To fit chaos into this framework it is necessary to remember that the transition schemes are generally probabilistic. Not all modifiability is connected with intrinsically random processes. For example, a compartment might be able to absorb disturbances by functioning in one of several dynamical regimes. In the absence of any information about the environment, switching from one regime to another may appear virtually random. In reality the situation is completely deterministic. However, it is also possible to consider situations in which the modifiability of a compartment is due to completely intrinsic factors. This is the case in genetic variability, in variability of immunoglobin molecules in the immune system, and in exploratory processes in the nervous system. This intrinsic modifiability also makes it possible to absorb disturbances and protect other compartments. The intrinsic variability of genes is the major form of adaptability in nature. It allows organisms to protect their essential dynamical properties in the face of environmental changes by varying less essential dynamical properties.

This intrinsic modifiability could be due to Brownian motion or it could be due to chaotic dynamics. As previously stated, distinguishing these two possibilities may be effectively undecidable. For the purpose of analysing the functional significance of chaos it is not necessary to make this distinction. If the dynamics appears chaotic, and this is not due to external forcing, it will make the same contribution to adaptability whatever the mechanism. However, there may be a significant advantage in structuring the system in a manner which fulfils the stretching, folding and rotating conditions required for dynamical chaos. Brownian motion is always present, but it is damped out in some dynamical systems and magnified in others. If a biological system obeys chaotic dynamics, this will ensure that its behaviour is chaotic. If the system obeys dynamics which are not sensitive to initial conditions, it is possible that the effects of the Brownian motion will be too negligible to make a significant contribution to adaptability.

1.4 Chaotic mechanisms of adaptability

Table 1.1 classifies chaotic mechanisms of adaptability. All the mechanisms involve the diversity generation in which chaotic dynamics could conceivably play a role.

The first category includes search processes in which an ensemble of possibilities is generated and tested. The most fundamental level of diversity generation is that of mutation, crossing over, recombination, and related genetic operations. These create a combinatorial explosion of possible genotypes on which natural selection acts. Conceivably some of the chemical dynamics underlying mutation are chaotic in nature or depend on intrinsic noise processes complemented by chaotic mechanisms. At the

Table 1.1 Possible functional roles of chaos.

---

(1) *Search* (diversity generation):
   genetic
   behavioural

(2) *Defence* (diversity preservation):
   immunological
   behavioural
   populational

(3) *Maintenance* (disentrainment processes):
   neural and other cellular networks
   age structure of populations

(4) *Cross-level effects:*
   interaction between population dynamics and gene structure

(5) *Dissipation of disturbance* (qualitative insensitivity to initial conditions)

---

present time, however, there is no evidence that dynamical chaos acts directly at the genetic level. Later we will argue that some indirect effects occur due to cross-level interactions with chaotic population dynamics. Another possible place of chaos in genetic diversity generation is in the origin of life. Nicolis *et al* [14] have recently presented a model for the origin of prebiological polymers in which the generation of sequence diversity is driven by chaotic reaction dynamics.

A second category of search processes is behavioural. For very simple organisms, such as micro-organisms, the behavioural mode is not as important as the genetic mode. This is due to the fact that the morphological simplicity of micro-organisms allows for a high degree of viable genetic diversity as well as for a short generation time. The use of genetic mechanisms of adaptation is restricted in higher plants by the longer generation time. However, the morphology is still relatively simple, therefore compatible with a large variety of viable genetic and morphological forms. The variety of morphological forms develops primarily as a growth response to light and moisture, though it is possible that some intrinsic diversity generation plays a role as well. Vertebrates have a much more intricate and sensitive morphology. The variety of genetic and morphological forms is severely restricted. The development of the immune system and the central nervous system provides for alternative mechanisms of diversity generation which compensate for these restrictions.

It is now known that physiologically plausible dynamic models of neurones can exhibit chaotic solutions. Chay [1] has shown that equations for membrane excitability can exhibit such behaviour in the absence of any forcing. It is also known that, in some central nervous system neurones, second messengers — especially cyclic AMP — control membrane electrical activity. The second-messenger system involves an elaborate set of biochemical reactions whose complexity allows for a rich set of dynamical behaviours [11]. Some evidence indicates that the dynamics of the cyto-skeleton may be involved [12]. We have modelled this system and have found that chaotic-appearing solutions occur. The key to chaos in these models is delicate threshold behaviour. Cyclic AMP builds up, either endogenously or due to presynaptic input. The internal state of the cell changes. How it responds to a given input pattern therefore changes. Rather complex endogenous rhythms can also be produced.

One can contemplate at least three roles for these chaotic regimes. First they can serve to generate diverse behaviour. Such behavioural diversity serves the same function as genetic diversity, and indeed compensates for restrictions on genetic diversity. The function here is to enhance exploratory behaviour.

The second function is defence. Rather than using the diversity of behaviour to explore, it can be used to avoid predators. An organism that moves in an unpredictable fashion is more difficult to catch than one which moves about in a highly determinate manner. The flitty behaviour of a butterfly makes it difficult to catch, and may have its basis in a chaotic neural mechanism [17].

A third possible role of neural chaos is the prevention of entrainment. It is conceivable that, in the absence of chaos, either very dull pacemaker activity would develop or highly explosive global neural firing patterns would emerge. Chaotic mechanisms would serve to maintain the functional independence of different parts of the nervous system. As already stated, a system whose parts vary in a more independent manner is more adaptable.

The immune system provides another example. Like the neuromuscular system the immune system compensates constraints on genetic, develop-mental and populational modes of adaptability in the vertebrates. High diversity of the immunoglobin molecules makes it possible for the organism to deal with the diversity of the microbial world. In addition the diversity of cell surface markers provides a means of privacy and defence. If the structures and dynamics of organisms were sufficiently regular to be predictable in detail, invasion by micro-organisms would become much easier. Whether or not the molecular and dynamic diversity of the immune system is a consequence of dynamical mechanisms of chaos is not known. At the level of control processes in the immune system, especially in so far as these processes involve dynamics of the endocrine system, there are ample possibilities for chaotic solutions which could serve to enhance the integrity of the organism by conferring a protective unpredictability on it.

Population dynamics provides a somewhat different example. The difference lies in the fact that neuromuscular, immunological and hormonal dynamics are a consequence of selection acting on individual organisms. Population dynamics is a more indirect consequence of selection acting on individuals, and to some extent a consequence of the logical requirement that the dynamics of a whole ecosystem be self-consistent. If the dynamics of the whole ecosystem fail in this respect, changes will inevitably occur, until finally self-consistency is achieved.

Many population dynamics models are known to exhibit chaos. These are probably the best-known examples of chaotic dynamics in biology. Admitting that function becomes a vague concept at the population level, it is nevertheless possible to look at population dynamics from a functional point of view to the extent that these dynamics must satisfy self-consistency. One role of chaos is again to preclude entrainment. If a population becomes highly entrained, its diversity is greatly reduced. For example, the age structure could become very narrow. Such reduction of diversity is adaptability reducing. Chaotic mechanisms would therefore serve to maintain the adaptability of populations.

Chaotic mechanisms could also serve to make a population less predictable to a predator. If the population oscillated in a very regular pattern, it would be easy for a predator or a disease vector to track it. The population dynamics of the predator could be co-ordinated to that of the host population in an anticipatory way. As in the immune system, chaotic mechanisms of population dynamics confer a defence-enhancing unpredictability.

Chaotic population dynamics has a surprising cross-level effect which has enormous significance for genetic structure and evolution. The effect was discovered in computational models of ecosystem dynamics in which modelled organisms feed and reproduce in a modelled environment [2,6]. Each organism consists of a set of genes (described at the nucleotide level) and a set of phenotypic traits determined by these genes. The population dynamics emerges from the interactions among the individual organisms. Since mutation and recombination are possible, and since space and energy are limited, organisms evolve by the Darwinian mechanism of variation and selection.

In some of our models the population dynamics is highly chaotic. The number of organisms increases and decreases in a highly irregular manner. In part this is due to the fact that the physiological states of the organisms are time dependent. An organism that has previously fed well may succeed in reproducing even in an environment in which food is scarce. On the other hand, organisms that have not already collected a lot of food may fail to reproduce even when a great deal of food becomes available. The availability of food is determined by the internal dynamics of the system. When organisms die, the matter they have accumulated is returned to the environment. The size of the population varies in an extremely chaotic manner, with the amplitude of the oscillations increasing as the time required to cycle matter between organisms and environment increases.

When the population size reaches a peak, few organisms succeed in reproducing. The advantage of being efficient and of having a 'streamlined' genome is slight. When the population size bottoms out, any organism whose traits fall into an accepted range will collect a full quota of food. The chance of reproducing is high even if the organism is not very efficient. Organisms that carry 'genetic junk' have a good chance of being injected into the population. In the intermediate zone between a population boom and a population bust, this superfluous genetic material is pruned out. As a consequence there is a continual injection of extraneous genetic material into the population when the population is small, and a flushing out of this material when it is of intermediate size. A certain amount of parasitic DNA will inevitably be present. The more violent the oscillations the greater the amount of parasitic genetic material.

This so-called parasitic genetic material turns out to serve a useful evolutionary function. If the genotypes could be maintained in a perfectly streamlined condition, the population would sit on top of an adaptive peak. The chance of discovering another adaptive peak by crossing over an adaptive valley would be negligibly small. The continuous injection of less than optimally fit organisms into the population means that many organisms will occupy positions well below the top of the peak. Technically these organisms can be described as carrying a genetic load, that is, a load of less than optimally fit genes. Since the load brings the organism closer to the valley between peaks, it increases the chance that some member of the population will make the transition to the unoccupied peak.

It is probably true that this effect is a consequence of population oscillations rather than chaos *per se*. But in fact in our artificial ecosystem models the oscillations were in all respects chaotic. Chaos at the level of population dynamics induced chaos at the level of gene structure. Chaos at the level of gene structure, as a form of genetic diversity, increased evolutionary adaptability and therefore facilitated the evolutionary process.

All the chaotic processes considered serve as a dynamic heat bath for dissipating disturbances. Disturbances from which biological systems are not physically isolated must either be absorbed by them in permanent changes or dissipated into a heat bath. Asymptotic orbital stability provides one route of dissipation. The biological system is initially modified, but in time the effect of the disturbance is completely absorbed in the random thermal activity of the environment. A chaotic system with a strange attractor can actually dissipate disturbance much more rapidly. Such systems are highly initial-condition sensitive, so it might seem that they cannot dissipate disturbance at all. But if the system possesses a strange attractor which makes all the trajectories acceptable from the functional point of view, the initial-condition sensitivity provides the most effective mechanism for dissipating disturbance.

1.5 Chaos and self-consistency

The possibility of interpreting biological chaos in functional terms is based on the fact that biological systems, to be interesting, must manage to stay in the game of life. To stay in the game of life it is necessary for the dynamics of the parts to be consistent with the dynamics of the whole, and for the dynamics of the whole to be consistent with that of the parts. The most important mechanism for achieving such self-consistency is natural selection. If the dynamics of individual organisms is inconsistent with the stability of the ecosystem as a whole, it is inevitable that natural selection will eliminate these organisms. Indirectly a dynamics for the whole eco-system is selected. If these dynamics are still inconsistent with the persis-tence of organisms, the characteristics of organisms will change through the mechanism of variation and selection until an acceptable global dynamics is achieved.

Can self-consistency considerations of this sort play any role in the abiological world? Ordinarily physicists do not think in functional terms. The so-called laws of nature are regarded as inexorable givens. In fact, however, history plays a role in inorganic nature. Geology, for example, is an historical subject. Whenever history enters into the picture, it is valid to ask why some structures and processes come to be and others do not. In history-dependent phenomena, such as those of geology and astronomy, it is valid to ask questions about stability. In order for these phenomena to have the 'right to persist', they must be stable. It is conceivable that turbulent and chaotic dynamics in such abiological systems could in some instances be interpreted as mechanisms for dissipating disturbance, and therefore as an essential contributor to the stability of the whole.

Whether the fundamental equations of physics are timeless or the product of a historical development is not known. Many physicists have seriously considered the possibility that, like the universe as a whole, the laws of physics are the product of an evolution. The consistency of microscopic laws with the macroscopic boundary conditions of the universe would then become as legitimate an object of inquiry as the consistency of organism dynamics with the macroscopic organisation of an ecosystem. Imagine, for example, a universe in which the microscopic dynamics were incapable of supporting an H-theorem. Macroscopic domains of the universe would never fall into a state of equilibrium. Continuous macro-scopic change would occur. If this macroscopic change altered the micro-scopic laws, these laws would eventually reach a form capable of supporting an H-theorem. The reason is simply that as soon as mechanisms of microphysical chaos are discovered, the universe as a whole will proceed much more rapidly to an equilibrium or at least to a slowly changing state, therefore to a state which allows the microphysical laws to retain a constant form.

Acknowledgements

This paper was written during a visiting professorship at the Molecular Biophysics Unit at the Indian Institute of Science, Bangalore, India. I am indebted to the Institute for its hospitality, and to the National Science Foundation Grant INT-83-11410 for supporting this visit. This research was also supported in part by NSF Grant MCS-82-05423.

References

[1] Chay, T.R. Chaos in a three-variable model of an excitable cell. *Physica-16D* 233–42 (1985).
[2] Conrad, M. Computer Experiments on the Evolution of Co-adaptation in a Primitive Ecosystem. PhD thesis, Biophysics Program, Stanford University, Stanford, California (1969).
[3] Conrad, M. Analysing ecosystem adaptability. *Math. Biosci.* **27**, 213–30 (1975).
[4] Conrad, M. Functional significance of biological variability. *Bull. Math. Biol.* **39**, 139–56 (1977).
[5] Conrad, M. Ecosystem stability and bifurcation in the light of adaptability theory. In *Bifurcation Theory and Applications in Scientific Disciplines,* ed. O. Gurel and O. E. Rössler, pp. 465–81. *Ann. NY Acad. Sci.* **316** (1979).
[6] Conrad, M. Algorithmic specification as a technique for computing with informal biological models. *BioSystems* **13**, 303–20 (1981).
[7] Conrad, M. *Adaptability*. Plenum Press, New York (1983).
[8] Dal Cin, M. Modifiable automata with tolerance: a model of learning. In *Physics and Mathematics of the Nervous System*, eds. M. Conrad, W. Guttinger and M. Dal Cin, pp. 442–58. Springer, Heidelberg (1974).
[9] Holden, A. V. and Muhamad, M.A. Chaotic activity in neural systems. In *Cybernetics and Systems Research 2*, ed. R. Trappl, pp. 245–50. Elsevier, Amsterdam (1984).
[10] Holden, A. V. and Muhamad, M.A. The identification of deterministic chaos in the activity of single neurones. *J. Electrophysiol. Tech.* **11**, 133–47 (1984).
[11] Kirby, K. and Conrad, M. The enzymatic neuron as a reaction-diffusion network of cyclic nucleotides. *Bull. Math. Biol.* **46**, 765–83 (1984).
[12] Liberman, E.A., Minina, S.V., Mjakotina, O.L., Shklovsky-Kordy, N.E. and Conrad, M. Neuron generator potentials evoked by intracellular injection of cyclic nucleotides and mechanical distension. *Brain Res.* **338**, 33–44 (1985).
[13] May, R.M. Simple mathematical models with very complicated dynamics. *Nature, Lond.* **261**, 459–67 (1967).
[14] Nicolis, G., Rao, J.S. and Rao, G.S. Chaos: a possible role in prebiotic evolution. Preprint, Department of Chemical Physics, Free University of Brussels (1984).
[15] Rössler, O.E. Continuous chaos—four prototype equations. In *Bifurcation Theory and Applications in Scientific Disciplines,* eds. O. Gurel and O.E. Rössler, p. 379. *Ann. NY Acad. Sci.* **316** (1979).
[16] Rössler, O. The chaotic hierarchy. *Z. Naturforsch.* **38a**, 788–801 (1983).
[17] I am indebted to O. Rössler for this example.

2

A graphical zoo of strange and peculiar attractors

A.V. Holden and M.A. Muhamad*

Department of Physiology,
The University, Leeds LS2 9NQ, UK

2.1 Trajectories and attractors

A large part of the interest in nonlinear dynamics arises from its applications: there is a strong belief that an understanding of the patterns of bifurcations in dynamical systems provides a means of understanding natural phenomena. If a variable measured in the course of an experiment settles down with time, to a constant, or a maintained oscillation, it seems reasonable to assume that it is approaching some stable, maintained course that corresponds to an equilibrium or periodic solution $\mathbf{x}(\mathbf{x}_0, t)$, that is obtained as $t \to \infty$, of some appropriate dynamical system.

$$(2.1) \qquad\qquad d\mathbf{x}(t)/dt = \mathbf{F}(\mathbf{x}(t))$$

$\mathbf{x}$ is a vector in R^n, with each of the n components representing a variable.

A system of equations that provides an accurate and physically satisfactory representation of an experimental system can be cumbersome, and even complicated and of high order, so solutions may often only be obtained as numerical approximations to solutions. Thus the numerical solutions themselves may be considered to be approaching an equilibrium or periodic solution.

The equations that represent physical systems that have an 'internal friction' are dissipative: the flow $t \to \mathbf{x}(t)$ on average contracts volumes in phase space, and so

$$\sum_{i=1}^{n} \partial F_i(\mathbf{x})/\partial x_i < 0$$

*Present address: Jabatan Fisiologi, Universiti Malaya, Kuala Lumpur 22-11, Malaysia.

Continued shrinking of volume in phase space means that stable, persistent motion in an n-dimensional dissipative system must be on a structure that has a dimension less than n: this structure is an attractor and occupies a subset X of phase space. Motions starting in some volume of phase space (the basin of attraction of the attractor) can follow complicated transients but, as $t \to \infty$, they finally approach the attractor. The attractor X has the properties that it is invariant under the flow and cannot be decomposed into nonoverlapping invariant pieces: any motion on the attractor is confined to the attractor, and not just part of the attractor. There is no transient motion on the attractor: all transients in a dissipative system are in the basin of attraction of an attractor on the approach to an attractor, and are not on attractors.

Virtually all possible initial states of a dissipative system are in the basin of an attractor, and so if the motion of a system is followed for long enough, it will end up on an attractor. Thus, just as an attractor may be imagined as the subset of phase space on which a motion starting in its basin of attraction is confined as $t \to \infty$, the attractor may be visualised by following a maintained trajectory after a time that is long enough for all transients to have decayed.

For a one-dimensional flow, the only possible attractor is a stable fixed point or sink. The bifurcations of fixed points are illustrated in Figs 3.1–3. A sink would appear as an equilibrium, or resting, state.

For a two-dimensional flow, there is also the possibility of a periodic solution or limit cycle, where the attractor is a simple closed curve. Trajectories leading to a sink and a limit cycle are illustrated in Fig. 2.1. If these trajectories were viewed only after they had reached the attractor, they would illustrate the attractors: the sink as a point in phase space, and the limit cycle attractor as a single closed loop in phase space. To construct

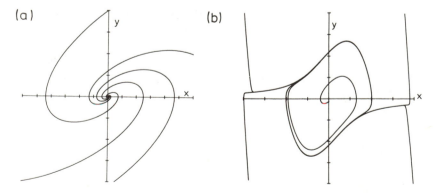

Fig. 2.1 Trajectories leading to (a) a sink, and (b) a limit cycle. (a) $dx/dt = -y$, $dy/dt = x - y$, with x-axis from -600 to 600, y-axis from -800 to 800; (b) the van der Pol oscillator $dx/dt = -y$, $dy/dt = x + y\mu\,(1 - x^2)$, with x-axis and y-axis from -4 to 4.

such a picture of the attractor, the values of the two variables $x(t)$ and $y(t)$ are required, when t is sufficiently large to be taken as if $t \rightarrow \infty$. Experimental measurements from an n-dimensional system are often only of one variable: however, it is possible to reconstruct an attractor from the values of a single variable, as in Figs 2.2–4 [13, 21, 33]. If a variable $x(t)$ is

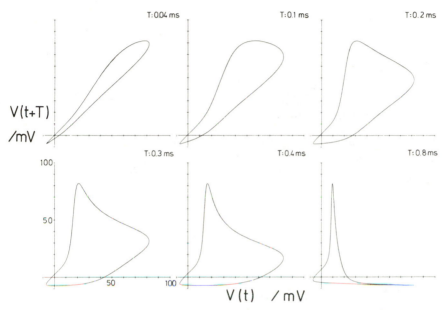

Fig. 2.2 Periodic attractor for the Hodgkin–Huxley [12] equations representing the membrane of the squid giant axon. This four-variable differential system responds to a constant membrane current density of 20 μA cm^{-2} by large-amplitude periodic solutions. A projection of this periodic attractor on to a plane may be constructed by plotting one variable, the experimentally observable membrane potential V, against itself delayed by a lag T. As T is changed from 0.04 to 0.8 ms, the shape of the reconstructed attractor changes; its topological properties do not change.

plotted against itself delayed by an appropriate time lag T, then $x(t)$ plotted against $x(t+T)$ will give a point for a sink, and a single closed loop for a limit cycle.

In a three-dimensional nonlinear system, or a periodically forced two-dimensional nonlinear system, there is also the possibility of changing motion that is not periodic, associated with a more complicated kind of attractor. If a dissipative system has only one attractor, its behaviour as $t \rightarrow \infty$ might not seem to depend very much on its initial conditions. Whatever the initial conditions, all trajectories will eventually arrive on the attractor. However, although phase space volume on average contracts in a dissipative system, so that trajectories passing through a large volume

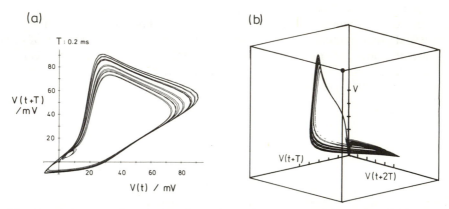

Fig. 2.3 Reconstruction, from values of membrane potential V only, of (a) two-
and (b) three-dimensional projections of a strange attractor for the Hodgkin–
Huxley equations driven by a sinusoidal forcing current density of $15(1 + 0.75 \cos (2\pi325\ t))$. The attractor is locally a flat, banded sheet that is twisted in a four-
dimensional state space. Experimental and numerical constructions of this attractor
are seen in Figs 12.5 and 12.6.

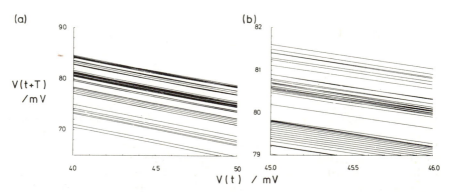

Fig. 2.4 A two-dimensional view of the same attractor, at increased (a) and further
increased (b) magnifications: the bands are composed of sub-bands, which are
themselves composed of yet further sub-sub-bands…. Such a self-similar structure,
which has a patterning that is independent of scale, has a fractal dimension.

of phase space end up passing through a smaller volume, this does not
mean that distance (or separation) is also contracting. If closely neighbouring points on an attractor give trajectories that, although confined to
the attractor, rapidly separate, then motion on the attractor shows a
sensitivity to initial conditions. This sensitivity to initial conditions
underlies chaos and can be produced by repeated stretching and folding
within the attractor. A measure of this sensitivity to initial conditions, the
maximal Lyapunov exponent, is discussed in Chapter 13.

An attractor that has a positive maximal Lyapunov exponent, and so has a high sensitivity to initial conditions, often has a very complicated structure, in that it may have a noninteger, fractal dimension, and have some of the properties of a Cantor set. The behaviour of trajectories on an attractor that exhibits sensitivity to initial conditions is strange and unexpected, and the geometry of such an attractor is often strange and complicated. Here we will use the term 'strange attractor' to refer to attractors that show sensitivity to initial conditions [4, 6, 14, 20, 27, 28, 29]. Such chaotic attractors need not have a fractal structure, and attractors with a fractal structure need not be chaotic [10].

This chapter illustrates some attractors: these static illustrations are a poor substitute for watching the trajectories approach and then wind around the attractor, and for following the sequence of bifurcations (changes in type of attractor) as a parameter is changed. All the figures in this chapter were produced using FORTRAN and GHOST-80 [7] or GINO-F [8] on the University of Leeds Amdahl 470/V7, but they may all be computed and displayed, and the bifurcation patterns followed, on a microcomputer.

2.2 Three-dimensional systems

2.2.1 *The Lorenz system*

The Lorenz system:

$$(2.2) \quad \begin{aligned} \mathrm{d}x/\mathrm{d}t &= -\sigma x + \sigma y \\ \mathrm{d}y/\mathrm{d}t &= -xz + rx - y \\ \mathrm{d}z/\mathrm{d}t &= xy - bz \end{aligned}$$

with σ, r and b positive parameters, was derived as a truncation of a partial differential equation for fluid convection [19, 32]. A flat fluid layer is heated from below and cooled from above: this represents the Earth's atmosphere heated by the ground's absorption of sunlight and losing heat into space. In the resultant convective motion, x represents the convective motion, y the horizontal temperature variation, and z the vertical temperature variation. The parameters σ, r and b are proportional to the Prandtl number, the Rayleigh number, and the size of the region whose behaviour is being approximated by the ordinary differential system (2.2): these parameters are positive. This system is considered in detail in Chapter 6, and is also formally equivalent to the equations for a single-mode, homogeneously broadened laser, treated in section 7.3. They may also be used to describe the behaviour of a variety of other physical systems.

Numerical approximations to solutions of the Lorenz equations are complicated for wide ranges of the parameters σ, b and r. Two-and three-dimensional views of a trajectory, when the integration time is sufficiently large for all transients to have decayed, are shown in Figs 2.5 and 2.6.

Fig. 2.5 Different faces of the Lorenz attractor. Projections of the attractor in the (a) zx–the mask, (b) yx, and (c) yz plane. ($\sigma = 10$, $r = 60$, $b = 8/3$.)

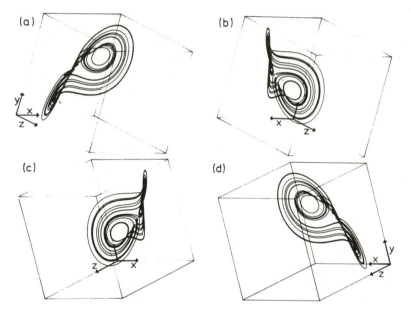

Fig. 2.6 Three-dimensional views of the Lorenz attractor of Fig. 2.5. The attractor is within the coordinate cube, which is tilted forward and then rotated around the y-axis by (a) 45°, (b) 135°, (c) −135°, and (d) −45°.

2.2.2 *Reversals of the Earth's magnetic field*

Throughout geological time there have been irregular changes and reversals of the Earth's magnetic field [1, 2]. If this behaviour is modelled by a modified disc dynamo [3], the appropriate magnetohydrodynamic partial differential equations may be truncated to give the simple system:

(2.3)
$$dx/dt = a(y - x)$$
$$dy/dt = zx - y$$
$$dz/dt = b - xy - cz$$

where a, b and c are positive parameters, and $b > ac(a + c + 3)/(a - 1 - c)$. The variables x and y relate to poloidal potentials and a toroidal magnetic field component, and z is related to the moment of angular momentum. The similarity of this system to the Lorenz equations is reflected in the shape of its strange attractor, shown in Figs 2.7 and 2.8.

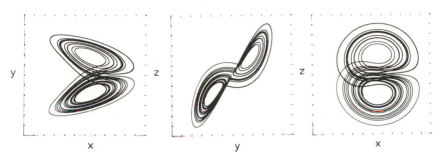

Fig. 2.7 Projections of the strange attractor of (2.3) into the (a) xy, (b) yz, and (c) xz planes, for $a = 14.625$, $b = 1.0$ and $c = 5.0$. The axes are all from -12 to $+12$.

2.2.3 *The Rössler attractor*

The Lorenz system (see eqn (2.2) and Chapter 6) has a complicated attractor, with trajectories spiralling around, and jumping between, two loops. In 1976 Rössler [23] introduced a simpler three-dimensional system that has only a single nonlinear cross-term, zx:

$$(2.4) \quad \begin{aligned} dx/dt &= -(y + z) \\ dy/dt &= x + ay \\ dz/dt &= b + z(x - c) \end{aligned}$$

where a, b and c are constants. This system may be considered to model the flow around one of the loops of the Lorenz attractor, and so is a model of a model. Here $a = b = 1/5$, and c is treated as a bifurcation parameter.

With $c = 5.7$ the flow is chaotic: the flow forms a single spiral embedded in a disc, with trajectories from the outer part of the spiral twisted, and folded back into the inner part of the spiral, forming a Möbius band. The construction of the attractor is shown in Fig. 2.9. This is similar to the experimental attractor seen in Fig. 8.6. A return map of a section through the attractor looks like a single-humped, thickened, one-dimensional map. Three-dimensional views of the attractor are shown in Fig. 2.10.

As c is increased from 2 to 4.2, there is a sequence of period-doubling bifurcations from a simple, period-one oscillation (Fig. 2.11). Chaos develops at the accumulation point of the period-doubling sequence, just above $c = 4.20$, with families of similar orbits confined to thin bands that grow from each of the period $2n$, $n \to \infty$, orbits. These attractor bands

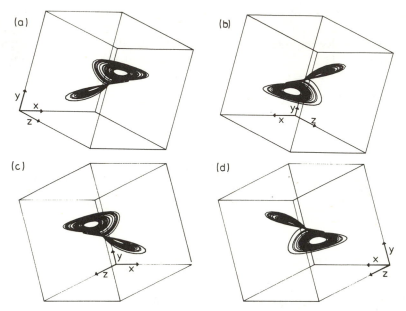

Fig. 2.8 Three-dimensional views of the strange attractor of Fig. 2.7. with the coordinate cube tilted forward and rotated about the y-axis by (a) 45°, (b) 135°, (c) –135°, and (d) –45°. The axes are all from –15 to +15.

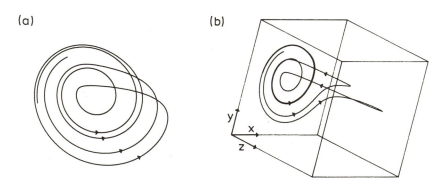

Fig. 2.9 Construction of the classical Rössler attractor, with $a = b = 0.2$, and $c = 5.7$ Trajectories diverge within the band ('stretching') and are folded back as they loop out and back in the z-direction. (a) Projection on x–y plane; (b) x and y axes from –14 to 14, z axis from 0 to 28.

are apparently separated by empty, repeller bands: a trajectory starting between the bands is rapidly drawn into one of the attractor bands. However, the attractor bands really form one, looped, attractor. Pairs of the attractor bands join in a sequence of reverse bifurcations until the bands in the strange attractor meet [23, 31].

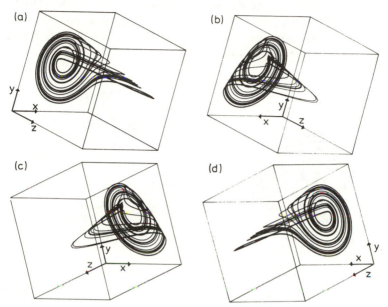

Fig. 2.10 Three-dimensional views of the attractor, with $a = b = 0.2$, and $c = 5.7$. The x-axis is from -14 to 14, the y-axis from -14 to 14, and the z-axis from 0 to 28. The axes form a cube, which has been rotated around the y-axis by (a) $45°$, (b) $135°$, (c) $-45°$, and (d) $-135°$.

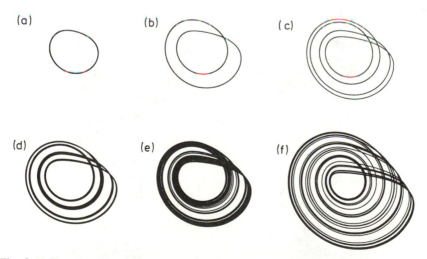

Fig. 2.11 Development of the classical Rössler attractor through a sequence of period doubling, and then reverse bifurcations. Projections in the x–y plane of the attractor for periodic and chaotic flows for the Rössler system with $a = b = 0.2$, and the bifurcation parameter c having the value of (a) 2.4, (b) 3.5, (c) 4.0, (d) 4.23, (e) 4.3, (f) 5.0.

A further increase in the parameter c gives a change in the shape of the attractor, which develops into a Rössler funnel as it expands with increasing c. The development and structure of the Rössler funnel are illustrated in Figs 2.12 and 2.13.

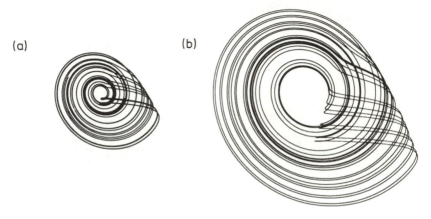

Fig. 2.12 Development of a Rössler funnel by a further increase in c, with $a = b = 0$, and (a) $c = 12$, (b) $c = 25$.

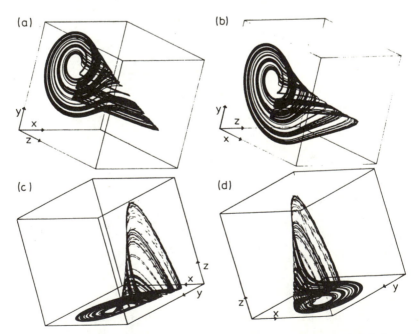

Fig. 2.13 Three-dimensional views of a Rössler funnel, with $a = 0.343$, $b = 1.82$ and $c = 9.75$.

2.2.4 *Abstract chemical kinetics*

All the systems considered above contain a cross-term, in which the rate of change of one variable is directly decreased by a term that is the product of two other variables. Such cross-terms cannot occur in a system that directly represents a chemical reaction, and the variables represent concentrations: the concentration of one chemical species cannot be directly decreased by a process in which that chemical does not take part [36]. Real chemical systems do show chaotic activity [5], but this could be due to spatial, hydrodynamic irregularities rather than true chemical chaos.

An abstract reaction mechanism, where the chemical reactions are at most second order and give a detailed mass balance when the system is closed, can be reduced to the system [40]:

$$
\begin{aligned}
dx/dt &= x(a_1 - k_1 x - z - y) + k_2 y^2 + a_3 \\
dy/dt &= y(x - k_2 y - a_5) + a_2 \\
dz/dt &= z(a_4 - x - k_5 z) + a_3
\end{aligned}
$$

(2.5)

The a_i are concentrations of reactants that are held constant, giving an

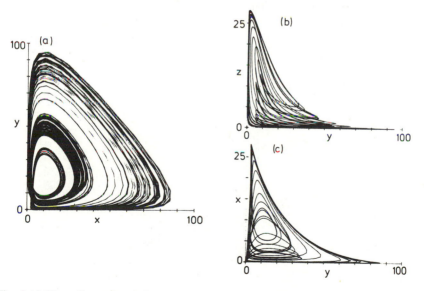

Fig. 2.14 Two-dimensional views of the trajectories at large times for eq (2.5) in the (a) *xy*, (b) *yz*, and (c) *zx* planes, with $k_1 = 0.25$, $k_2 = 0.001$, $k_5 = 0.5$, $a_1 = 30$, $a_2 = a_3 = 0.01$, $a_4 = 16.5$, and $a_5 = 10$.

open system, and the k_i are rate constants: thus the a_i and the k_i are positive. Numerical integrations of eqn (2.5) show irregular, apparently chaotic, activity (Figs 2.14 and 2.15).

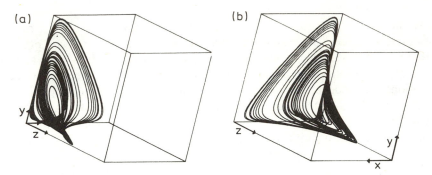

Fig. 2.15 Three-dimensional views of the strange attractor of Fig. 2.14, with the axes z from 0 to 50, and x and y from 0 to 100.

2.3 Four-dimensional systems

In the three-dimensional systems treated in section 2.2, the strange attractors are locally planar: a small displacement perpendicular to this sheet will decay, as the trajectory returns to the attractor; a small displacement along the sheet will remain, as a trajectory is effectively pushed forward in time; and a small lateral displacement will grow in time. This sensitivity to initial conditions (or instability for small lateral displacements) is reflected by the single positive Lyapunov exponent of a strange attractor of a three-dimensional system. A strange attractor of a four-dimensional system can have two positive Lyapunov exponents, and so can have solutions that are more irregular than chaos.

A simple system introduced by Rössler [24–26] that exhibits such hyperchaos is:

(2.6)
$$dx/dt = -y - z$$
$$dy/dt = x + 0.25y + w$$
$$dz/dt = 3 + xz$$
$$dw/dt = -0.5z + 0.05w$$

Two-dimensional views of the strange attractor are shown in Fig. 2.16: the xy plane view is reminiscent of the Rössler funnel. The system (2.6) has been obtained from a system similar to (2.4) by adding a linear variable, w. The variable z is activated whenever a threshold value of x is exceeded, and the activation of z leads to a reinjection of the trajectory to a new region in xyw space. The motion in xyz space is similar to that of (2.4): compare Fig. 2.17 with Fig. 2.9b. Three-dimensional outline views of the strange attractor in xyz, xyw, zyw and zxw space are shown in Fig. 2.18.

Although this is a very simple four-dimensional nonlinear system, it is difficult to visualise its attractor. One method of representing motion in 4-space on a plane is by representing a point in 4-space by a line on the plane, where each end of the line represents a pair of coordinates [17]: this

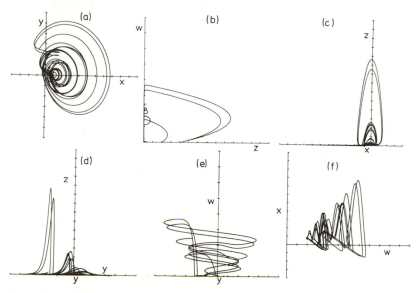

Fig. 2.16 Two-dimensional views of the irregular motion of (2.6) in (a) the xy, (b) zw, (c) xz, (d) yz, (e) yw and (f) wx planes. The axes are: x from −110 to 40, y from −60 to 60, z from 0 to 280, and w from 0 to 170.

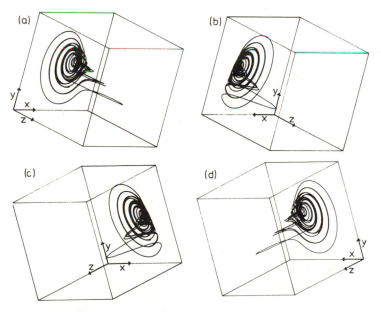

Fig. 2.17 Three-dimensional views of the attractor of 2.16 in xyz space. The attractor is within the coordinate cube, which is tilted forward and then rotated around the y axis by (a) 45°, (b) 135°, (c) −135°, and (d) −45°. Same axis range as in Fig. 2.16.

is used in Fig. 2.19 to illustrate the strange attractor of Fig. 2.16.

2.4 Forced nonlinear oscillators

2.4.1 *The forced Duffing's equation*

Ueda [37, 38] has presented an extensive gallery of periodic and chaotic motions of a forced oscillator with a cubic term:

$$(2.7) \qquad d^2x/dt^2 + adx/dt + x^3 = b\cos(t)$$

where the parameters a and b are positive and $a < 1$ and $b < 25$. There is a rich variety of 2π-periodic (harmonic and higher harmonic) and subharmonic solutions that map out periodic attractors: a few periodic attractors are illustrated in Fig. 2.20, where the system is rewritten as

$$(2.8) \qquad \begin{aligned} dx/dt &= y \\ dy/dt &= -ay - x^3 + b\cos(t) \end{aligned}$$

and the solutions (after a long integration time, so all transients have decayed) are plotted in the xy plane. For some combinations of a and b there is more than one possible stable solution: two different periodic solutions, obtained with the same a and b, but different initial conditions, are seen in Figs. 2.20d and e. Each stable solution is associated with a different attractor, with its own basin of attraction.

Since, for a given pair of parameters (a, b), there can be more than one attractor, a small change in the initial condition or one of the parameters can switch the solution between attractors.

A few of the changes from periodic to chaotic, and between periodic, attractors produced by small increases in b are shown in Fig. 2.21. The bifurcation patterns in the (a, b) parameter space have been plotted for $0 < b < 250$ [30], and as b is increased the same pattern of bifurcations into non-2π-periodic solutions (period doubling, chaos) is repeatedly seen. For small values of a, as b is increased, a periodic solution bifurcates into a pair of symmetric solutions, which then undergo a period-doubling cascade into a chaotic solution. A further increase in b gives a return to a symmetric pair of periodic solutions that undergo period doubling into chaos, which then leads into a new periodic solution. This process is then repeated. At large values of a, the chaotic solution returns directly to a single periodic solution, which then splits into a symmetric pair of solutions that undergo a period-doubling cascade back into chaos.

A related system is:

$$(2.9) \qquad d^2x/dt^2 + adx/dt + a_1x + x^3 = b\cos(t)$$

When a_1 is positive, this can represent the behaviour of a charged particle in a periodic field [16], and when a_1 is negative, a buckled beam undergoing forced lateral vibrations [15].

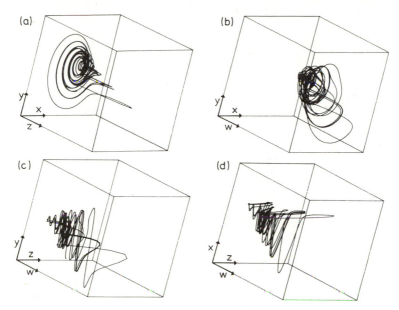

Fig. 2.18 Three-dimensional views of the attractor of Figure 2.16 in (a) *xyz* (b) *xyw* (c) *zyw* and (d) *zxw* space. Same axis range as in Figure 2.16.

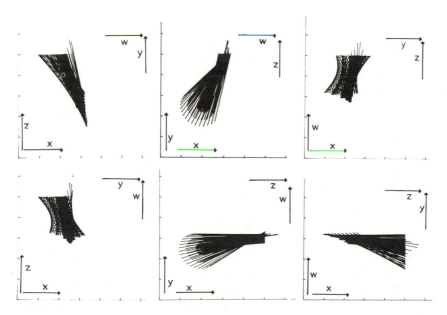

Fig. 2.19 Different views of attractor in 4-space: each line represents a point, joining two pairs of coordinates. The axes ranges have been normalised, so that the motion is within a square representing a hypercube.

2.4.2 *Forced negative resistance oscillator*

The system

(2.10)
$$dx/dt = y$$
$$dy/dt = a(1 - x^2)y - x^3 + b\cos(ft)$$

was derived by Ueda and Akamatsu in 1981 [39] from the differential equation representing a sinusoidally forced nonlinear electronic oscillator. Unlike the Duffing system, this system is oscillatory in the absence of any forcing. When $b > 0$ the oscillations may be entrained by the driving sinusoid, with a periodic pattern of **m** oscillations occurring every **n** cycles of the driving sinusoid, where **m** and **n** are simple integers.

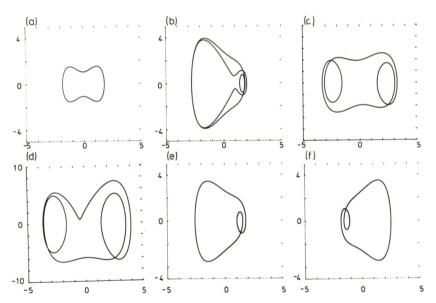

Fig. 2.20 Periodic solutions of eqn (2.8) with (a) $a = 0.3$, $b = 2.0$, (b) $a = 0.2$, $b = 5.0$, (c) $a = 0.2$, $b = 16.5$, (d) $a = 0.2$, $b = 23.5$, and (e) and (f) $a = 0.10$, $b = 3.5$, but with different initial conditions. These figures are plotted after $t = 200$ to allow all transients to decay.

Some combinations of b and f fail to entrain the system; the response is either quasi-periodic, or is irregular and chaotic. Quasi-periodic oscillations may be produced for all f when b is sufficiently small. For some pairs of b and f the motion is irregular and chaotic. Figure 2.22 shows periodic, quasi-periodic and chaotic motions in the y–x plane; the motion is trapped on the periodic, quasi-periodic and strange attractors. Ueda and Akamatsu present a different view of the attractors by using stroboscopic portraits: the trajectory is sampled at a fixed phase of each driving cycle. In such a stroboscopic portrait a periodic attractor appears as a small number of points, a quasi-periodic attractor appears as a closed curve, and a strange attractor appears as an intricate structure.

2.4.3 *The forced Brusselator*

The Brusselator is a formal set of chemical reactions:

$$A \rightarrow X$$
$$B + X \rightarrow Y + D$$
$$2X + Y \rightarrow 3X$$
$$X \rightarrow E$$

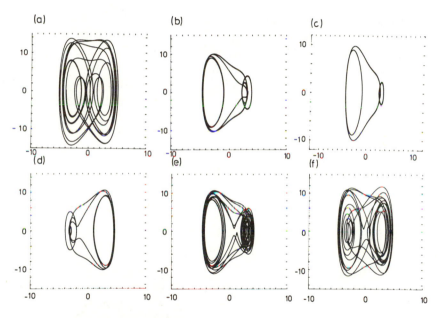

Fig. 2.21 Chaotic and periodic solutions, with $x_0 = 0.3$, $y_0 = 0.5$, $a = 0.3$, and $b =$ (a) 30.75, (b) 31.0, (c) 31.0, (d) 32, (e) 33, (f) 34.0.

in a spatially homogeneous system, where the inverse reactions are ignored and the initial and final reactant concentrations A, B, D and E are maintained at set values. This system was introduced by Prigogine and Lefever in 1968 [22] as an abstract model of an autocatalytic, none-quilibrium system: the trimolecular mechanism is an implausible but convenient way of introducing nonlinearity. When all the kinetic constants are equal to one, the chemical kinetic equations are:

(2.11)
$$dx/dt = A + x^2 y - Bx - x$$
$$dy/dt = Bx - x^2 y$$

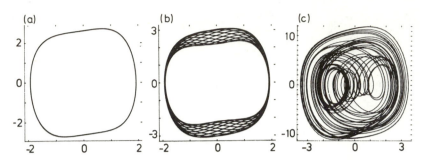

Fig. 2.22 Attractors for the forced, negative-resistance oscillator. (a) A simple periodic attractor: $b = 0, f = 1.617$; (b) a quasi-periodic attractor: $b = 1.0, f = 4.0$; and (c) a strange attractor, associated with chaotic motion: $b = 17, f = 4.0$.

which has a single steady-state solution that is unstable when $B > A^2 + 1$, when there is a stable limit cycle. Thus this system may be considered to represent a formal chemical oscillator, with x and y representing the concentrations of X and Y [18].

Tomita and Kai [35] added a sinusoidal forcing term of amplitude a and frequency f:

$$(2.12) \qquad \begin{aligned} dx/dt &= A + x^2 y - Bx - x + a\cos(ft) \\ dy/dt &= Bx - x^2 y \end{aligned}$$

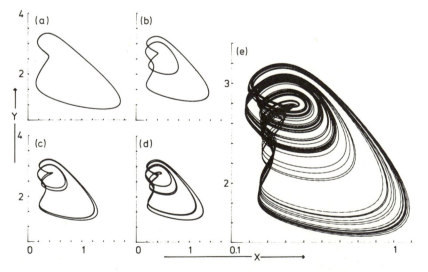

Fig. 2.23 Periodic and chaotic attractors for the sinusoidally forced Brusellator, with $A = 0.4$, $B = 1.2$, and $a = 0.05$. A sequence of period doublings with $f =$ (a) 0.6, (b) 0.8, (c) 0.83, (d) 0.84; (e) a chaotic attractor, with $f = 0.95$.

to give the possibility of quasi-periodic and chaotic solutions. Quasi-periodic solutions were found at sufficiently small a, and chaotic solutions developed after a sequence of period doublings as f is increased when $a = 0.05$ (see Fig. 2.23).

2.4.4 *The glycolytic oscillator*

Oscillations can occur in all the metabolites of the glycolytic pathway, in which the oxidation of one molecule of glucose leads to the production of two molecules of ATP, by a complicated sequence of enzyme-catalysed reactions. Although a realistic model is quite complicated [9], a simple model of the form:

(2.13)
$$dx/dt = -xy^2$$
$$dy/dt = xy^2 - y$$

reproduces many of the features of the glycolytic oscillations [11].

The addition of a sinusoidal forcing term $f(t) = a + b \cos (\omega t)$, to give

(2.14)
$$dx/dt = f(t) - xy^2$$
$$dy/dt = xy^2 - y$$

gives, with forcing ($a < 1$ but close to 1, here $a = 0.999$, $b = 0.42$), periodic, quasi-periodic and chaotic solutions as the frequency ω is changed [34]—see Fig. 2.24.

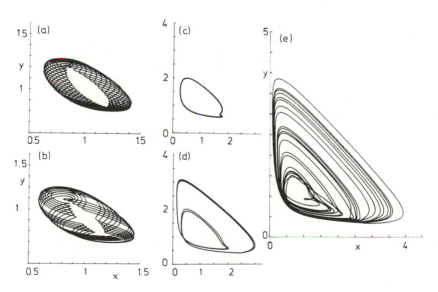

Fig. 2.24 Quasiperiodicity (a,b), period-doubling (c,d), and chaos (e) in eqns (2.14) for the frequency ω of the forcing function (a) 4.5, (b) 3.5, (c) 2.0, (d) 1.8, and (e) 1.75.

References

[1] Bullard, E. C. Reversals of the earth's magnetic field. *Phil. Trans. Roy. Soc. A*, **263**, 481–524 (1968).
[2] Bullard, E. C. The disk dynamo. In *Topics in Nonlinear Dynamics, a Tribute to Sir Edward Bullard*, ed. S. Jorna. *American Institute of Physics Conference Proceedings* **46**, 373–89 (1978).
[3] Chillingworth, D. R. J. and Holmes, R. J. Dynamical systems and models for reversals of the Earth's magnetic field. *Math. Geol.* **12**, 41–59 (1980).
[4] Eckman, J. -P. Roads to turbulence in dissipative dynamical systems. *Rev. Mod. Phys.* **53**, 643–54 (1981).
[5] Epstein, I. R. Oscillations and chaos in chemical systems. *Physica 7-D* 47–56 (1983).
[6] Garrido, I. and Simo, C. Some ideas about strange attractors. *Lecture Notes in Physics* **179**, ed. L. Garrido, pp. 1–18. Springer, Berlin (1983).
[7] GHOST graphical output system. United Kingdom Atomic Energy Authority Culham Laboratory, Oxford OX14 3DB (1983).
[8] GINO-F. CAD Centre Ltd, Cambridge CB3 0HB (1976).
[9] Goldbeter, A. and Lefever, R. Dissipative structures for an allosteric model. *Biophys. J.* **12**, 1302–15 (1972).
[10] Grebogi, C., Ott, E., Pelikan, S. and Yorke, J. A. Strange attractors that are not chaotic. *Physica-D* **13**, 261–8 (1984).
[11] Higgins, J. A chemical mechanism for oscillation of glycolytic intermediates in yeast cells. *Proc. natl Acad. Sci. (USA)* **51**, 989–94 (1964).
[12] Hodgkin, A. L. and Huxley, A. F. A quantitative description of membrane current and its application to conduction and excitation in nerve. *J.Physiol.* **117**, 500–44 (1952).
[13] Holden, A. V. and Muhamad, M. A. The identification of deterministic chaos in the activity of single neurones. *J. Electrophysiol. Tech.* **11**, 135–47 (1984).
[14] Holmes, P. 'Strange' phenomena in dynamical systems and their physical implications. *Appl. Math. Modelling* **1**, 362–6 (1977).
[15] Holmes, P. A nonlinear oscillator with a strange attractor. *Phil. Trans. Roy. Soc. (London)* **292A**, 419–48 (1979).
[16] Huberman, B. A. and Crutchfield, J. P. Chaotic states of anharmonic systems in periodic fields. *Phys. Rev. Lett.* **43**, 1743–7 (1979).
[17] Lábos, E. and Turcsányi, B. On the reversible and irreversible representations of motions in R_n to R_2. *Physica-16D*, 124–32 (1985).
[18] Lefever, R. and Nicolis, G. Chemical instabilities and sustained oscillations. *J. Theor. Biol.* **30**, 267–84 (1971).
[19] Lorenz, E. N. Deterministic nonperiodic flow. *J. Atmos. Sci.* **20**, 130–41 (1963).
[20] Ott, E. Strange attractors and chaotic motions of dynamical systems. *Rev. Mod. Phys.* **53**, 655–71 (1981).
[21] Packhard, N. H., Crutchfield, J. P., Farmer, J. D. and Shaw, R. S. Geometry from a time series. *Phys. Rev. Lett.* **45**, 712–16 (1980).
[22] Prigogine, I. and Lefever, R. Symmetry breaking instabilities in dissipative systems. *J. Chem. Phys.* 1695–1700 (1968).
[23] Rössler, O. E. An equation for continuous chaos. *Phys. Lett.* **57A**, 397 (1976).
[24] Rössler, O. E. An equation for hyperchaos. *Phys. Lett.* **71A**, 155–7 (1979).
[25] Rössler, O. E. Chaotic oscillations: an example of hyperchaos. *AMS Lectures in Applied Maths* **17**, 141–55 (1979).
[26] Rössler, O. E. The chaotic hierarchy. *Z. Naturforsch.* **38a**, 788–801 (1983).
[27] Ruelle, D. Strange attractors. *Math. Intell.* **2**, 126–37 (1980).

[28] Ruelle, D. Small random perturbations of dynamical systems and the definition of attractors. *Comm. Math. Phys.* **82**, 137–51 (1981).

[29] Ruelle, D. and Takens, F. On the nature of turbulence. *Comm. Math. Phys.* **20**, 167–92 (1971); see also: Note concerning our paper 'On the nature of turbulence' *Comm. Math. Phys.* **23**, 343–4 (1971).

[30] Sato, S-I. Sano, M. and Sawada, Y. Universal scaling property in bifurcation structure of Duffing's and of generalized Duffing's equation. *Phys. Rev. A* **28**, 1654–8 (1983).

[31] Shaw, R. Strange attractors, chaos and information flows. *Z. Naturforsch.* **36a**, 80 (1981).

[32] Sparrow, C. *The Lorenz Equations: Bifurcations, Chaos and Strange Attractors.* Springer, Berlin, 269 pp. (1982).

[33] Takens, F. Detecting strange attractors in turbulence. *Lecture Notes in Mathematics* **898**, eds D. A. Rand and L. S. Young, pp. 366–81, Springer, Berlin (1981).

[34] Tomita, K. and Daido, H. Possibility of chaotic behaviour and multi-basins in forced glycolytic oscillator. *Phys. Lett.* **79A**, 133–7 (1980).

[35] Tomita, K. and Kai, T. Stroboscopic phase portrait and strange attractors. *Phys. Lett.* **66A**, 91–3 (1978).

[36] Toth, J. and Hars, V. Orthogonal transforms of the Lorenz and Rössler equations. *Physica-D* in press (1985).

[37] Ueda, Y. Steady motions exhibited by Duffing's equation: a picture book of regular and chaotic motions. In *New Approaches to Nonlinear Problems in Dynamics,* ed. P. Holmes, pp. 311–22. SIAM, Philadelphia, Pa. (1979).

[38] Ueda, Y. Explosion of strange attractors exhibited by Duffing's equation. *Ann. NY Acad. Sci.* **357**, 422–34 (1980).

[39] Ueda, Y. and Akamatsu, N. Chaotically transitional phenomena in the forced negative resistance oscillator. *IEEE Trans. CS* **28**, 217–24 (1981).

[40] Willamowski, K.-D. and Rössler, O. E. Irregular oscillations in a realistic abstract quadratic mass action system. *Z. Naturforsch.* **35a**, 317–18 (1980).

Part II

Iterations

3
One-dimensional iterative maps

H. A. Lauwerier

Mathematical Institute, University of Amsterdam,
Roetersstraat 15, 1018 WB Amsterdam, The Netherlands

3.1 Introduction

In this chapter a survey is given of the main properties of one-dimensional maps. My approach is that of an applied mathematician and I have adopted a somewhat informal style. I have tried to give the reader a better understanding of the sometimes very complicated regular and irregular behaviour of discrete dynamical systems, not by stating and proving theorems in endless succession, but by illustrating the fundamental ideas using simple worked-out cases. In the next chapter two-dimensional systems are considered in a similar way. Both chapters share a common bibliography of selected books and papers. Our models are mainly drawn from population dynamics. Much attention is given to one-dimensional and two-dimensional quadratic mappings, as they exhibit almost all the interesting properties a map can have. The most important papers with a similar approach are those by Hénon [24–26] and May [44–46]. A good survey with an emphasis on theoretical aspects was given recently by Whitley [55]. The book by Gumowski and Mira [22] is mainly concerned with area-preserving two-dimensional maps and contains a wealth of experimental results. The book by Iooss [27] is more directed to the theoretical and technical aspects of the bifurcation of two-dimensional maps. The monograph of Collet and Eckmann [11] offers a variety of interesting details, theoretical as well as experimental. Perhaps the best source is the proceedings of the Les Houches Summer School edited by Iooss *et al.* and published in 1983 [28].

I consider iterations of the one-dimensional map

(3.1) $$T : x \rightarrow f(x)$$

sometimes written as

$$x' = f(x)$$

or in an iterative way as

$$x_{n+1} = f(x_n)$$

Unless stated otherwise, $f(x)$ is considered sufficiently smooth and it is assumed that the range of $f(x)$ is contained in its domain. The m-fold iterate of $f(x)$ is denoted by $f^m(x)$ to be distinguished from the mth power $(f(x))^m$.

A point ξ for which $T\xi = \xi$, i.e

(3.2) $f(\xi) = \xi$

is called a *fixed point* of T. From any starting point x_0 we may form a (forward) orbit by taking the sequence $\{x_n\}$ of its iterates $x_n = T^n x_0$. The orbit of a fixed point consists of just this single point. The next possibility is a *cycle* or a *periodic orbit* formed by periodic points. A *periodic point* ξ of order m is a fixed point of $f^m(x)$ if m is the lowest natural number for which this is true. Then $T\xi$, $T^2\xi$,..., $T^{m-1}\xi$ are similar periodic points of order m.

The local behaviour in a neighbourhood of a fixed point ξ is determined by its *multiplier*:

(3.3) $\lambda = f'(\xi)$

where f' is the derivative of f. This gives the following possibilities:

$$\begin{aligned}
|\lambda| < 1 \quad &\text{attracting or stable} \\
\lambda = 0 \quad &\text{superstable} \\
|\lambda| > 1 \quad &\text{repelling or unstable} \\
|\lambda| = 1 \quad &\text{neutral}
\end{aligned}$$

For a periodic orbit we have a similar classification. Let $x_0, x_1 ,..., x_{m-1}$ be the members of an m-cycle. Then all those points are fixed points of $f^m(x)$ and with respect to that map they have the same multiplier

(3.4) $\lambda = f'(x_0)f'(x_1)f'(x_2)...f'(x_{m-1})$

It is important to note that a periodic orbit which contains a critical point, i.e. a point x_0 for which $Df(x_0) = 0$, is superstable.

For the orbit of a starting point x_0 we have the following possibilities:

(a) x_0 is a fixed point.

(b) x_0 is a periodic point.

(c) x_0 is an *eventually periodic point*. This means that x_0 is a pre-image of some order of a periodic point, or in other words that, for some integer m, $T^m x_0$ is a periodic point.

(d) x_0 is an *asymptotically periodic point*. This means that the orbit contains a subsequence converging to a stable periodic point.

(e) x_0 is an *aperiodic point* if it is not of the previous types. The orbit is called aperiodic, stochastic or chaotic.

3.2 The elementary bifurcations

We often consider a family of maps:

(3.5) $$T : x \to f(\mu, x)$$

where μ is a real parameter. Then we can study how the properties of T change when μ varies. When we fix our attention to a particular fixed point $\xi(\mu)$, its nature may change from stable to unstable and this may happen in two ways according to the sign of its multiplier $\lambda(\mu)$. Further fixed points may suddenly appear or disappear when μ passes a certain value. This gives three types of bifurcation, which will be described below. Each type is illustrated by a representative example.

3.2.1 *Transcritical bifurcation*

(3.6) $$x_{n+1} = (1 + \mu)x_n + x_n^2$$

The fixed point of interest is always $x = 0$. The multiplier $\lambda = 1 + \mu$ is close to one, so μ is considered a small bifurcation parameter. For $\mu > 0$ the fixed point is unstable, for $\mu < 0$ (of course $\mu > -2$) it is stable. There is a second fixed point, $x = -\mu$. Its multiplier is $\lambda = 1 - \mu$. It is unstable for $\mu < 0$ and stable for $\mu > 0$ (with $\mu < 2$). This can be pictured in the so-called bifurcation diagram of Fig. 3.1.

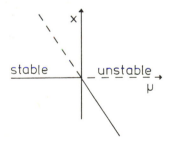

Fig. 3.1 Transcritical bifurcation.

3.2.2 *Flip bifurcation*

(3.7) $$x_{n+1} = -(1 + \mu)x_n + x_n^3$$

The trivial fixed point $x = 0$ loses stability at $\mu = 0$. There are two other fixed points but since they are not close to $x = 0$ they are of no concern to us. However, there is the 2-cycle $\pm \sqrt{\mu}$ which comes into existence for $\mu > 0$. Its multiplier is $\lambda = f'(\sqrt{\mu})f'(-\sqrt{\mu}) = (1 - 2\mu)^2$, which means stability if μ is sufficiently small. The bifurcation diagram is shown in Fig. 3.2.

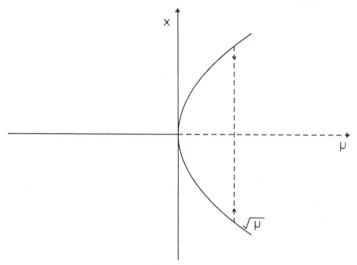

Fig. 3.2 Flip bifurcation.

3.2.3 *Fold bifurcation*

(3.8) $$x_{n+1} = \mu + x_n - x_n^2$$

For $\mu < 0$ there is no fixed point at all. For $\mu > 0$ we have the two fixed points $\pm \sqrt{\mu}$. The corresponding multipliers are $1 \mp 2\sqrt{\mu}$. This means that when μ passes the bifurcation point $\mu = 0$, a pair of fixed points are created, one stable and the other unstable. The bifurcation diagram is given in Fig. 3.3.

The three examples illustrating the various types of bifurcation are generic and can be considered as normal forms with only the lowest nonlinear term being present. Any map showing flip bifurcation for $x = 0$ and $\mu = 0$ can be written as

$$x_{n+1} = -(1 + \mu)x_n + \sum_{k=2}^{\infty} a_k x_n^k$$

According to the theory of normal forms, there exists a coordinate transformation which removes all even powers. The odd powers are in resonance with $\lambda = -1$ and cannot be removed.

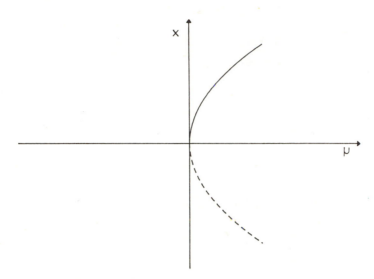

Fig. 3.3 Fold bifurcation.

3.3 The logistic map

In 1976, R. M. May wrote an exciting review article [44] in which he showed that the simplest possible nonlinear iterative process

$$(3.9) \qquad x_{n+1} = ax_n(1 - x_n) \, , \, 0 < a \leqslant 4$$

already has a very complicated dynamical behaviour. Many mathematicians were inspired by this, and since then a continuous stream of interesting papers on iterative maps is flooding the mathematical world. If (3.9) is considered as a model in population dynamics where x_n measures the relative number of individuals of the nth generation, a should be restricted to the real interval $(0,4]$. But pure mathematicians prefer with good reason the generalisation of (3.9) to the complex domain: see section 4.5. A very good recent survey is given in Blanchard [9].

The elementary bifurcation behaviour of (3.9) has been illustrated in Fig. 3.4. The fixed points and their multipliers are $x = 0$, $\lambda = a$ and $x = 1 - 1/a$, $\lambda = 2 - a$. At $a = 1$ we observe transcritical bifurcation with exchange of stability. At $a = 3$ the stable branch loses stability according to flip bifurcation. A 2-cycle p,q is born for which

$$q = ap(1 - p) \, , \, p = aq(1 - q)$$

An elementary calculation shows that

$$p + q = (a + 1)/a \, , \, pq = (a + 1)/a^2$$

so that p and q are the roots of

$$a^2x^2 - a(a + 1)x + (a + 1) = 0$$

Their multiplier follows from (3.4) as

$$\lambda = a^2(1-2p)(1-2q) = -a^2 + 2a + 4$$

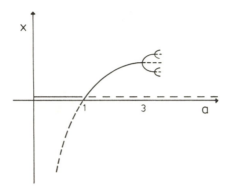

Fig. 3.4 Bifurcation diagram of logistic map up to period doubling.

We note that $\lambda = 1$ for $a = 3$ and $\lambda = -1$ for $a = 1 + \sqrt{6} = 3.449490$. This again means flip bifurcation of both the p-branch and the q-branch. The 2-cycle loses stability and a stable 4-cycle is created. This is merely the beginning of an infinite sequence of flip bifurcations and period doublings. The first few values a_k where a 2^k-cycle is born are collected in the following table:

a_1	3	a_5	3.568759
a_2	3.449499	a_6	3.569692
a_3	3.544090	a_7	3.569891
a_4	3.564407	a_8	3.569934

These values appear to converge to a limit a_∞ in a geometric progression as:

(3.10) $$a_k \approx a_\infty - c\mathscr{F}^{-k}$$

We have

$$a_\infty = 3.569946\ldots$$
$$c = 2.6327\ldots$$
$$\mathscr{F} = 4.669202\ldots$$

Feigenbaum noted in 1975 that this pattern of period doubling is a quite universal phenomenon and — what is most important — that for a very large class of maps the constant $\mathscr{F}$ has the same value. This constant is now named the *Feigenbaum constant*. For practical purposes it gives us the possibility of making a prediction of a_∞ as soon as the first few period-doubling values are known. This holds for any map with a single parameter a. For the present map it works as follows. Take $a_2 = 3.4495$ and $a_3 = 3.5541$ as bifurcation values found by using a pocket calculator. Then, using (3.10), we have

$$a_\infty \approx \frac{\mathscr{F}a_3 - a_2}{\mathscr{F} - 1} = 3.5699$$

which is correct to four decimals.

The interval $(a_\infty, 4)$ contains an infinite number of small windows of a-values for which there exists a stable m-cycle. The first such cycles to appear beyond a_∞ are of even period. Next, odd cycles appear in descending order. The period 3-cycle first appears for $a = 3.828427$ and stays stable up to $a = 3.841499$. These values can easily be obtained by means of a pocket calculator by using the property that the multiplier of the 3-cycle runs from $+ 1$ to $- 1$ in the corresponding a-window. At the end of the 3-window we have flip bifurcation, i.e. the beginning of a stable 6-cycle and of further period doublings. The same is true of course for all other stable cycles of odd order.

Outside the windows there are no stable periodic orbits although there is an infinite number of unstable cycles. The dynamic behaviour of the map is then called chaotic. The most chaotic case, $a = 4$, deserves special attention since in that case the iterative map can be parametrised by an elementary function.

If

(3.11)
$$x_{n+1} = 4x_n(1 - x_n)$$

we may write

$$x_n = \sin^2 (2^n \beta \pi)$$

where $0 \leqslant \beta < 1$. It is helpful to write β in 2-adic form as

$$\beta = 0 \cdot b_1 b_2 b_3 b_4 \ldots$$

Then at each iteration step the foremost binary digit is lost. If x_0 is an arbitrary starting point, then as a rule β is an irrational number with an infinite string of zeros and ones like the tossing of a coin. This means that as a rule the orbits are aperiodic. Periodic orbits, always unstable, are produced by rational β. The first few interesting cases are

$$\beta = \tfrac{1}{3} \qquad \text{fixed point } \tfrac{3}{4}$$
$$\beta = \tfrac{1}{5} \qquad \text{2-cycle } (5 \pm \sqrt{5})/8$$
$$\beta = \tfrac{1}{7} \qquad \text{3-cycle } 0.188, 0.611, 0.950$$
$$\beta = \tfrac{1}{9} \qquad \text{3-cycle } 0.117, 0.413, 0.970$$

These special cases make clear in an almost dramatic way what we may expect from computer experiments. Let us assume that we are working with a personal computer where fractions are given with a precision of 40 binary digits; then, after 40 iteration steps, all memory of the start has been lost completely. Much depends on the way the computer is instructed to fill in vacant binary positions! Although in theory aperiodic behaviour is the rule, in practice all orbits are eventually periodic. By way of illustration we consider the discrete version of (3.11) as produced by a computer with an accuracy of d decimals with round-off

$$x_{n+1} = 10^{-d} \, \mathrm{IP}(4 \, . \, 10^d x_n(1 - x_n) + 0.5)$$

or

$$(3.12) \qquad\qquad y_{n+1} = \mathrm{IP}(4y_n(1 - 10^{-d}y_n) + 0.5)$$

where y_n is an integer and IP means the integer part of a number. A simple computer experiment with $d = 3$ shows that 709 orbits end up in the fixed point $y = 0$, that 216 orbits converge to the 2-cycle $250 \longleftrightarrow 750$, and that the remaining 76 orbits have a common 13-cycle! Yet the computer can be extremely helpful in analysing the dynamic behaviour of an iterative map. One of the first things one should do is to make a global bifurcation map as shown in Fig. 3.5 for the logistic map (3.9). The interval $2.9 \leqslant a \leqslant 3.9$ has been divided in 220 equally spaced intervals, and for each a-value an approximation of the limit set of the orbit of $x_0 = \frac{1}{2}$ is given by plotting the iterates of order 250 up to 400 (cf. [11]).

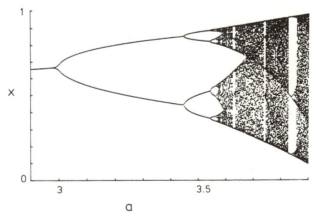

Fig. 3.5 Bifurcation diagram of logistic map for $2.9 < a < 3.9$

3.4 Parametrisation

Let

$$(3.13) \qquad\qquad x_{n+1} = f(x_n)$$

be an iterative map with a repelling fixed point at $x = 0$. If $f(x)$ is an analytic function which is holomorphic at $x = 0$, then the orbit can be parametrised by

$$(3.14) \qquad\qquad x_n = F(a^n c)$$

where $F(z)$ is an analytic function of its complex argument z and where a is the multiplier of the fixed point. The constant c is determined by the start x_0. The existence of such a parametrising function is guaranteed by the following theorem that goes back to Poincaré and has been redis-covered many times thereafter.

Theorem

The functional equation

$$(3.15) \qquad\qquad F(az) = f(f(z))$$

where $f(z)$ is holomorphic at $z = 0$ with $f(0) = 0$, $f'(0) = a$, $|a| > 1$, has a solution $F(z)$ holomorphic at $z = 0$ with $F(0) = 0$. With the additional condition $F'(0) = 1$, the solution is unique. If $f(z)$ is an entire function, then $F(z)$ also is entire.

The equation (3.15) is called the *Poincaré functional equation*, and the unique solution of (3.15) with $F'(0) = 1$ is called the *Poincaré function*. Only in some very special cases is $F(z)$ an elementary function. We list a few cases.

$x \to ax$	$F(z) = z$	$x_n = a^n c$
$x \to 4x(1 - x)$	$F(z) = \sin^2\sqrt{z}$	$x_n = \sin^2(2^n c)$
$x \to 2x(1 - x)$	$F(z) = (e^{2z} - 1)/2$	$x_n = (\exp(2^n c) - 1)/2$
$x \to -2x(1 - x)$	$F(z) = \frac{1}{2} - \cos(\frac{\pi}{3} + \frac{2z}{\sqrt{3}})$	$x_n = \frac{1}{2} - \cos(\frac{\pi}{3} + (-2)^n c)$
$x \to \dfrac{ax}{1 + x}$	$F(z) = \dfrac{(a-1)z}{a - 1 + z}$	$x_n = \dfrac{(a-1)a^n}{a^n + c}$
$x \to \dfrac{2x}{1 + x^2}$	$F(z) = \dfrac{e^{2z} - 1}{e^{2z} + 1}$	$x_n = \tanh(2^n c)$

What can be said in the general situation will be discussed by considering the logistic map with $2 \le a \le 4$. Then the Poincaré function satisfies the equation

$$(3.16) \qquad\qquad F(az) = aF(z)(1 - F(z))$$

It is not difficult to determine the power-series expansion

$$(3.17) \qquad\qquad F(z) = z - c_2 z^2 + c_3 z^3 - c_4 z^4 + \ldots$$

Substitution gives

$$c_2 = \frac{1}{a-1}, c_3 = \frac{2}{(a-1)(a^2-1)}, c_4 = \frac{a+5}{(a-1)(a^2-1)(a^3-1)}$$

In practice $F(z)$ can be calculated for quite large values of z by a clever combination of (3.16) and a few terms of the power-series expansion (3.17). If the power series is sufficiently accurate, say up to $|z| = 1$, then for determining $F(z)$ with $|z| \geqslant 1$ with equal accuracy, we first compute $F(z/a^m)$ where $m \approx \log |z|/a$, and we take m iterations of (3.16). In this case $F(z)$ is an entire function which means that its power-series expansion converges for all z. On the positive real axis, where we write $z = x$, $F(x)$ can be interpreted as the infinitely iterated function $F^\infty(x)$ of the original map $f(x) = ax(1 - x)$. As x increases, $F(x)$ takes all values of $f(x)$ and its iterates in progressive order. Meanwhile x runs to infinity whereas the domain of $f^\infty(x)$ is still the unit interval. The projection of $F(x)$ on the F-axis can be visualised as a rope of infinite length with an infinite number of folds with turning points at the extrema of $F(x)$. The dynamic behaviour of the one-dimensional map is translated into an expanding similarity map $x \rightarrow ax$ along the rope. The resulting regular or chaotic behaviour of the original map is then caused by the interplay of the similarity map along the rope and the pattern of the folds. The special case $a = 4$ discussed before has shown that the two patterns can be in resonance for a countable set of 'rational' orbits, the unstable periodic points of all orders. If a is slightly less than 4, we are still in the chaotic regime of the logistic map and the overall situation will be not much different. However, $F(x)$ is no longer periodic but almost periodic with a not very regular pattern of critical points. In order to give an idea of what $F(x)$ looks like, we give two illustrations. In Fig. 3.6 the regular case $a = 3.56$ is shown, where there is a stable 8-cycle. In Fig. 3.7 the chaotic case $a = 3.9$ is illustrated.

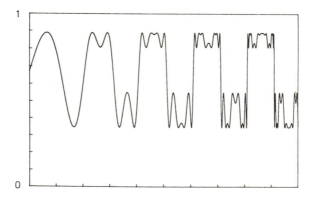

Fig. 3.6 $F(A^T)$ for $0 < T < 10$, with $A = 3.56$. This gives an 8-cycle.

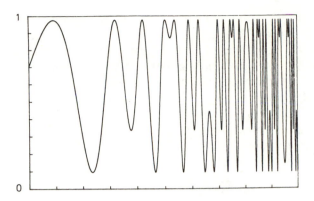

Fig. 3.7 $F(A^T)$ for $0 < T < 7$, with $A = 3.9$. This gives chaos.

The attracting domain of a stable fixed point with a nonvanishing multiplier can also be parametrised by the Poincaré function which satisfies (3.15). In that case the theorem has to be applied to the inverse map. However, there are complications since the inverse map is generally non-unique so that $F(x)$ may exist only in some neighbourhood of $x = 0$. The attracting domain of a superstable fixed point, say $f(0) = f'(0) = 0$ and $f''(0) \neq 0$, can be parametrised by a so-called *Boettcher equation* satisfying the following functional equation

(3.18) $$H(z^2) = f(H(z))$$

with $H(0) = 0$. Again $H(z)$ is holomorphic at $z = 0$. The proof of (3.18) is based on the property that

$$(f^n(x))^{2^{-n}} \to G(x)$$

where $G(x)$ is analytic. A simple observation shows that $G(x)$ satisfies the functional equation

(3.19) $$G(f(z)) = G^2(z)$$

The inverse of $G(z)$ is the *Boettcher function* satisfying (3.18). Using $H(x)$ we obtain a parametrisation of the orbits converging to the superstable fixed point $x = 0$ as

(3.20) $$x_n = H(c^{2n})$$

Example

For the map $x \to x^2/(1 - 2x^2)$ the Boettcher function is $H(z) = z/(1 + z^2)$.

3.5 Period doubling

Period doubling is a quite common phenomenon in one-dimensional and more-dimensional iterative maps. What has been observed for the logistic map is almost true for any other similar map. Feigenbaum [16–18] discovered the astonishing fact that maps of a very large class show the same pattern of period doubling with the same constants of scaling. It is hoped that the following qualitative description will give a better under-standing of the universality of the period doubling. It is based on the original papers of Feigenbaum, and contributions by Helleman and May.

We consider an arbitrary map $x \to f(\mu, x)$ where the fixed point $x = 0$ is subjected to flip bifurcation for $\mu = 0$. If all terms of the third and higher orders are omitted, we may write the map as (cf. 3.7):

$$(3.21) \qquad x_{n+1} = -(1 + \mu)x_n + x_n^2$$

The existence of a 2-cycle p, q requires that $q = -(1 + \mu)p + p^2$, $p = -(1 + \mu)q + q^2$. A simple calculation shows that $p + q = \mu$, $pq = -\mu$, so that

$$(3.22) \qquad 2p = \mu + \sqrt{\mu^2 + 4\mu}, \ 2q = \mu - \sqrt{\mu^2 + 4\mu}$$

The multiplier of p is

$$(3.23) \qquad (2p - 1 - \mu)(2q - 1 - \mu) = 1 - 4\mu - \mu^2$$

The iterated map is explicitly

$$(3.24) \qquad x_{n+1} = (1 + \mu)^2 x_n + (\mu + \mu^2)x_n^2 - 2(1 + \mu)x_n^3 + x_n^4$$

For this map $x = p$ is an ordinary fixed point. The idea is to shift the origin to $x = p$ and to 'renormalise' the resulting map into the form (3.21). The substitution $x_n = p + y_n$ gives

$$y_{n+1} = (1 - 4\mu - \mu^2)y_n + Cy_n^2 + \ldots$$

with

$$C = 4\mu + \mu^2 - 3\sqrt{\mu^2 + 4\mu}$$

Using C as a rescaling factor, we bring this map in the form (3.21) as

$$x_{n+1} = -(1 + \mu_{n+1})x_n + x_n^2 \ldots$$

with

$$x = Cy \text{ and}$$

$$(3.25) \qquad \mu_{n+1} = \mu_n^2 + 4\mu_n - 2$$

If the higher-order terms are omitted, the same argument can be repeated *ad infinitum*.

Each time, μ is stepped up a little according to the map (3.25). The limit value is

$$\mu_\infty = \tfrac{1}{2}(-3 + \sqrt{17}) = 0.56$$

Its multiplier is $2\mu_\infty + 4$ or $1 + \sqrt{17} = 5.12$, which is an analytic approximation to Feigenbaum's constant, 4.67. If this value is substituted in the expression of C, we find

$$\frac{1 + \sqrt{17}}{2} - 3\sqrt{\frac{1 + \sqrt{17}}{2}} = -2.24$$

as an approximation to Feigenbaum's universal scaling factor $\alpha = -2.50$.

The mechanism of period doubling can also be made clear in a graphical way by starting from the map $T : x \rightarrow f(\mu, x)$ for which $x = 0$ is a superstable periodic point of order 2 as shown in Fig. 3.8a for the bifurcation value $\mu = \mu_1$. Geometrically this means that OABC is a square. The next illustration shows the iterate T^2. If μ is increased a bit,

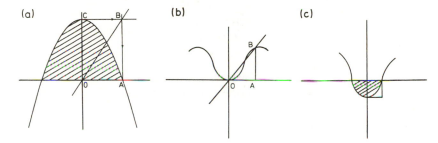

(a)   (b)   (c)

Fig. 3.8 Three stages of period doubling and renormalisation.

it can be imagined that the graph of T^2 is shifted downwards, and slightly deformed, until the negative part of T^2 is similar to the positive part of T in Fig. 3.8a. We assume that, for the value $\mu = \mu_2$, after a flip bifurcation, $x = 0$ is again superstable and of period 2. With a scaling factor α, the latter map is 'renormalised' into the shape of Fig. 3.8a:

$$\alpha T^2(\mu_2, x/\alpha) \approx T(\mu_1, x)$$

Repetition of the argument gives a sequence of μ-values for which $x = 0$ is a superstable fixed point of order 2^n at $\mu = \mu_n$. The limit value $\mu = \mu_\infty$ is again approached in a geometric progression with the Feigenbaum ratio of $\mathcal{F} = 4.66920\ldots$, and the universal scaling factor $\alpha = -2.50291\ldots$. At the limit $T(\mu, x)$ has become a self-similar map satisfying $\alpha T^2(x/\alpha) = T(x)$, without any change in μ. This relation can be written as

(3.26) $$g(x) = \alpha g(g(x/\alpha)) , \; g(0) = 1$$

a quite remarkable functional equation discovered by Feigenbaum. It has been the object of extensive studies in the last few years.

This relation shows that $g(x)$ may be considered as a fixed point in function space. Feigenbaum's constant appears to be an eigenvalue of its local linearisation, which gives an independent way of determining its numerical value. It enabled Feigenbaum to obtain the following approximations (in November, 1975)

$$\mathcal{F} = 4.669\ 201\ 609\ 102\ 990\ 9\ldots$$
$$\alpha = -\ 2.502\ 907\ 875\ 509\ 589\ 284\ldots$$

It turns out that (3.26) has the following analytical solution

(3.27) $g(z) = 1 + c_1z^2 + c_2z^4 + c_3z^6 + c_4z^8 + c_5z^{10} + c_6z^{12} + \ldots$

with

$$
\begin{aligned}
c_1 &= -\ 1.527\ 633\ldots \\
c_2 &= \ 0.104\ 815\ldots \\
c_3 &= \ 0.026\ 706\ldots \\
c_4 &= -\ 0.003\ 527\ldots \\
c_5 &= \ 0.000\ 082\ldots \\
c_6 &= \ 0.000\ 025\ldots
\end{aligned}
$$

This approximation is easy to find with the aid of a computer. It gives $\alpha = 1/g(1)$ correct to 4 decimals after the decimal point.

3.6 Stochastic properties

When the iterative map $T : x \rightarrow f(x)$ has no stable periodic orbits, at least not in the relevant domain, the dynamic behaviour is loosely called chaotic. Chaotic motion can be characterised in many, sometimes very subtle, ways. In practical situations the invariant distribution $P(x)$ and the Lyapunov exponent σ are the most useful concepts. In order to simplify the discussion we consider the case of a unimodal map with $0 \leqslant x \leqslant 1$ as shown in Fig. 3.9. Then for x there are at most two pre-images, y and z. The *invariant distribution* $P(x)$ can be interpreted as the probability $P(x)dx$ of finding a value $f(x)$ in the interval $x, x + dx$. Invariance means that

(3.28) $$P(x) = P(T^{-1}x)$$

with

$$\int P(x)dx = 1$$

The probability of finding x in $x, x + dx$ equals the sum of the probabilities of finding the pre-images y and z in corresponding intervals as shown in Fig. 3.9. Thus we have

$$P(x)dx = P(y)dy + P(z)dz$$

with

$$x = f(y) = f(z)$$

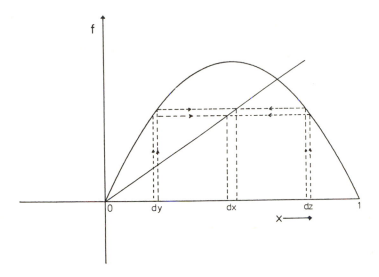

Fig. 3.9 The construction of the invariant distribution.

This gives a functional equation, sometimes called the Perron–Frobenius equation:

$$(3.29) \qquad P(x) = \frac{P(y)}{|f'(y)|} + \frac{P(z)}{|f'(z)|}$$

This equation can rarely be solved analytically. However, in practice it can be solved numerically in an iterative way by using an approximation of $P(x)$ on the right-hand side. This should then lead to a better approximation. The invariant distribution is not always a classical function but it can be a generalised function or a superposition of a classical function and Dirac functions. According to a theorem by Lasota and Yorke, $P(x)$ is absolutely continuous if $f(x)$ is everywhere expanding, i.e. $|P'(x)| > \alpha > 1$ with a few exceptions. The simplest case is the so-called tent map of Fig. 3.10.

$$(3.30) \qquad f(x) = \begin{cases} 2x & 0 \leqslant x \leqslant \tfrac{1}{2} \\ 2 - 2x & \tfrac{1}{2} < x \leqslant 1 \end{cases}$$

The functional equation (3.29) is here

$$2P(x) = P(x/2) + P(1-x/2)$$

Its solution is simply the usual Lebesgue measure: $P(x) = 1$.

The apparent regularity of the dynamic behaviour of the tent map is also observed when one tries to determine the pre-images of an arbitrary point, $x = 0 \cdot b_1b_2b_3 \ldots$, in binary notation. If $\bar{b}_k = 1 - b_k$, the

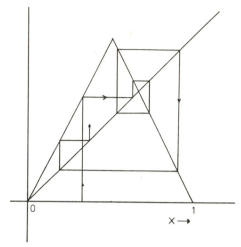

Fig. 3.10 The tent map with a stochastic orbit.

complementary binary digit, the two pre-images are

$$T^{-1}x \quad \begin{cases} 0 \cdot 0\, b_1 b_2 b_3 \ldots \\ 0 \cdot 1\, \bar{b}_1 \bar{b}_2 \bar{b}_3 \ldots \end{cases}$$

and next

$$T^{-2}x \quad \begin{cases} 0 \cdot 00 b_1 b_2 b_3 \ldots \\ 0 \cdot 01 \bar{b}_1 \bar{b}_2 \bar{b}_3 \ldots \\ 0 \cdot 10 b_1 b_2 b_3 \ldots \\ 0 \cdot 11 \bar{b}_1 \bar{b}_2 \bar{b}_3 \ldots \end{cases}$$

and so on. Clearly the 2^n pre-images $T^{-n}x$ are distributed uniformly on the unit interval in a very regular way.

The next simple case is the logistic map with $a = 4$:

$$x_{n+1} = 4x_n(1 - x_n)$$

The functional equation (3.29) becomes

$$P(x) = \frac{P(\tfrac{1}{2} - \tfrac{1}{2}\sqrt{1-x}) + P(\tfrac{1}{2} + \tfrac{1}{2}\sqrt{1-x})}{4\sqrt{1-x}}$$

This equation already shows some aspects of what can be expected in the general case. The right-hand side becomes infinite at $x = 1$ so that $P(x) \approx c(1 - x)^{-\frac{1}{2}}$ close to $x = 1$. In view of the symmetry $x \longleftrightarrow 1 - x$, we also have $P(x) \approx cx^{-\frac{1}{2}}$ close to $x = 0$. However, we need not worry since there is a better way of finding the invariant distribution. The transformation $x = \sin^2(\pi\xi/2)$ relates the logistic map to the tent map (3.30) with ξ written for x. Thus we have

$$P(x)dx = P(\xi)d\xi = d\xi$$

so that

$$P(x) = d\xi/dx = \frac{2}{\pi \sin \pi\xi}$$

or

(3.31)
$$P(x) = \frac{1}{\pi \sqrt{x(1-x)}}$$

An attempt to determine numerically the invariant distribution of the general logistic map $x_{n+1} = ax_n(1 - x_n)$ in the chaotic regime of a runs into difficulties since $P(x)$ may have Dirac-function components. However, if we are satisfied with a low precision, a good approximation can readily be obtained in a Monte Carlo way by dividing the unit interval in, say, a hundred equal subintervals and counting the number of times the elements of a few random orbits fall into a specific subinterval. The Perron–Frobenius equation can be used in an iterative way to determine the cumulative distribution function. Figure 3.11 shows the result for $a = 3.7$ after 20 iterations.

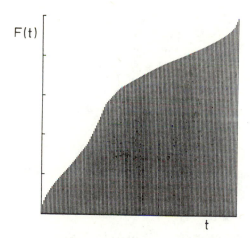

Fig. 3.11 The invariant distribution of the logistic map for $a = 3.7$.

Let $T : x \to f(x)$ be a given map. In order to facilitate the discussion, we assume that T transforms the unit interval into itself and that normally the orbits are aperiodic. Then for a given $\varphi(x)$ we may define the time mean

(3.32)
$$\langle \varphi(x) \rangle = \lim_{n \to \infty} \frac{1}{n} \sum_{k=0}^{n-1} f(Tx)$$

and the space mean

(3.33)
$$\langle \varphi(x) \rangle_s = \int f(x) d\mu$$

where

(3.34)
$$d\mu = P(x)dx$$

is the invariant measure.

For so-called ergodic systems the space mean almost always equals the time mean. In principle the time mean depends on the special choice of the starting point and is an invariant function for which

$$\langle \varphi(Tx) \rangle = \langle \varphi(x) \rangle$$

Ergodicity means that the general orbit of T is aperiodic and gives a dense covering of the basic interval.

Perhaps the most important application is the time mean of the logarithmic slope, i.e. of

$$\varphi(x) = \log \left| \frac{df}{dx} \right|$$

the so-called *Lyapunov exponent*, σ:

(3.35)
$$\sigma = \lim_{n\to\infty} \frac{1}{n} \sum_{k=0}^{n-1} \log \left| \frac{df}{dx} \right|$$

Except for a set of measure zero, σ is independent of the initial value. For $\sigma > 0$ we have a chaotic orbit; for $\sigma < 0$ we have an eventually periodic orbit. Assuming ergodicity, the Lyapunov exponent can also be found from

(3.36)
$$\sigma = \int \log | df/dx | \, d\mu$$

if the invariant measure $d\mu = P(x)dx$ is known. As an illustration we consider again the logistic map

$$x_{n+1} = 4x_n(1 - x_n)$$

for which the invariant measure is given by (3.31).

Using (3.36) we find

$$\sigma = \frac{1}{\pi} \int_0^1 \frac{\log |4(1 - 2x)|}{\sqrt{x(1 - x)}}$$

$$= \int_0^1 \log (4 \cos(\pi\xi/2)) d\xi = \log 2$$

Figure 3.12 gives an idea of the Lyapunov exponent as a function of a in the interval (3.6,4). However, due to the restricted degree of accuracy, the fine details seen in Figure 13.2, such as the presence of small parameter windows with a stable period of low order, are missed. The most striking feature is the dip at $a = 3.83$ for which there is a stable 3-cycle.

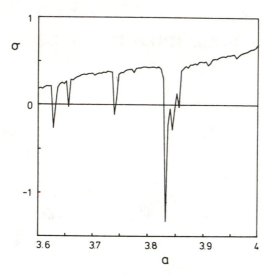

Fig. 3.12 Graph of the Lyapunov exponents for the logistic map $x \to ax\,(1-x)$.

References

Full details of the references cited in this chapter can be found at the end of Chapter 4.

4
Two-dimensional iterative maps

H. A. Lauwerier

Mathematical Institute, University of Amsterdam,
Roetersstraat 15, 1018 WB Amsterdam, The Netherlands

4.1 Introduction

This chapter is a continuation of the previous one. The same notations and definitions will be used, with obvious generalisations. Also, almost all interesting features here are illustrated by quadratic maps. Most examples considered in this chapter have their origin in population dynamics such as the logistic-delay equation, a predator–prey model and models of host–parasitoid interaction. Particular attention is given to the phenomena of Hopf bifurcation and Arnold tongues. The calculus of normal forms is presented here in a way leading to quantitative results such as the size of the Hopf circle (ellipse) and the position of the Arnold tongue in the case of weak resonance. The theory is given in sections 4.3 and 4.4. Applications are given in section 4.5. In section 4.6 we give a brief survey of one-dimensional maps in the complex plane. This piece of classical analysis of the great French school headed by Poincaré, Picard, Montel and Fatou, and by Julia [30] is undergoing a revival. What are now called Julia sets are the prototypes of the strange attractors. Only a brief discussion is given. A good survey containing many new results is given by Blanchard [9]. The next section on area-preserving maps is also kept short. This topic is extensively treated within the framework of Hamiltonian systems. The following publications may be recommended: Berry [6, 7], Hénon [24, 26], Gumowski and Mira [22], and Lichtenberg and Lieberman [38]. The analysis of stable and unstable manifolds in a mapping is of great value for the understanding of its possible chaotic behaviour. In section 4.2 we discuss the possible parametrisation of the unstable manifold of a saddle. For a wide class of mappings this can be done by means of a holomorphic or perhaps entire analytic function. The intersection of an unstable and a stable manifold, the occurrence of homoclinic or heteroclinic points, is an indication of chaotic aspects. This subject is considered in the last section. Rather than giving definitions and theorems we have preferred to explain the basic notions by simple examples. A strange attractor can be considered as the limit set of an unstable manifold which is a curve of infinite length with an infinite number of loops without self-intersection and with a Cantor set as a cross-section. This is illustrated by a simple geometrical construction, the essence of which goes back to the founder of modern topology, L. E. J. Brouwer.

4.2 Fixed points and invariant manifolds

The basic notations are as for one-dimensional maps with obvious extensions. More explicitly we may write

$$(4.1) \qquad T \begin{cases} x_{n+1} = f(x_n, y_n) \\ y_{n+1} = g(x_n, y_n) \end{cases}$$

The local behaviour of the map at a fixed point P is governed by its local linearisation for which

$$(4.2) \qquad J \begin{pmatrix} f_x & f_y \\ g_x & g_y \end{pmatrix}$$

taken at P is the corresponding matrix. The eigenvalues λ_1 and λ_2 of J are called the *multipliers* of the fixed point. Orbits $P_0 P_1 P_2 \ldots$ in the neighbourhood of the fixed point can generally be described by

$$P_n \approx C_1 \lambda_1^n + C_2 \lambda_2^n$$

which implies a classification of fixed points according to the positions of the fixed points in the complex plane.

If $|\lambda_1| < 1$ and $|\lambda_2| < 1$, then all orbits are locally attracted by P and the fixed point is described as *stable*. If at least one eigenvalue is outside the unit circle, the fixed point is *unstable*.

If λ_1 and λ_2 are real, then coordinates can be chosen such that locally:

$$(4.3) \qquad x' = \lambda_1 x \,, \; y' = \lambda_2 y$$

(a) If $0 < \lambda_1 < 1$, $0 < \lambda_2 < 1$, successive points of an attracting orbit are situated on an invariant curve

$$\frac{\log |x|}{\log \lambda_1} - \frac{\log |y|}{\log \lambda_2} = \text{constant}$$

which looks like a parabolic arc. The fixed point is called a *node* (Fig. 4.1).

(b) If $\lambda_1 > 1$, $\lambda_2 > 1$, we have a similar situation for orbits in backward sense. The fixed point is a repelling node.

Two-dimensional iterative maps

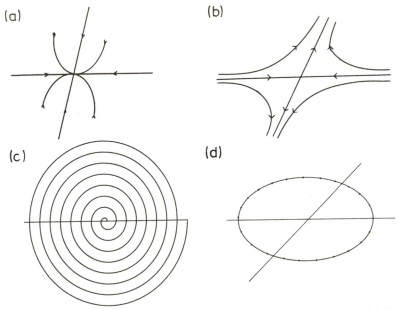

Fig. 4.1 (a) Stable node; invariant curves. (b) Saddle; invariant curves. (c) Focus; invariant curve. (d) Centre; periodic orbit on invariant curve.

(c) If $0 < \lambda_1 < 1$ and $\lambda_2 > 1$, the invariant curves are hyperbolic arcs with the exception of the coordinate axes. Along the x-axis on both sides of the origin the orbits are attracted by the fixed point. Along the y-axis the orbits are repelled by the fixed point. In a more general setting the x-axis would be called the stable manifold of the fixed point. The y-axis is then called the unstable manifold. The fixed point is called a *saddle*.

(d) If at least one multiplier is negative, the situation is as for the squared map T^2. Successive points of an orbit on an invariant curve now lie alternately on two distinct branches.

(e) If λ_1 and λ_2 are conjugate complex, we sometimes write $\lambda_1 = \lambda$, $\lambda_2 = \bar{\lambda}$ with $\lambda = a \exp(i\alpha)$ $0 < \alpha < \pi$. The locally linearised map can be written after a suitable affine transformation of coordinates very conveniently in polar coordinates as $r_{n+1} = ar_n$, $\theta_{n+1} = \theta_n + \alpha$. The invariant curves are described by $r \exp(-\theta/\alpha \log a) = $ constant.

(f) If $a \neq 1$, the invariant curves are logarithmic spirals. The fixed point is called a *focus*. For $a < 1$ it is stable and attracting. For $a > 1$ it is unstable and repelling.

(g) If $a = 1$, the invariant curves are ellipses. The fixed point is called a *centre*. If α/π is rational and if the map is linear, all orbits are periodic. If α/π is irrational, then almost all orbits on the same ellipse are covering it densely.

There are a few degenerate cases. We mention only the so-called *star-node* which is described by (4.3) with $\lambda_1 = \lambda_2$.

If the multipliers of a fixed point are not on the unit circle the local linearisation at the fixed point gives a reliable picture of what to expect there for the nonlinear map. In many cases the original map (4.1) contains a bifurcation parameter μ such that for $\mu < 0$ the multipliers are inside the unit circle, and that for $\mu = 0$ one or both multipliers are 1 in absolute value. Let us imagine that for $\mu < 0$ both multipliers are real and in the interval $(-1, 1)$, and that $\lambda_1 = 1$ for $\mu = 0$ whereas λ_2 stays inside $(-1, 1)$. The behaviour of the map close to the bifurcation point $\mu = 0$ is essentially that of the one-dimensional map for the single variable corresponding to the multiplier λ_1. Thus we may expect transcritical bifurcation as described in Chapter 3. In a similar way we may expect flip bifurcation if λ_1 crosses the unit circle at -1 whereas λ_2 is somewhere between -1 and 1. If λ_1 and λ_2 are both complex and inside the unit circle for $\mu < 0$, and if they cross the unit circle at the points $\exp \pm i\alpha$, we have a *Hopf bifurcation*, a phenomenon which has no counterpart in one-dimensional maps.

For a linear map the unstable manifold of a saddle is a straight line. For a nonlinear map it is generally a curved line tangent to the eigenvector at the saddle for the largest multiplier. Further away from the saddle the unstable manifold can be a highly complicated curve with an infinity of loops and self-intersections.

Example

$$x_{n+1} = 2x_n, y_{n+1} = y_n/2 + 7x_n^2$$

The origin is a saddle with multipliers 2 and $\frac{1}{2}$. The stable manifold is the y-axis. The unstable manifold is the parabola $y = 2x^2$. The invariant curves are $y = 2x^2 + C/x$

In many cases the unstable manifold can be determined analytically. We consider the map (4.1) in the form

(4.4) $$x_{n+1} = y_n, y_{n+1} = g(x_n, y_n)$$

where the origin is a saddle with a multiplier $a > 1$. Sometimes we are starting from a two-step recurrent relation

(4.5) $$x_{n+1} = g(x_{n-1}, x_n)$$

and then (4.5) is the corresponding two-dimensional map. If the map originally is of the form (3.1), it suffices to take $f(x, y)$ as a new y-coordinate. The parametrisation of the unstable manifold is postulated as

(4.6) $$x = F(t), y = F(at)$$

where $F(t)$ is supposed to be a solution of the functional equation

(4.7) $$F(a^2 z) = g[F(z), F(az)]$$

where $F(z)$ has power-series expansion

(4.8) $$F(z) = \sum_{k=0} c_k z^k, c_1 = 1$$

This equation is the two-dimensional analogue of the Poincaré equation (3.15). Again, if $g(x, y)$ is analytic $F(z)$ is also analytic, and if $g(x, y)$ is entire, e.g. a polynomial, $F(z)$ is an entire function.

The technique will be illustrated by the logistic-delay equation considered by Pounder and Rogers [53] and by Aronson *et al.* [3] The model is

(4.9) $$x_{n+1} = y_n, y_{n+1} = ay_n(1 - x_n)$$

The fixed point $(0, 0)$ is stable for $0 < a \leqslant 1$ and becomes a saddle for $a > 1$. For $1 < a \leqslant 2$ the fixed point $(1 - 1/a, 1 - 1/a)$ is stable but is subjected to Hopf bifurcation at $a = 2$. For $a > 1$ the saddle at the origin has the multipliers 0 and a.

The x-axis is, in a trivial way, its stable manifold. The unstable manifold can be parametrised by an entire function $F(z)$ satisfying

$$F(a^2 z) = aF(az)(1 - F(z))$$

Substitution of the expansion (4.8) gives

$$a^k(a^k - 1)c_{k+1} = -\sum_{j=1}^{k} a^{j-1} c_j c_{k+1-j}$$

for $k = 1, 2, 3 \ldots$ from which the coefficients can be determined. The first few are

$$c_2 = \frac{-1}{a^2 - a}, c_3 = \frac{1}{a^3(a-1)^2} \ldots$$

Using the power-series expansion in combination with the Poincaré equation, a considerable length of the unstable manifold emanating from the origin can be determined quite accurately. An attempt to determine the unstable manifold in the form $y = \varphi(x)$ would give only a small initial segment since $\varphi(x)$ is bound to diverge at the first maximum of $F(t)$. If x is sufficiently small, we find, by elimination of t,

$$y = ax - x^2 + \frac{3a + 1}{a^2(a-1)} x^3 + O(x^4)$$

showing that the unstable manifold leaves the origin as a parabolic arc. In this case the unstable manifold can be expected to end at the fixed point $x = y = 1 - 1/a$ if $1 < a \leqslant 2$. If $a > 2$, the unstable manifold converges to a Hopf curve or perhaps to a strange attractor. In Fig. 4.2a an illustration is given of the case $a = 2.27$ where the unstable manifold is tangent to the stable manifold. In Fig. 4.2b a blow-up of the folded part at the origin is given.

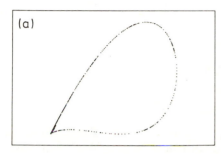
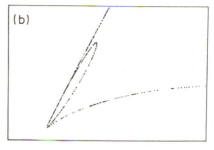

Fig. 4.2 The unstable manifold of the origin in the logistic-delay model close to homoclinic tangency. $a = 2.27$. (a) Horizontal axis from -0.3 to 1.3, vertical axis from -0.1 to 1.1. (b) Horizontal axis from -0.03 to 0.13, vertical axis from -0.01 to 0.11.

4.3 Normal form

We consider a map

$$(4.10) \qquad x_{n+1} = f(x_n, y_n), y_{n+1} = g(x_n, y_n)$$

for which the origin is a fixed point with multipliers λ, $\bar{\lambda}$ close to the unit circle. We write

$$(4.11) \qquad \lambda = (1 + \mu) \exp i\alpha, \, 0 < \alpha < \pi$$

and consider μ as a bifurcation parameter and α as a constant. If α/π is rational, we write

$$(4.12) \qquad \alpha = 2\pi l/m$$

where m is the lowest possible integer.

If $\mu < 0$, the origin is a locally stable fixed point, but as μ increases to positive values, all kinds of bifurcations are possible. If α/π is irrational, or if α/π is rational with $m \geqslant 5$, we have Hopf bifurcation. However, if $m = 3$ or $m = 4$, the situation is much more complicated. The appropriate technique to study the bifurcation behaviour is to reduce the map to its normal form. Since we are interested in explicit formulae which can be used in actual cases, we bring the original map in the form

$$(4.13) \qquad \begin{cases} x_{n+1} = y_n \\ y_{n+1} = Ax_n + By_n + \sum g_{jk}x_n^j y_n^k, \, j + k \geqslant 2 \end{cases}$$

by taking $f(x, y)$ as a new y-variable. The constants A,B follow from the multipliers as

(4.14) $A = -\lambda\bar\lambda , B = \lambda + \bar\lambda$

The next step is the introduction of conjugate complex coordinates z, $\bar z$ by means of

(4.15) $z = ix\bar\lambda - iy , \bar z = -ix\lambda + iy$

Locally they have the meaning of eigenvector coordinates at the origin. Substitution gives

(4.16) $z_{n+1} = \lambda z_n + \sum a_{jk} z_n^j \bar z_n^k , j + k \geqslant 2$

and a similar expression in conjugate complex form. The coefficients a_{jk} can easily be derived from the coefficients g_{jk}. We need them only in the approximation of lowest order with respect to μ. Since

(4.17) $2x \sin \alpha \approx z + \bar z , 2y \sin \alpha \approx z\,e^{i\alpha} + \bar z\,e^{-i\alpha}$

we have

(4.18) $\begin{cases} 4i\,a_{20} \sin^2\alpha \approx g_{20} + g_{11}e^{i\alpha} + g_{02}e^{2i\alpha} \\ 2i\,a_{11} \sin^2\alpha \approx g_{20} + g_{11}\cos \alpha + g_{02} \\ 4i\,a_{02} \sin^2\alpha \approx g_{20} + g_{11}e^{-i\alpha} + g_{02}e^{-2i\alpha} \end{cases}$

The map written in the form (4.16) will be called the pre-normal form. According to Arnold [1] it is possible to define new complex coordinates w, $\bar w$ such that most nonlinear terms in (4.16) can be removed. The terms that cannot be removed without the coordinate transformation becoming singular at $\mu = 0$ are called resonating terms. The terms $z^2\bar z$, $z^3\bar z^2$, ... are always in resonance. If α/π is irrational they are the only resonating terms. Then

(4.19) $z_{n+1} = \lambda z_n + Q z_n^2 \bar z_n + O(z_n^5)$

is the corresponding normal form.

If α/π is rational, there are more resonating terms starting with $\bar z^{m-1}$. If $m \geqslant 5$, cases of so-called weak resonance, we have the same normal form (4.17) but with $O(z^4)$ for $m = 5$. The cases $m = 3,4$ give a so-called strong resonance. The corresponding normal forms are

(4.20) $z_{n+1} = \lambda z_n + Q z_n^2 \bar z_n + R\bar z_n^3 + O(z_n^5)$

for resonance 1: 4, and

(4.21) $z_{n+1} = \lambda z_n + P\bar z_n^2 + Q z_n^2 \bar z_n + O(z_n^4)$

for resonance 1: 3.

The first problem is to obtain an explicit expression for Q in the general case (4.19). The coordinate transformation which brings (4.16) into the normal form (4.19) is written as

(4.22) $$w = z + p_{20}z^2 + p_{11}z\bar{z} + p_{02}\bar{z}^2 + \dots$$

where w is the new coordinate. Substitution in (4.19) gives

$$z_{n+1} + p_{20}z_{n+1}^2 + p_{11}z_{n+1}\bar{z}_{n+1} + p_{02}\bar{z}_{n+1}^2 + \dots$$
$$= \lambda(z_n + p_{20}z_n^2 + p_{11}z_n\bar{z}_n + p_{02}\bar{z}_n^2 + \dots)$$
$$+ Qz_n^2\bar{z}_n + \dots$$

Substitution of (4.16) on the left-hand side turns this into an identity. Equating coefficients of z^2, $z\bar{z}$ and $\bar{z}^2$ we find

$$\begin{cases} a_{20} + \lambda^2 p_{20} = \lambda p_{20} \\ a_{11} + \lambda\bar{\lambda} p_{11} = \lambda p_{11} \\ a_{02} + \bar{\lambda}^2 p_{02} = \lambda p_{02} \end{cases}$$

Since we have excluded the cases of strong resonance, it is possible to determine coefficients p_{20}, p_{11}, p_{02}, which guarantee the vanishing of the quadratic terms in (4.16).

Again in the lowest-order approximation to $\mu = 0$ we obtain

(4.23) $$p_{20} = \frac{a_{20}}{\lambda - \lambda^2}, p_{11} = \frac{a_{11}}{\lambda - 1}, p_{02} = \frac{a_{02}}{\lambda - \bar{\lambda}^2}$$

with $\lambda = \exp(i\alpha)$.

Equating the coefficients of $z^2\bar{z}$ we find from the same identity

(4.24) $$Q = 2\lambda a_{11}p_{20} + (\lambda\bar{a}_{11} + \bar{\lambda}a_{20})p_{11} + 2\bar{\lambda}\bar{a}_{02}\bar{p}_{02} + a_{21}$$

If the values of p_{20}, p_{11}, p_{02} found above are substituted, we obtain the following useful expression

(4.25) $$Q = \frac{|a_{11}|^2}{1 - \bar{\lambda}} + \frac{2|a_{02}|^2}{\lambda^2 - \bar{\lambda}} + \frac{2\lambda - 1}{\lambda(1 - \lambda)}a_{11}a_{20} + a_{21}$$

with $\lambda = \exp(i\alpha)$.

The technique of the calculation of the intrinsic parameter Q will be demonstrated for the logistic delay map (4.9). Translation of the coordinates to the fixed point $(1 - 1/a, 1 - 1/a)$ gives, in shifted coordinates,

$$\begin{cases} x_{n+1} = y_n \\ y_{n+1} = -(a-1)x_n + y_n - ax_ny_n \end{cases}$$

The eigenvalue equation is $\lambda^2 - \lambda + (a-1) = 0$, which gives $\alpha = \pi/3$, $\mu = -1 + \sqrt{a-1}$. The transformation (4.17) gives the pre-normal form (4.16) with $3a_{20} = 2i\exp(i\pi/3)$, $3a_{11} = 2i$, $3a_{02} = 2i\exp(-i\pi/3)$. Finally, (4.25) gives $Q = \frac{2}{3}(1 - i\sqrt{3})$.

In the case of resonance 1:4, the normal form (4.20) requires the calculation of two intrinsic parameters. Starting from the pre-normal form (4.16) we find as before

$$(4.26) \quad Q = \tfrac{1}{2}(1 - i)|a_{11}|^2 - (1 + i)|a_{02}|^2 + \tfrac{1}{2}(1 + 3i)a_{11}a_{20} + a_{21}$$

$$(4.27) \qquad\qquad R = -\tfrac{1}{2}(1 - i)(a_{11} + 2i\bar{a}_{20})a_{02} + a_{03}$$

By way of illustration we consider the following variant of the logistic delay map

$$\begin{cases} x_{n+1} = y_n \\ y_{n+1} = ay_n(1 - \tfrac{1}{2}x_n - \tfrac{1}{2}y_n) \end{cases}$$

The fixed point $x = y = 1 - 1/a$ bifurcates for $a = 3$. The eigenvalue equation is $2\lambda^2 - (3 - a)\lambda + (a - 1) = 0$, giving $\alpha = \pi/2$, $\mu = -1 + \sqrt{(a-1)/2}$. For the coefficients at $\mu = 0$ of the pre-normal form we find $8a_{20} = 3(-1-i)$, $4a_{11} = 3i$, $8a_{02} = 3(-1 + i)$. The parameters of the normal form are then: $32Q = 9(2-i)$, $32R = -9i$.

4.4 Hopf bifurcation and Arnold tongues

We consider a two-dimensional map with two parameters for which the origin is a bifurcating fixed point. Its multipliers are a complex conjugate pair λ, $\bar{\lambda}$ with

$$(4.28) \qquad\qquad \lambda = (1 + \mu)e^{i\alpha}$$

The parameter μ is the small bifurcation parameter and α is a free constant in $(0, \pi)$. For $\mu < 0$ the origin is stable, but if μ increases and becomes positive, we may have a Hopf bifurcation. This means that orbits close to the origin which acts as a repeller are attracted either by an elliptical invariant curve or by a periodic cycle. We shall give no proofs but show how in actual cases the nature of the Hopf bifurcation can be determined in a very explicit way.

As in the previous section our starting point is the map

$$(4.29) \qquad \begin{cases} x_{n+1} = y_n \\ y_{n+1} = Ax_n + By_n + \sum g_{jk}x_n^j y_n^k \end{cases}$$

The linear transformation

$$(4.30) \qquad\qquad z = ix\bar{\lambda} - iy, \ \bar{z} = -ix\lambda + iy$$

brings this in the pre-normal form

$$(4.31) \qquad\qquad z_{n+1} = \lambda z_n + \sum a_{jk}z_n^j \bar{z}_n^k$$

It is important to note that the nonlinear coefficients a_{jk} can be taken at the bifurcation point $\mu = 0$.

Postponing the cases of strong resonance we may reduce (4.31) to its normal form

(4.32) $$z_{n+1} = (1 + \mu)e^{i\alpha}z_n + Q z_n^2 \bar{z}_n + \ldots$$

where Q is given by (4.25).

Introducing polar coordinates

(4.33) $$z = re^{i\theta}$$

we obtain

$$r_{n+1} = r_n |(1 + \mu)e^{i\alpha} + Q r_n^2| + \ldots$$

or with

(4.34) $$Q = -q e^{i\gamma}$$

in the lowest-order approximation

(4.35) $$r_{n+1}^2 = (1 + 2\mu) r_n^2 - 2q \cos(\alpha - \gamma) r_n^4$$

This can be interpreted as a one-dimensional map with a possible non trivial fixed point $r = R_H$ satisfying

(4.36) $$R_H^2 = \frac{\mu}{q \cos(\alpha - \gamma)}$$

If $\cos(\alpha - \gamma) > 0$, this gives a Hopf circle which is an attracting invariant curve for $\mu > 0$. Of course, in the original x, y-coordinates, it would be an invariant ellipse. If $\cos(\alpha - \gamma) < 0$, we have a similar invariant curve for $\mu < 0$ but this time it is unstable and repelling. The bifurcation diagram is given in Figs 4.3a and b. The value R_H will be called the Hopf radius.

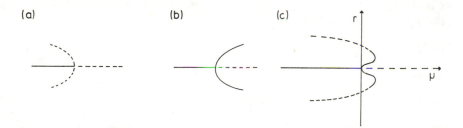

(a)   (b)   (c)

Fig. 4.3 (a) Backward Hopf bifurcation. (b) Forward Hopf bifurcation. (c) A singular form of Hopf bifurcation called a crater bifurcation.

Using the expression (4.25) we have

(4.37) $$\frac{\mu}{R_H^2} = \tfrac{1}{2}|a_{11}|^2 + |a_{02}|^2 - Re \frac{(2-\bar{\lambda})a_{11}a_{20}}{\lambda(1-\lambda)} - Re\bar{\lambda}a_{21}$$

In the original coordinates of (4.29) the Hopf circle $z\bar{z} = R_{\mathrm{H}}^2$ is transformed into the elllipse

(4.38) $x^2 - 2xy \cos\alpha + y^2 = R_{\mathrm{H}}^2$

If $\cos(\alpha - \gamma) \approx 0$, then the description of the bifurcation behaviour needs a further term of the normal form. Instead of (4.35) we obtain the approximation

(4.39) $r_{n+1}^2 = (1 + 2\mu)r_n^2 - 2Ar_n^4 + 2Br_n^6$

where $A = q \cos(\alpha - \gamma)$. For $r_{n+1} = r_n$ we have

(4.40) $\mu = Ar^2 - Br^4$

For $A > 0$, $B > 0$, for example, the corresponding bifurcation diagram is shown in Fig. 4.3c.

For $\mu > 0$ we obtain a pair of Hopf circles. The radii are given by

(4.41) $\begin{cases} 2BR_+ = A + \sqrt{A^2 - 4\mu B} \\ 2BR_- = A - \sqrt{A^2 - 4\mu B} \end{cases}$

The inner Hopf circle $|z| = R_-$ turns out to be stable. The outer Hopf circle $|z| = R_+$ is unstable. The two circles coalesce for $\mu = A^2/(4B)$, a point of fold bifurcation. For $\mu < 0$ a single unstable Hopf curve remains.

If $A>0$, $B<0$ the situation is very much like the ordinary Hopf bifurcation. If $A<0$ the bifurcation behaviour is as for $A>0$ but with $\mu \rightarrow -\mu$.

Next we consider the weak resonance when α/π is a rational number. Assuming the existence of a stable Hopf circle we have two possibilities. Either the Hopf circle is densely covered by aperiodic orbits or it contains an attracting cycle of order m when α is given by (4.12). Starting from the normal form

(4.42) $z_{n+1} = \lambda z_n + Q z_n^2 \bar{z}_n + \ldots + c\bar{z}_n^{m-1} + \ldots$

we take

(4.43) $\lambda = (1 + \mu e^{i\varphi})e^{i\alpha}, \quad \alpha = 2\pi l/m$

which means that the neighbourhood of $\exp(i\alpha)$ is described by local polar coordinates μ and φ. This gives for the Hopf radius

(4.44) $qR_H^2 \cos(\alpha - \gamma) = -\mu \cos\varphi$

as a little extension of (4.36).

The action of (4.42) on the Hopf circle is described by

$\theta_{n+1} = \theta_n + \alpha + \arg(1 + \mu e^{i\varphi} + e^{-i\alpha}QR^2 + \ldots + c_1 R^{m-2} e^{-im\theta_n} + \ldots)$

where $\theta = \arg z$ and c_1 is a constant. Substitution of (4.44) and a little trigonometry gives

(4.45)

$$\theta_{n+1} = \theta_n + \alpha + \arg\left(1 + i\mu\,\frac{\sin(\varphi + \alpha -\gamma)}{\cos(\alpha - \gamma)} + \ldots + c_2\mu^{(m-2)/2}\,e^{-im\theta_n} + \ldots\right)$$

where c_2 is a constant.

The existence of a periodic m-cycle implies the existence of a θ-value such that, in a first-order approximation,

$$\mu\,\frac{\sin(\varphi + \alpha -\gamma)}{\cos(\alpha - \gamma)} + \mu^{(m-2)/2}\,Im(c_2 e^{-im\theta}) = 0$$

Such a value of θ can be found if there exists the inequality

(4.46) $$|\sin(\varphi + \alpha - \gamma)| \leqslant C\mu^{(m-4)/2}$$

where the constant C depends on the coefficient of $\bar{z}^{m-1}$ in (4.42). The region determined by (4.46) has the shape of two horns in a symmetric position. However, since we have assumed forward Hopf bifurcation, only the part outside the unit circle has a meaning. A typical case has been illustrated in Fig. 4.4a for $\gamma = \alpha = \pi/3$. Then the boundary of (4.46) consists of two circular arcs.

(a)　　　　　　　　　　　　　(b)

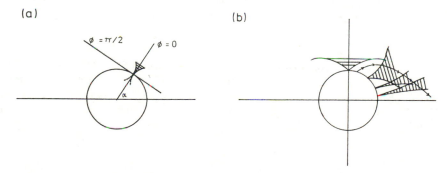

Fig. 4.4 (a) A case of 1:6 resonance with an Arnold tongue. (b) A one-parameter route through Arnold tongues.

At each point with a rational value of α/π we may expect the cuspal beginning of such a so-called *Arnold tongue* or *Arnold horn*. Close to the unit circle they are symmetric with respect to the line

(4.47) $$\varphi = \gamma - \alpha \pmod{\pi}$$

called the axis of the Arnold tongue.

As a rule, for higher rational values of α/π, the tongues have sharper cusps and are narrower. If $\cos(\alpha - \gamma) \approx 0$ the analysis breaks down. In such a case the Arnold tongues are almost tangent to the unit circle and they may give the impression of combed hair.

The theory of normal forms is local by nature. However, as a rule, the Arnold tongues may extend much further in the unstable region of the λ-parameter plane, and Arnold tongues from different resonances may overlap.

In practical cases one often has an iterative map containing a single parameter, a. If, for some critical value, the map shows Hopf bifurcation, its behaviour can be understood by embedding the map in some two-parameter map. Then the two-parameter map may exhibit the whole show of weak resonances and Arnold tongues. Its restriction to the original map means that in the λ-parameter plane for increasing values of a we follow a single curve as sketched in Fig. 4.4b. This curve may cross various Arnold tongues in a more or less systematic way. Aronson *et al.* [3] have studied such behaviour in detail.

In the case of strong resonance 1:4, the bifurcation behaviour is rather complicated. The appropriate normal form is (4.20), but in view of scaling the only intrinsic parameter is $Q/|R|^2$. The various possibilities can be summarised as follows.

If $|Q| < |R|$, there exists a single 4-cycle which is always unstable. There is the possibility of a stable 'Hopf circle' looking like a rounded square. If $|Q| > |R|$, there exists a sector of Q-values, the equivalent of an Arnold horn, for which there are two 4-cycles. The inner cycle is always unstable. The outer cycle is always stable. Outside the sector there may be a stable 'Hopf circle'. If, in the latter case, $ImQ < -|R|^2$, there exists a Hopf circle, but if $ImQ > |R|^2$, no such invariant curve exists.

In the case of strong resonance 1:3, the bifurcation behaviour is less complicated. The normal form obtained from (4.21) by suitable scaling in its lowest-order approximation is

$$(4.48) \qquad z_{n+1} = e^{2\pi i/3}((1 + \mu e^{i\varphi})z_n - e^{i\varphi}\bar{z}_n^2)$$

There exists the 3-cycle

$$\mu \to \mu \exp\frac{2\pi i}{3} \to \mu \exp -\frac{2\pi i}{3}$$

but this cycle is always unstable.

4.5 Some special maps

4.5.1 The following model is a more general version of the logistic delay equation (cf. [33–36])

$$(4.49) \qquad x_{n+1} = ax_n(1 - (1 - b)x_n - bx_{n-l})$$

with $a > 0$ and $0 \le b \le 1$. For $b = 1$ it is the model studied by Pounder and Rogers [52] and by Aronson *et al* [3]. We write (4.49) in the form of a two-dimensional map

$$(4.50) \qquad \begin{cases} x_{n+1} = y_n \\ y_{n+1} = a y_n (1 - b x_n - (1-b) y_n) \end{cases}$$

The trivial fixed point $(0,0)$ is stable for $a \le 1$ but becomes a saddle for $a > 1$. Its stable manifold is the line $y = 0$. The unstable manifold can be parametrised by entire analytic functions as shown in section 4.2. The nontrivial fixed point $x = y = 1 - 1/a$ is stable in the interval $1 \le a < a_0$ where

$$a_0 = \frac{3 - 2b}{1 - 2b} \text{ for b} \le \tfrac{1}{4}$$

and

$$a_0 = 1 + 1/b \text{ for } b \ge \tfrac{1}{4}$$

The transition at $a = 1$ is an example of transcritical bifurcation. At the line $a = (3 - 2b)/(1 - 2b)$ with $b < \tfrac{1}{4}$, we have flip bifurcation and at the line $a = 1 + 1/b$ with $b > \tfrac{1}{4}$ a case of Hopf bifurcation. At this Hopf line we have $\cos \alpha = (3 - a)/2$ for $2 \le a < 5$. This means that α varies between $\pi/3$ and π. This includes the strong resonance cases 1:4 for $a = 3, b = \tfrac{1}{2}$ and 1:3 for $a = 4, b = \tfrac{1}{3}$.

The Hopf radius is

$$(4.51) \qquad R_H = \frac{(a-1)(5-a)}{a} \sqrt{32\mu/3}$$

where $\mu = -1 + \sqrt{b(a-1)}$. The axis of the Arnold horn is determined by $\varphi = \gamma - \alpha = \arg((1 + \cos \alpha) + i(3 \sin \alpha + \cot 3\alpha/2))$.

A few special cases are collected in Table 4.1

Table 4.1   Special cases of the logistic delay equation.

a	b	$\alpha°$	Resonance	$R_H/\sqrt{\mu}$	$\gamma° - \alpha°$
2	1	60	1:6	1.225	60
2.382	0.724	72	1:5	1.240	62.6
2.653	0.605	80	2:9	1.194	63.7
3.618	0.382	108	3:10	0.816	-18.0
4.414	0.293	135	3:8	0.370	86.3
4.618	0.276	144	2:5	0.244	86.5

As a model in population dynamics the variables x, y have only a meaning inside the triangle bounded by $x = 0$, $y = 0$ and $bx + (1-b)y = 1$.

For some values of a and b the triangle is not mapped inside itself and so there may be an escape region such that orbits starting there leave the triangle after a number of iterations. For increasing values of a, chaotic phenomena predominate gradually. The Hopf circle may deform into a complicated attractor with an infinity of folds with a Cantor-like cross-section, a strange attractor. The escape region may take up almost all the area of the triangle, leaving perhaps a two-dimensional Cantor-like pattern of bounded aperiodic orbits.

$$(4.52) \qquad \begin{cases} x_{n+1} = y_n \\ y_{n+1} = ay_n(1-x_n/2-y_t/2), \, 1 < a \leqslant 4 \end{cases}$$

This is a very interesting special case of the map (4.50). The nontrivial fixed point $x = y = 1-1/a$ is stable up to $a = 3$. At this point we have a case of Hopf bifurcation with resonance $1:4$. Two 4-cycles are born. The unstable cycle follows the pattern $\sigma_1 \rightarrow 2/a \rightarrow \sigma_2 \rightarrow 2/a$, with $\sigma_1 + \sigma_2 = 1-1/a$, $\sigma_1 \sigma_2 = 4/a^2$. The stable cycle stays stable up to $a_2 = 3.627630$, which is the beginning of a period-doubling sequence with $a_3 = 3.666598$, $a_4 = 3.672522$ and ending with $a_\infty = 3.673990$. It follows the well-known Feigenbaum pattern with the Feigenbaum constant 4.6692. For such large values of a it appears to be very difficult to find bounded orbits when doing computer experiments. In Fig. 4.5 a plot is given of a few escape regions of low order, sets of initial points of orbits leaving the triangular domain $x > 0$, $y > 0$, $x + y < 2$ after a few iterations. The boundaries of these escape regions are the successive pre-images of the line $y = 0$. However, the numerical determination of boundaries of higher order very soon runs into technical difficulties requiring the plotter mechanism to act with the speed of light. It is not difficult to determine the behaviour on a line intersecting the escape regions transversally. In Fig. 4.6 a plot is given of the escape numbers of 100 points distributed uniformly on the line $x = 0$, $0 < y < 2$. Such plots can easily be obtained on a personal computer. Theoretically there is an infinite number of points with the structure of a Cantor set for which the orbits are bounded forever. Numerically they may be hard to find (cf. [33]).

4.5.2   Maps of the kind

$$(4.53) \qquad \begin{cases} x_{n+1} = y_n \\ y_{n+1} = Ax_n + By_n + Cx_n^2 + Dx_ny_n + Ey_n^2 \end{cases}$$

may serve to demonstrate all sorts of bifurcation phenomena with respect to the fixed point $x = y = 0$. If $C = 0$ the map is also invertible. Even with this restriction there are still many possibilities. Of course

$$(4.54) \qquad A = -\lambda_1\lambda_2, \, B = \lambda_1 + \lambda_2$$

where λ_1 and λ_2 are the multipliers. If $C = 0$, the remaining coefficients count as a single degree of freedom in view of a possible scaling.

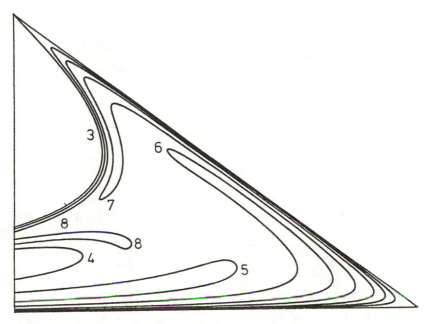

Fig. 4.5 Some escape regions for $a = 3.5$.

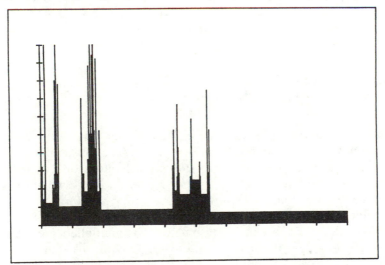

Fig. 4.6 Escape numbers along $x = 0, 0 < y < 2$ for $a = 3.7$.

The case $C = 0$, $D = 3$, $E = -2$, is a case of resonance $1:4$. The parameters of the normal form are $16Q = 19 - 3i$, $16R = -5 + 27i$. In Fig. 4.7 we have plotted the case $\mu = 0.005$, $\varphi = 4\pi/9$, with $\lambda_1 = (1 + \mu e^{i\varphi})i$. More details can be found in Lauwerier [33].

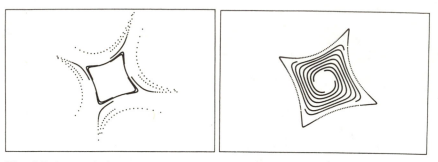

Fig. 4.7 A rounded square as a limit cycle in a 1:4 resonance class.

4.5.3 The following map belongs to the class of predator–prey models

$$(4.55) \qquad \begin{cases} x_{n+1} = ax_n(1 - x_n - y_n) & 2 < a \leq 4 \\ y_{n+1} = bx_n y_n & 2 < b \leq 4 \end{cases}$$

The nontrivial fixed point, $x = 1/b$, $y = 1 - 1/a - 1/b$, becomes unstable for $b > 2a/(a-1)$. At the line $b = 2a/(a-1)$ we have Hopf bifurcation with $\cos \alpha = (5-a)/4$. In particular, for $a = 3$, $b = 3$, we have a case of 1:6 resonance. The parameter of the normal form is $Q = -3i\sqrt{3}$. This gives the Hopf radius and the position of the Arnold horn as $\mu/R_H^2 = 9/2$, $\varphi = 30°$, with $\mu = -1 + \sqrt{a(b-2)/b}$.

$$(4.56) \qquad \begin{cases} x_{n+1} = ax_n(1 - x_n - y_n) \\ y_{n+1} = ax_n y_n \end{cases}$$

This is a very interesting special case of the previous map with a very rich bifurcation behaviour. Starting from $a = 0$, the trivial fixed point $(0, 0)$ is stable up to $a = 1$, where we have transcritical bifurcation. Stability is taken over by the fixed point $x = 1 - 1/a$, $y = 0$. At $a = 2$ we have again transcritical bifurcation. The nontrivial equilibrium $x = 1/a$, $y = 1 - 2/a$ is then stable up to $a = 3$, where we get Hopf bifurcation in the form of a stable 6-cycle, a clear case of 1:6 resonance. The multipliers are the roots of $\lambda^2 - \lambda + (a - 2) = 0$, so that $Re \, \lambda = \frac{1}{2}$. This means that for $a > 3$ the multiplier leaves the unit circle along the line $\varphi = 30°$, i.e. along the axis of the 1:6 Arnold horn. The periodic points of the 6-cycle follow the pattern $(p, q) \rightarrow (p, r) \rightarrow (r, p) \rightarrow (q, p) \rightarrow (q, r) \rightarrow (r, q)$, where $p + q = 1 - 1/a$, $pq = 1/a^2$, $r = 1/a$. At $a = (11 + \sqrt{85})/6 = 3.37$ the cycle is subjected to a secondary Hopf bifurcation. The relevant angle is $120°9'$ so that the behaviour is very close to 1:3 resonance. Each periodic point gives birth to three other periodic points and we obtain a stable 18-cycle as a prelude to a sequence of period doublings. Experimentally for $a = 3.4$ a stable 72-cycle is found. The bifurcation diagram is sketched in Fig. 4.8. In Fig. 4.9 we have illustrated the map in the case $a = 3.43$ at the transition to chaos. The reticular pattern apparently has to do with the unstable manifolds of the 6-cycle each having three branches.

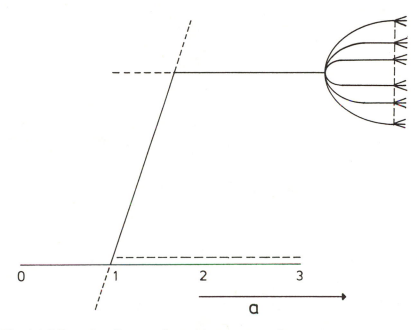

Fig. 4.8 Bifurcation diagram of a one-parameter predator-prey map.

4.5.4 Maps of the kind

$$(4.57) \qquad \begin{cases} x_{n+1} = ax_n\varphi(y_n) \\ y_{n+1} = ax_n - x_{n+1} \end{cases}$$

with $x > 0$, $y > 0$, $a > 1$ and where $\varphi(y)$ is decreasing monotonously from $\varphi(0) = 1$ to $\varphi(\infty) = 0$, are models of parasitoid–host interaction. The origin is always a saddle with multipliers 0 and a. There exists a nontrivial fixed point

$$(4.58) \qquad (c/(a-1),c) \text{ with } \varphi(c) = 1/a$$

As a rule $\varphi(y)$ contains a second parameter b so that the bifurcation behaviour can be studied in the a,b-parameter plane. Generally there are regions of stability and instability with Hopf bifurcation at the line

$$(4.59) \qquad a^2 c \varphi'(c) + a = 1$$

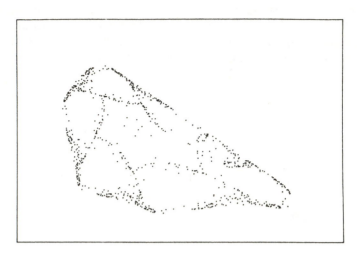

Fig. 4.9 A predator-prey map with a strange attractor, $a = 3.43$.

The corresponding angle α is given by $2 \cos \alpha = 1 - 1/a$ This means that $60° < \alpha < 90°$. The interesting conclusion is that a possible stable periodic cycle has 7 as its lowest order. We list the following three subcases.

(1) $$\varphi(y) = 1/(1 + y^b)$$

The nontrivial fixed point is stable for $b < 1$. However, at the Hopf line $b = 1$, we have $\cos(\alpha - \gamma) = 0$ for all a. Thus there is no Hopf bifurcation.

(2) $$\varphi(y) = a \exp(-y^b)$$

This model is due to Hassell and Varley. The nontrivial fixed point is stable for $a\, b \log a < a - 1$. The radius of the Hopf circle is given by

(4.60) $$\frac{\mu}{R_H^2} = \frac{(a - 1)^2 - a \log^2 a}{4a^2 \log^2 a}$$

which is positive for all values of a.

(3) $$\varphi(y) = \exp \frac{-\sqrt{1 + y} + 1}{b}$$

a model suggested by J.A.J. Metz. The nontrivial fixed point is stable for

$$b > \frac{a}{2(a - 1) - a \log a} - \frac{1}{\log a}$$

The model requires that $a < 4.9216$. The radius of the Hopf circle is given by

(4.61)
$$\frac{\mu}{R_H^2} = \frac{2(c + 1)^{3/2} - b(2c + 3)}{16b(c + 1)^2}$$

However, the numerator changes sign at $a = 1.6304$ corresponding to $\alpha = 36°.23'$. This means forward Hopf bifurcation for $1.6304 < a < 4.9216$. Close to the lower value we have the kind of Hopf bifurcation as described by Fig. 4.3c. Metz and his co-workers have obtained a wealth of experimental computer results which will be made available in reports and preprints jointly with this author.

A typical case is shown in Fig. 4.10. There is a stable inner Hopf cycle and an outer unstable one.

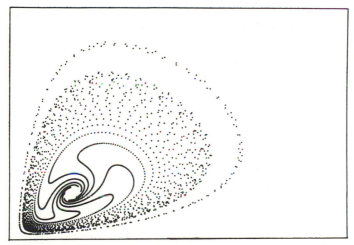

Fig. 4.10 A parasitoid-host map showing singular Hopf bifurcation, $a = 4.2$, $b = 10$. Horizontal axis from 0 to 400, vertical axis from 0 to 1200.

4.6 Rational complex maps

We consider a map of the kind

(4.62)
$$T: z \rightarrow F(z)$$

in the complex plane with $z = x + iy$ where $F(z)$ is a holomorphic function. The most important properties of such maps were described by Gaston Julia in a large prize essay [30] continuing earlier work by Fatou. Interest in the subject remained dormant for half a century. Recently there has been a reawakening of interest following the availability of computers.

The study of such complex maps is very rewarding. It offers the pure mathematician a subject of great intrinsic beauty and the numerical analyst a better understanding of Newton's algorithm in the complex plane. It yields a number of strange attractors and repellers, some of which have already gained some popularity, such as Mandelbrot's San Marco attractor or Douady's rabbit. A good recent survey is given by Blanchard [9]. A modern version of Julia's work containing many new results is given by Thurston [54]. Illustration of nice fractals can be found in Mandelbrot [42] and Peitgen *et al.* [52].

In this section we restrict ourselves to a few salient points. It is assumed that $F(z)$ is a rational function. Then there are periodic points of all periods. The most important notion in the theory is that of the so-called *Julia set* $J(F)$, which is defined as the closure of the set of all unstable periodic points.

Julia proved the following properties.

(1) It is a perfect set.
(2) It is invariant with respect to T and its inverses.
(3) It is an attractor of the inverse mapping.
(4) It is densely covered by the pre-images of one of its points.

There are only three kinds of Julia set. $J(F)$ can be totally disconnected as a Cantor set. $J(F)$ can be linearly connected. Then it is a continuous Jordan curve that, as a rule, is nonsmooth, with a fractal dimension. $J(F)$ can be all of $\mathbb{C}$, but this is a rather rare case.

Most features of the theory are illustrated by the complex version of the logistic map

$$(4.63) \qquad\qquad z \rightarrow az(1-z)$$

where a can be any complex number. Since all quadratic maps are equivalent, sometimes the version

$$(4.64) \qquad\qquad z \rightarrow z^2 + c$$

is used, where $c = (2a-a^2)/4$. For $a = 4$ the map can be parametrised as $z_n = \sin^2(2^n\gamma)$. This shows that all orbits starting outside the closed unit interval I on the real axis are attracted by $z = \infty$. Orbits on I are almost always aperiodic. Obviously I is the Julia set in this case.

For $a = 2$ the map (4.64) will be used with $c = 0$. In this case the orbits are either attracted by $z = 0$ or by $z = \infty$, with the exception of those starting on the unit circle, which is the Julia set. Again on this circle the dynamic behaviour is chaotic. These cases are very exceptional in the sense that their Julia sets are smooth curves. If, for example, a is real and $2 < a < 4$, the Julia set is a fractal curve. A typical case is illustrated in Fig. 4.11 for $a = 3$, the so-called San Marco atttractor. Such an illustration can be obtained from the inverse of (4.64)

$$(4.65) \qquad\qquad z_{n+1} = \sigma_n \sqrt{z_n - 3/4}$$

where (σ_n) is a random sequence of ± 1. Starting from $z_0 = 3/2$, a known point of $J(F)$, all further points give a dense covering of the Julia curve. However, $J(F)$ appears to be covered in a nonuniform way. Some parts of it appear to be visited very rarely. Yet this method is very efficient on a personal computer for obtaining a first impression of the possible shape of the Julia set. However, more accurate plots require special computer techniques. The Julia set of (4.63) can also be derived from the Boettcher function which describes the attracting domain of $z = \infty$ (cf.). [36]). With a slight adaption the Boettcher equation of (4.64) is $H(z^2) = H^2(z) + c$, where $zH(z)$ is holomorphic for $|z| < 1$. Explicitly,

$$H(z) = \sum_{k=0}^{\infty} h_k z^{2k-1}$$

with $h_0 = 1$, $h_1 = -c/2$, $h_2 = c(c-2)/8,\ldots$ The unit circle $|z| = 1$ is the natural boundary of $H(z)$. The Julia set is obtained as its image, i.e. all complex values of the Fourier series

$$\sum_{k=0}^{\infty} h_k \exp((2k-1)\, 2\pi\theta i)$$

In Fig. 4.12 a picture is given of Douady's rabbit, the Julia set of (4.64) with $c = -0.1226 + i0.7449$, a value for which the origin is an element of a superstable 3-cycle. Again the Julia set is connected. It is a continuous curve with a dense covering of multiple points. It separates the plane in an infinite number of simply connected domains.

In Fig. 4.13 we illustrate a totally disconnected Julia set for (4.64) with $c = 0.5$.

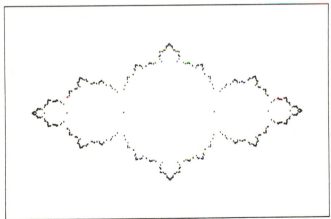

Fig. 4.11 The San Marco attractor.

4.7 Area-preserving maps

Most area-preserving maps in actual applications result from Hamiltonian systems in three or more dimensions. Let us consider such a Hamiltonian system as a set of three differential equations in an x,y,z-space. As a rule an orbit is a continuous closed or nonclosed curve that is uniquely determined by an initial condition. The limit set can be a fixed point, a

Fig. 4.12 Approximation of Douady's rabbit obtained as a seven times iterated conformal map of a circle.

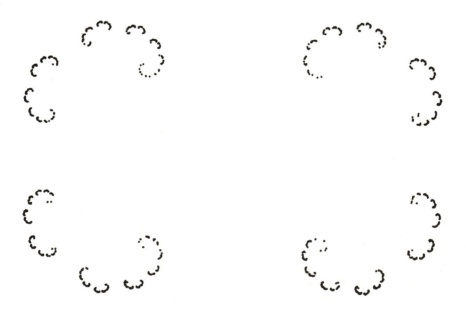

Fig. 4.13 A disjoint Julia set as the unstable invariant set of $z \to z^2 + 1/2$.

limit cycle or something more complicated. The intersection of the orbits with a surface, say the x,y-plane, defines a two-dimensional diffeomorphic map called a *Poincaré map*. This inherits all kinds of properties from the spatial behaviour. A limit cycle gives a periodic point, an invariant surface in space, e.g. a torus, gives an invariant curve, etc. The most important property is that the Poincaré map of a Hamiltonian system is area-preserving. If the Hamiltonian system is integrable, the dynamic behaviour is that of regular motion characterised by invariant tori in phase space. The corresponding Poincaré map can be modelled in polar coordinates r,θ as the so-called *twist map*

$$(4.66) \qquad \begin{cases} r_{n+1} = r_n \\ \theta_{n+1} = \theta_n + 2\pi\varphi(r_n) \end{cases}$$

where $\theta(r)$ is the r-dependent rotation number. The circles $r = $ constant are invariant curves corresponding to tori in space. On a circle where $\varphi(r)$ is irrational an orbit is filled densely for any initial condition. If $\varphi(r)$ is rational, all orbits are periodic with the same period. If the Hamiltonian system is nonintegrable we have irregular motion. If the system is very close to an integrable one, we have the celebrated KAM theorem. A very clear and elementary description is given in Hénon [26].

Briefly the KAM theorem, the combined work of Kolmogorov, Arnold and Moser, says that in a slightly perturbed integrable system most invariant tori of the unperturbed system remain invariant. Such invariant tori are characterised by a rotation number that is sufficiently far from all rational values. This means φ must satisfy the inequality

$$|\varphi - p/q| > \epsilon q^{-5/2}$$

for all rationals p/q with the same small constant ϵ. The consequences of the KAM theorem can be observed in the following so-called *standard map*, a perturbed twist map studied by Chirikov and Taylor (cf. [38]).

$$(4.67) \qquad \begin{cases} r_{n+1} = r_n + a \sin \theta_n \\ \theta_{n+1} = \theta_n + r_{n+1} \end{cases}$$

and in many other maps.

In x, y-coordinates an area-preserving map is described by

$$(4.68) \qquad T \begin{cases} x_{n+1} = f(x_n, y_n) \\ y_{n+1} = g(x_n, y_n) \end{cases}$$

where

$$(4.69) \qquad \frac{\partial f}{\partial x} \frac{\partial g}{\partial y} - \frac{\partial f}{\partial y} \frac{\partial g}{\partial x} = 1$$

for all x, y. As a consequence the multipliers λ_1 and λ_2 of a fixed point satisfy the condition $\lambda_1 \lambda_2 = 1$. This means that the fixed point is either a centre or a saddle. The behaviour of an area-preserving map in the neighbourhood of a centre, also called an *elliptic fixed point*, can be studied by considering the following quadratic map

(4.70)
$$T \begin{cases} x_{n+1} = x_n \cos \alpha - y_n \sin \alpha + x_n^2 \sin \alpha \\ y_{n+1} = x_n \sin \alpha + y_n \cos \alpha + x_n^2 \cos \alpha \end{cases}$$

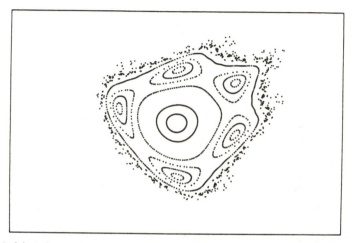

Fig. 4.14 (a) A few orbits of Henon's quadratic map, $\cos \alpha = 0.4$. Horizontal axis from -1.6 to 1.6, vertical axis from -1.2 to 1.2. (b) Blow-up of the previous map at the saddle $0.5696, 0.1622$. Horizontal axis from 0.50 to 0.64, vertical axis from 0.10 to 0.20.

as did Hénon [24]. In Fig. 4.14 a number of orbits are given for $\cos \alpha = 0.4$. Close to the origin there are a number of closed invariant curves, the idealisation perhaps of cycles with a high period. This is in perfect agreement with the predictions of the KAM theorem. Further away we see that an invariant curve of the unperturbed (i.e. linear) system with a low rotation number has been broken up into a cycle of centres and saddles of higher periodicity in alternating order. Each secondary centre is an ordinary fixed point of some power of T and we may repeat the whole argument. Thus each secondary centre is surrounded by secondary invariant curves or rings of centres and saddles of higher order. The conspicuous island structure caused by a broken invariant curve has already been described in great detail by Poincaré and Birkhoff, but their beauty and intricacy have been revealed by modern computers. For there is still more. Each saddle or *hyperbolic fixed point* is surrounded by a chaotic region, and chaotic regions belonging to different hyperbolic periodic points may coalesce. However, a chaotic orbit cannot pass an invariant curve. This means that a map such as the one shown in Fig. 4.14 is always a composition of elliptic and hyperbolic fixed points of any order, invariant curves and stochastic rings. Each island is a microcosm with the same features as the whole map. Such an area-preserving map can be said to be self-similar. In Fig. 4.14b a few orbits are sketched in the neighbourhood of a hyperbolic point in Hénon's map of Fig. 4.14a. We notice secondary island structures but the aperiodic orbit with its cloudlike structure is the most conspicuous phenomenon. It is very instructive to watch the forming of such a stochastic orbit on the screen of a personal computer. Each time, the orbit leaves the plotting area but returns after a few iteration steps. We may imagine that the orbit is leaving along a branch of the unstable manifold and returning along a branch of the stable manifold. In fact the analysis of the structure of these two invariant curves gives considerable insight into what is going on. As a rule there exists at least one *homoclinic point*, an intersection of the stable and the unstable manifold. However, all its images and preimages are also homoclinic points. This, combined with the area-preserving property, means that the overall structure is of an almost unbelievable complexity. More about this will be said in the next section. In Fig. 4.15 we reproduce a sketch by Birkhoff illustrating the generic behaviour of a perturbed twist map with the formation of island chains.

Fig. 4.15 Birkhoff's illustration of the formation of self-similar island structures.

Illustrations of such situations can be found in the book by Gumowski and Mira [22] and in Hénon [26]. Gumowski and Mira gave a detailed study of maps like

(4.71) $$x_{n+1} = y_n, \; y_{n+1} = -x_n + 2F(y_n)$$

where

(4.72) $F(y) = cy + 2(1-c)y^2/(1 + y^2)$

The parameter c measures the nonlinear perturbation. For c close to 1 we may expect the phenomena predicted by Birkhoff and the KAM theorem. As c decreases, more and more KAM curves are destroyed, which gives more space to stochastic behaviour. A typical plot is given in Fig. 4.16 for the case $c = 0.25$. Shown is a single aperiodic orbit of 4000 points starting from $x = y = 0.2$

MacKay [39,40] has given an interesting and very detailed analysis of the effects of an increasing amount of nonlinear perturbation. According to his findings the KAM curve with the golden ratio as its rotation number is the last one to disappear before full chaos sets in. This has to do with the fact that $(-1 + \sqrt{5})/2$ has the 'simplest possible' continued fraction $(1,1,1,1, \ldots)$. MacKay and many others have considered area-preserving maps on a toroidal surface. This is equivalent to a periodic map on a square grid. The following map due to Arnold and Avez [2] is one of the simplest examples

(4.73) $\begin{cases} x_{n+1} = x_n + y_n \\ y_{n+1} = x_n + 2y_n, \text{ mod } 1 \end{cases}$

Arnold showed the action of this map by the distortions of a cat's face after a few iterations. The map is now nicknamed *Arnold's cat map*. It can be analysed in an explicit way. The origin is a saddle with multipliers $a = (3 + \sqrt{5})/2$ and $1/a$. The unstable manifold is given by

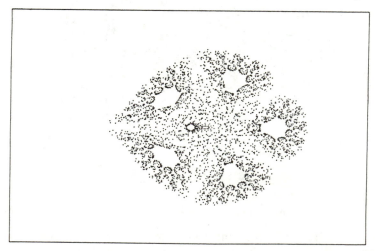

Fig. 4.16 A stochastic orbit of an area-preserving map as a model of the interaction of elementary particles. Horizontal axis from −12 to 12, vertical axis from −9 to 9. $c = 0.25$.

$2y = (1 + \sqrt{5})x$, mod 1. The stable manifold is $2y = (1 - \sqrt{5})x$, mod 1. Each invariant line gives a uniform covering of the unit square (or the torus) which is accordingly densely covered by homoclinic points. Further analysis shows that the map is also ergodic and mixing, meaning as chaotic as possible.

4.8 Stochastic aspects, strange attractors

As for one-dimensional maps, an *invariant measure* $d\mu$ or an *invariant distribution* $P(x,y)$ can be defined. We write

(4.74) $$d\mu = P(x,y)dxdy$$

   In practical applications one may try to determine $P(x,y)$ for a stochastic region by means of a Monte Carlo technique.

Example
   The map $T : x \to y$, $y \to 4x(1-x)$, is chaotic in the square $0 \leqslant x, y \leqslant 1$. It can be parametrised by $x_{2k} = \sin^2(2^k\alpha)$, $x_{2k+1} = \sin^2(2^k\beta)$. The square map T^2 degenerates into two one-dimensional chaotic maps, $x \to 4x(1 - x)$, $y \to 4y(1 - y)$. The invariant distribution is:

$$\frac{1}{\pi^2 \sqrt{xy(1 - x)(1 - y)}}$$

   This example is a rare case of a two-dimensional map for which the invariant distribution can be found in the form of a simple explicit formula. An approximation of the invariant distribution can also be found when one tries to solve the so-called *Perron–Frobenius equation*

(4.75) $$P(x,y) = \sum \frac{P(\xi, \eta)}{|J(\xi,\eta)|}$$

summed over all inverses (ξ,η) of (x,y) and where J is the Jacobi determinant of $DT(\xi,\eta)$. A trial solution on the right-hand side of (4.75) may then lead to a better approximation on the left-hand side.
   In actual applications it is convenient to have a bounded mapping. Of course this can always be arranged if necessary by introducing new coordinates by means of an inversion: $x \to x/(1 + x^2 + y^2)$, $y \to y/(1 + x^2 + y^2)$, or something similar. For area-preserving maps P is a constant in agreement with (4.75). In fact an area-preserving map has a single inverse and $J = 1$ at all points. Thus (4.75) is trivially solved by $P = 1$. If the map has a simple attractor, a set of periodic points or a one-dimensional curve, then $P(x,y)$ is a generalised function with delta-function components. If $P(x,y)$ exists in some region as a classical nonvanishing function, then this means that the map has a stochastic

or chaotic behaviour in that region. A thorough discussion of such behaviour would imply an amount of ergodic theory for which there is no place here. There are a number of books and survey papers in which the interested reader may find the necessary details (e.g. [2,8,32]). For an ergodic dynamical system, time averages and space averages are equal. This means that for almost all measurable functions $\varphi(x,y)$

$$(4.76) \qquad \lim \frac{1}{N} \sum_{k=0}^{N-1} \varphi(x_k, y_k) = \int \varphi(x, y) \, d\mu$$

It is, however, very difficult in practice to decide whether a given map is ergodic or not. Usually the relation (4.76) is accepted as being valid. Perhaps the most important application of (4.76) is the calculation of the *Lyapunov exponents*. They are the generalisation of the corresponding concept for one-dimensional map (cf. (3.35), (3.36)).

Let (x_0, y_0) be the starting point of an aperiodic orbit (x_k, y_k), then we define the geometrical means of the Jacobian matrix after n steps by

$$(4.77) \qquad \{DT(x_{n-1}, y_{n-1}) \cdot DT(x_{n-2}, y_{n-2}) \dots DT(x_0, y_0)\}^{1/n}$$

Its eigenvalues are written as $\lambda_1(n)$ and $\lambda_2(n)$. Then the Lyapunov exponents σ_1 and σ_2 are defined as

$$(4.78) \qquad \sigma_j = \lim_{n \to \infty} \log |\lambda_j(n)|$$

Example

In Arnold's cat map (4.77)

$$\begin{cases} x_{n+1} = x_n + y_n \\ y_{n+1} = x_n + 2y_n \end{cases}, \text{mod } 1$$

we always have

$$DT(x, y) = \begin{pmatrix} 1 & 1 \\ 1 & 2 \end{pmatrix}$$

with $\lambda_1 = (3 + \sqrt{5})/2$, $\lambda_2 = (3 - \sqrt{5})/2$. This gives $\sigma_1 = 0.962$, $\sigma_2 = -0.962$.

The Lyapunov exponents measure the mean distortion of an infinitesimal circle. Such a circle will be transformed on the average into an ellipse with the principal deformation ratios $\exp \sigma_1$ and $\exp \sigma_2$.

The matrix product in (4.77) is not commutative. However, by taking the determinant value, we obtain the following result

$$(4.79) \qquad \sigma_1 + \sigma_2 = \lim_{n \to \infty} \frac{1}{n} \log |J(x_n, y_n)|$$

where $J(x_n, y_n)$ is the determinant value of the Jacobian matrix at the nth point of an aperiodic orbit. In particular, for an area-preserving map we have $\sigma_1 + \sigma_2 = 0$ as in the example above. Assuming the validity of (4.76) the relation (4.79) can be replaced by

(4.80) $$\sigma_1 + \sigma_2 = \int \log |J(x, y)| d\mu$$

If both Lyapunov exponents are negative, all areas are contracting, which is the indication of the existence of a simple attractor. If the Lyapunov exponents differ in sign, we may have a strange attractor. Such a behaviour may occur in a map where the dynamics can be described in terms of a stable and an unstable manifold of the same saddle.

Many maps combine regular and stochastic aspects. It is hardly possible to give a systematic account of all kinds of behaviour. We restrict ourselves here to some highlights. A basic concept is that of a *Cantor set* or a Cantor discontinuum. The traditional example is that of the 'middle thirds'. Starting from the closed unit interval [0, 1] we delete the open middle third (1/3, 2/3). From the remaining two closed intervals we also delete the open middle third (1/9, 2/9) and (7/9, 8/9) and so on *ad infinitum*. The remainder, the classical Cantor set, is a totally disconnected perfect set. It is uncountable and self-similar and it is of zero measure. The elements of this Cantor set can be represented as real numbers x between 0 and 1 in a base-3 expansion

$$x = a_1(\tfrac{1}{3}) + a_2(\tfrac{1}{3})^2 + \ldots + a_n(\tfrac{1}{3})^n + \ldots$$

where a_j is 0, 1 or 2. The removal of the middle thirds is equivalent to deleting all fractions x which have a 1 in their expansion. Thus the Cantor

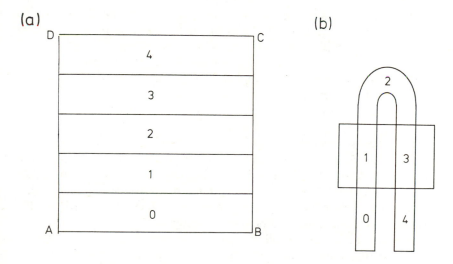

Fig. 4.17 A horseshoe mapping.

set is simply the collection of all numbers written in ternary expansion with
0s and 2s. There are many variations and generalisations, all called Cantor
sets. For example, all numbers x between 0 and 1 in a base-5 expansion
with only 1s and 3s form such a Cantor set. Its geometrical construction is
obvious. At the first step from the unit interval the three open intervals (0,
1/5) (2/5, 3/5) and (4/5, 1) are removed, etc.

The next important concept is *Smale's horseshoe*. It can be introduced in
the following way (cf. Fig. 4.17). The unit square 0<x, y<1 is shortened
in the x-direction by the factor 5 and stretched in the y-direction by the
same factor. The resulting vertical strip is then curved as a horseshoe and
placed upon the original square as shown in Fig. 4.17b. The points in the
horizontal strips labelled 1 and 3 are then mapped inside the square but
the points in the strips numbered 0, 2, 4 fall outside the square. After a
great many iterations most orbits will leave the square. However, all orbits
which stay forever form a so-called horseshoe. If the coordinates of a point
of the square are written in a base-5 expansion as

$$\begin{cases} x = 0 \cdot x_1 x_2 x_3 x_4 \ldots \\ y = 0 \cdot y_1 y_2 y_3 y_4 \ldots \end{cases}$$

then the mapping can be described as follows. If $y_1 = 0, 2, 4$, the image is
outside the unit square, but if $y_1 = 1, 3$, we have the image

$$\begin{cases} x' = 0 \cdot y_1 x x x \ldots \\ y' = 0 \cdot y_2 y_3 y_4 y_5 \ldots \end{cases}$$

All points for which the x_j and y_i are either 1 and 3 belong to orbits
staying permanently in the square. Thus they are the elements of the
horseshoe. The projection on either axis forms a Cantor set of the type
described above. Therefore the horseshoe appears to be the Cartesian
product of two Cantor sets. The dynamics of the horseshoe can be
described as a shift of the double-infinite sequence formed by the
expansions of x and y, . . . $y_4 y_3 y_2 y_1 x_1 x_2 x_3 x_4$ . . . . An iteration step is then
translated into a single shift to the right. In abstract language a horseshoe
is a two-symbol shift. This construction also shows that there exists a
countable set of periodic orbits and an uncountable set of aperiodic orbits.

The horseshoe as described above can be generalised in a variety of
ways but all horseshoes share the same characteristics. The occurrence of a
horseshoe in a map is an indication of chaos. According to a theorem of
Smale and Birkhoff, a horseshoe is generated when the unstable manifold
and the stable manifold of a hyperbolic fixed point intersect transversally.
Such an intersection is called a homoclinic point. Thus we have the
pattern: homoclinic point → horseshoe → chaos.

The next topic is the so-called *strange attractor*. There is no generally
accepted mathematical definition. However, as a rule, a strange attractor
is a nontrivial subset of a line of infinite length with an infinite number of

loops with a Cantor set as a cross-section. The basic idea is given in Fig. 4.18, which is almost self-explanatory. Starting from the unit square, a middle-third rectangular region is deleted, its boundary staying away from the previous boundary at a distance 1/3 in horizontal or vertical direction. From the remaining area a region is deleted, the boundary of which stays away at a distance 1/9 from the previous boundary, etc.

When, in this way, an infinite number of open, simply connected regions have been deleted, the remainder is a continuous polygon of infinite length. The cross-section, say at $y = \frac{1}{2}$, is the classical Cantor set. The initial part of this polygon can easily be followed in Fig. 4.18. Points of the polygon can be parametrised by their Euclidean distance s to the origin taken along the polygon. The polygon can be interpreted as the unstable manifold of the origin. The action of the two-dimensional mapping on this invariant line may be described as $s \to as$, $a > 1$. In this way orbits can be constructed both geometrically and numerically. Along the polygon the one-dimensional dynamic behaviour is uninteresting. However, the combined action of this most simple one-dimensional map and the infinite number of loops inside the square produces a complicated strange attractor as the two-dimensional limit set of the orbits along the polygon. This model could be analysed further in a quantitative way. We confine ourselves, however, to remarking that the nature of the limit set may be dependent on a. It could happen that, due to some resonance

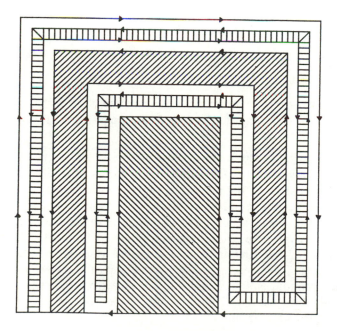

Fig. 4.18 A geometrical construction of a strange attractor.

effect, the limit set reduces to a few attracting points, a periodic cycle. It could also happen that some parts of the polygon are densely covered by the limit set. The model described here is based upon a similar model invented by L.E.J. Brouwer, the founder of modern topology, as an illustration of a common boundary of three neighbouring countries. So we propose to call it the 'Brouwer attractor'.

A simple example of a strange attractor was given in 1976 by Hénon [25], who studied quadratic maps, in the standard representation

(4.81)
$$\begin{cases} x_{n+1} = y_n \\ y_{n+1} = bx_n + ay_n - y_n^2 \end{cases}$$

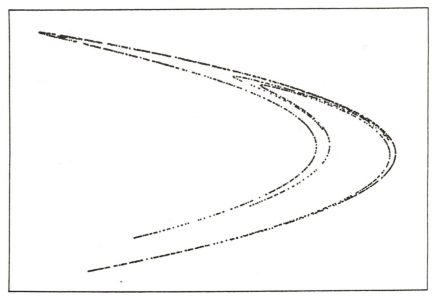

Fig. 4.19. The Hénon attractor. Horizontal axis from −1.5 to 1.5, vertical axis from −0.45 to 0.45.

For $a = 3.1678$, $b = 0.3$, he discovered the strange attractor shown in Fig. 4.19. This attractor shows all the features attributed to a strange attractor. In particular the numerical experiments revealed a self-similarity and a Cantor-like cross-section. The strange attractor can be considered as the limit set of the invariant unstable manifold of the saddle at the origin. For the given values of a and b its multipliers are −0.0920 and 3.2598. The small value of the first multiplier suggests a strong lateral attraction. Indeed, numerical experiments show that, apart from an initial arc, the unstable manifold is almost indistinguishable from the strange attractor. We recall from section 4.2 the parametrisation of the unstable manifold emanating from the origin

(4.82) $x = F(t), y = F(\lambda t)$

where λ is the largest multiplier and where $F(z)$ is an entire analytic function satisfying

(4.83) $$F(\lambda^2 z) = bF(z) + aF(\lambda z) - F^2(\lambda z)$$

with $F(z) = z + c_2 z^2 + c_3 z^3 + \ldots$

As a final example of a strange attractor we present the following model, details of which can be found in ref. [37].

(4.84) $$\begin{cases} x_{n+1} = \tfrac{1}{3} x_n(1-2y_n) + y_n \\ y_{n+1} = 4 y_n(1-y_n) \end{cases}$$

The unit square $0 \leqslant x, y \leqslant 1$ is mapped like a pinched horseshoe into itself, cf. Fig. 4.20. There are two fixed points on both saddles. The saddle $(0, 0)$ has multipliers $1/3$, 4. The saddle $(9/14, 3/4)$ has multipliers $-1/6$, -2. The horizontal lines $y = 0$ and $y = 3/4$ are the corresponding stable manifolds. Both unstable manifolds can be parametrised in a very explicit way. The unstable manifold of the origin is determined by

(4.85) $$\begin{cases} x = \tfrac{1}{2} - \sum_{k=1}^{\infty} 3^{-k} \varphi_k(t) \\ y = \tfrac{1}{2}(1 - \cos t) \end{cases}$$

where

$$\varphi_k(t) = \frac{\sin t}{2^k \sin (2^{-k} t)} = \cos \frac{t}{2} \cos \frac{t}{2^2} \ldots \cos \frac{t}{2^k}$$

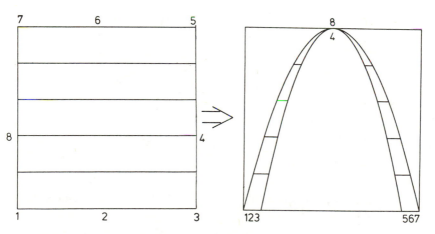

Fig. 4.20 A pinched horseshoe map.

According to (4.85) it is like a sine curve folded up an infinite number of times so that it fits inside the unit square. It appears that in this case the strange attractor consists of this unstable manifold together with its

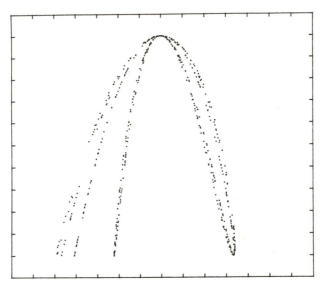

Fig. 4.21 An orbit on the unstable manifold of the origin. Horizontal axis from −0.1 to 1.1, vertical axis from −0.1 to 1.1

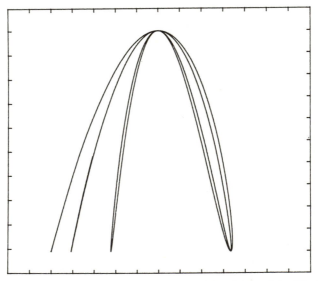

Fig. 4.22 The first few loops of the unstable manifold of the origin. Horizontal axis from −0.1 to 1.1, vertical axis from −0.1 to 1.1.

closure. Again cross-sections with horizontal lines are Cantor sets. Of particular interest is the cross-section at $y = 3/4$, the stable manifold of the second fixed point. These points of intersection are the so-called heteroclinic points. They are obtained from (4.85) by all parameter values $t = \pm 2\pi/3 + 2m\pi$ where m is an integer. The homoclinic points are given by all parameter values $t = 2^m \pi$. At those points the two manifolds are tangent in a multiple way. At each point of such homoclinic tangency an infinite number of folds come together. The unstable manifold of the second fixed point can be parametrised in a similar way. However, the closures of both unstable manifolds are identical. In Figs 4.21 and 4.22 a point plot and a corresponding line plot of the unstable manifold (4.85) are given.

A strange attractor has a fractal dimension as a rule. Lack of space prevents us from discussing this interesting topic. A good survey is given by Farmer *et al.* [15], and methods for estimating the dimension of attractors associated with experimental data are reviewed in Chapters 13 and 14.

References and selected books and papers

[1] Arnold, V.I. *Geometrical Methods in the Theory of Ordinary Differential Equations,* Springer, New York (1983).
[2] Arnold, V.I. and Avez, A. *Ergodic Problems of Classical Mechanics.* Benjamin, London (1968).
[3] Aronson, D.G., Chory, M.A., Hall, G.R. and McGehee, R.P. Bifurcations from an invariant circle for two-parameter families of maps of the plane: a computer assisted study. *Comm. Math. Phys.* **83**, 303–54 (1982).
[4] Aronson, D.G., Chory, M.A., Hall, G.R. and McGehee, R.P. Resonance phenomena for two-parameter families of maps of the plane: uniqueness and non-uniqueness of rotation numbers. In *Nonlinear Dynamics and Turbulence,* eds. Barenblatt *et al.,* pp. 35–47. Pitman, London.
[5] Barnsley, M.F., Geronimo, J.S. and Harrington, A.N. On the invariant sets of a family of quadratic maps. *Comm. Math. Phys.* **88**, 479–501 (1983).
[6] Berry, M. Regular and irregular motion. In *Topics in Nonlinear Mechanics,* ed. S. Jorna, pp. 16–120, New York (1978).
[7] Berry, M. Semiclassical mechanics of regular and irregular motion. In *Chaotic Behaviour of Deterministic Systems,* eds. Iooss *et al.,* pp. 171–271. Elsevier North-Holland, Amsterdam (1983).
[8] Billingsley, P. *Ergodic Theory and Information.* Wiley, New York (1965).
[9] Blanchard, P. Complex analytic dynamics on the Riemann sphere. *BAMS* **11**, 85–141 (1984).
[10] Campbell, D. K., Farmer, J.D. and Rose, H. (eds.) *Order in Chaos, Los Alamos Conference,* May 1982. *Physica D-7* (1983).
[11] Collet, P. and Eckmann, J.P. *Iterated Maps on the Interval as Dynamical Systems.* Birkhäuser, Boston (1980).
[12] Curry, J. H. and Yorke, J.A. A transition from Hopf bifurcation to chaos: computer experiments with maps on R^2. In *The Structure of Attractors in Dynamical Systems.* Springer, Berlin, pp. 48–68 (1978).
[13] Curry, J.H., Garnett, L. and Sullivan, D. On the iteration of a rational function: computer experiments with Newton's method. *Comm. Math. Phys.* **91**, 267–77 (1983).
[14] Eckmann, J.P. Routes to chaos with special emphasis on period doubling. In *Chaotic Behaviour of Deterministic Systems,* eds. Iooss *et al.,* pp. 455–510. Elsevier North-Holland, Amsterdam (1983).

[15] Farmer, J.D., Ott, E. and Yorke, J.A. The dimensions of chaotic attractors. In *Order in Chaos, Los Alamos Conference,* May, 1982, eds Campbell *et al., Physica D-7,* 153–80 (1983).

[16] Feigenbaum, M.J. The metric universal properties of period doubling bifurcations and the spectrum for a route to turbulence. In *Nonlinear Dynamics,* ed. R.H.G. Helleman, pp. 330–6. *Ann. NY Acad. Sci.* **357,** 330–6 (1980).

[17] Feigenbaum, M.J. Universal behaviour in nonlinear systems. In *Order in Chaos, Los Alamos Conference,* May 1982, eds. Campbell *et al., Physica D-7,* 16–39 (1983).

[18] Feigenbaum, M.J. Universal behaviour in nonlinear systems. In *Nonlinear Dynamics and Turbulence,* eds. Barenblatt *et al.,* pp. 101–38. Pitman, New York (1983).

[19] Garrido, L. (ed.) Dynamical systems and chaos. *Proc. Conf. Stat. Mech. Barcelona,* Springer, Berlin, 179 (1983).

[20] Guckenheimer, J., Oster, G.F. and Ipatchki, A. The dynamics of density-dependent population models. *J. Math. Biol.* **4,** 101–47 (1977).

[21] Guckenheimer, J. and Holmes, P. *Nonlinear Oscillations, Dynamical Systems and Bifurcations of Vector Fields.* Springer, Berlin (1983).

[22] Gumowski, I. and Mira, C. *Recurrences and Discrete Dynamical Systems.* Springer, Berlin, p. 809 (1980).

[23] Helleman, R.H.G. (ed.) Nonlinear dynamics. *Ann. NY Acad. Sci.* **357** (1980).

[24] Hénon, M. Numerical study of quadratic area-preserving mappings. *Q. Appl. Math.* **27,** 219–312 (1969).

[25] Hénon, M. A two-dimensional mapping with a strange attractor. *Comm. Math. Phys.* **50,** 69–78 (1976).

[26] Hénon, M. Numerical exploration of Hamiltonian systems. In *Chaotic Behaviour of Deterministic Systems,* eds. Iooss *et al.,* pp. 53–170. Elsevier North-Holland, Amsterdam (1983).

[27] Iooss, G. *Bifurcation of Maps and Applications,* Elsevier North-Holland, Amsterdam (1979).

[28] Iooss, G., Helleman, R.H.G. and Stora, R. (eds.) *Chaotic Behaviour of Deterministic Systems.* Elsevier North-Holland, Amsterdam (1983).

[29] Jorna, S. (ed.) *Topics in Nonlinear Dynamics.* American Institute of Physics, New York (1978).

[30] Julia, G. Mémoire sur l'itération des fonctions rationelles. *J. Math. Pure Appl.* **4,** 47–245 (1918).

[31] Lanford, O.E. Smooth transformation of intervals. In *Sém. Bourbaki* vol. 1980/1981. Springer, Berlin, pp. 36–54 (1980).

[32] Lanford, O.E. Dynamical systems. In *Chaotic Behaviour of Deterministic Systems,* eds. Iooss *et al.,* pp. 3–51. Elsevier North-Holland, Amsterdam (1983).

[33] Lauwerier, H.A. *Bifurcation of a map at resonance 1:4.* Report Amst. Math. Centre TW 245 (1983).

[34] Lauwerier, H.A. *Local bifurcation of a logistic delay map.* Report Amst. Math. Centre TW 246 (1983).

[35] Lauwerier, H.A. *Global bifurcation of a logistic delay map.* Report Amst. Centre Math. Comp. Sci. AM-R8402 (1984).

[36] Lauwerier, H.A. *Entire functions for the logistic map.* Report Amst. Centre Math. Comp. Sci. AM-R8404 (1984).

[37] Lauwerier, H.A. *A case of a not so strange strange attractor.* Report Amst. Centre Math. Comp. Sci. AM-R8406 (1984).

[38] Lichtenberg, A. J. and Lieberman, M.A. *Regular and Stochastic Motion.* Springer, Berlin (1982).

[39] MacKay, R.S. *Renormalisation in area-preserving maps*. Thesis, Princeton University (1982).

[40] MacKay, R.S. A renormalisation approach to invariant circles in area-preserving maps. In *Order in Chaos, Los Alamos Conference*, May 1982, eds. Campbell *et al., Physica D-7*, 283–300 (1983).

[41] Mandelbrot, B.B. Fractal aspects of the iteration of $z \to \lambda z(1-z)$ for complex λ and z. In *Nonlinear Dynamics*, ed. R.H.G. Helleman, pp. 249–59. *Ann. NY Acad. Sci.* **357**, 249–59.

[42] Mandelbrot, B.B. *The Fractal Geometry of Nature*. W. H. Freeman, San Francisco (1982).

[43] Mandelbrot, B.B. On the quadratic mapping $z \to z^2 - \mu$ for complex μ and z. In *Order in Chaos, Los Alamos Conference*, May 1982, eds. Campbell *et al., Physica D-7*, 224–39 (1983).

[44] May, R.M. Simple mathematical models with very complicated dynamics. *Nature, Lond.* **261**, 459–67 (1976).

[45] May, R.M. Nonlinear phenomena in ecology and epidemiology. In *Nonlinear Dynamics*, ed. R.H.G. Helleman, pp. 267–81. *Ann. NY Acad. Sci.* **357**, 267–81 (1980).

[46] May, R.M. Nonlinear problems in ecology and resource management. In *Chaotic Behaviour of Deterministic Systems*, eds. Iooss *et al.*, pp. 514–63. Elsevier North-Holland, Amsterdam (1983).

[47] Mira, C. Complex dynamics in two-dimensional endomorphisms. *Nonlinear Analysis* **4**, 1167–87 (1980).

[48] Misiurewicz, M. Strange attractors for the Lozi mappings. In *Nonlinear Dynamics*, ed. R.H.G. Helleman, pp. 348–58. *Ann. NY Acad. Sci.* **357**, 348–58.

[49] Misiurewicz, M. Maps of an interval. In *Chaotic Behaviour of Deterministic Systems*, eds. Iooss *et al.*, pp. 565–90. Elsevier North-Holland, Amsterdam (1983).

[50] Morris, H.C., Ryan, E.E. and Dodd, R.K. Snap-back repellers and chaos in a discrete population model with delayed recruitment. *Nonlinear Analysis* **7**, 571–621 (1983).

[51] Nusse, H.E. Complicated dynamical behaviour in discrete population models. *Nw Arch. v. Wisk.* **2** (4), 43–81 (1984).

[52] Peitgen, H.O., Saupe, D. and Haeseler, F.V. Cayley's problem and Julia sets. *Math. Intell.* **6**, 11–20 (1984).

[53] Pounder, J.R. and Rogers, T.D. The geometry of chaos: dynamics of a nonlinear second-order difference equation. *Bull. Math. Biol.* **42**, 551–97 (1980).

[54] Thurston, H.E. On the dynamics of iterated rational maps. Preprint.

[55] Whitley, D. Discrete dynamical systems in dimensions one and two. *Bull. Lond. Math. Soc.* **15**, 177–217 (1983).

Part III

Endogenous chaos

5
Chaos in feedback systems

A. Mees

Department of Mathematics, University of Western Australia,
Nedlands, W.A. 6009, Australia

5.1 Summary

Nonlinear systems theory is of great importance to anyone interested in
feedback systems. It is also true that the theory of feedback systems has
made important contributions to nonlinear systems theory. This chapter
discusses some chaotic feedback systems, drawn from electronic circuit
theory and elsewhere, and shows how they may be analysed. So far, most
of the techniques required have been taken directly from the usual
differential and difference equation theory, but some results with a more
control system-theoretic flavour are now available.

5.2 Feedback systems may be chaotic

Any dynamical system described, for example, by difference or differential
equations may be regarded as a feedback system. Conversely, any finite
dimensional feedback system may be described by difference or differential
equations, which we shall call a state-space description, but many physical
and biological systems lend themselves most naturally to a description
which highlights feedback rather than state [22].

We think of a *system* as a causal connection between a set of time
functions v called *inputs* and a set of functions y called *outputs:*

$$(5.1) \qquad\qquad y = sv$$

The notation used is not to be construed as suggesting linearity or even
finite dimensionality! Even in the simplest cases, y depends on the whole
past of v. We may connect systems together by making some or all of the
outputs of one system inputs to another, and we can represent the
connections by a directed graph. A *feedback system* is then a set of

interconnected systems for which the graph contains a cycle. More simply put, a feedback system has its inputs affected by its outputs.

We can write $y_k = s_k v_k$ for systems $s_1, \ldots, s_n$. Each input v_k may depend on outputs y_j. If there is a chain such that some v_k ultimately depends on its own y_k, we have a feedback system. The simplest feedback system will be

$$y_1 = s_1 v_1$$
$$y_2 = s_2 v_2$$
$$v_1 = y_2 + u$$
$$v_2 = y_1$$

Here we regard u as an input from 'the environment' and y_1 or y_2 or both as the output. Assuming y_1 is what we are interested in, we can write

(5.2) $y_1 = s_1 v_1$
(5.3) $v_1 = s_2 s_1 v_1 + u$

where we regard (5.2) as describing the 'readout map' which shows the environment what the system is doing, and (5.3) as being the system equations. Notice that (5.3) is a fixed point problem: find v_1 to satisfy the equation for given s_1, s_2 and u. For detailed discussions of the definition of a system, the meaning of feedback, and other issues we have glossed over, consult the control-theory literature [3, 22].

Feedback systems are common in both the natural world and the works of man. It is reasonable to model predator–prey systems as feedback systems, for example. Here s_1 might be the prey population system and s_2 the predator population system, with v_1 being the entire history of the prey population. The external input u would probably be zero, but might represent colonisation from another part of the environment. One can also use more abstract feedback models even for a single species in which, say, v_1 is a population birthrate over time, s_1 outputs the survival rate to maturity, and s_2 outputs the birthrate for a given mature population. The point here is that the number of births affects future generations of breeding adults, and the size of the breeding population affects the number of births.

At the simplest model level there is little to choose between represent-ation as a feedback system as against representation as a difference or differential equation, but it is well known that for larger system models there are both conceptual and technical advantages to the feedback system representation wherever it is a natural one. Since it is widely felt that large systems may be likely to have chaotic modes of behaviour, it is a little surprising that there has been so little fuss about chaos in the feedback systems literature [23]. If anything, it would seem more likely that chaos would have first come to notice in feedback systems, yet the reverse is true: in thinking about feedback systems one tends to imagine globally stable equilibria, limit cycles, and precious little else. Why is this?

The main reason appears to be that feedback systems have been studied mainly by engineers (especially control engineers) who are interested in *synthesis* more than in *analysis*. Artificial systems are usually carefully constructed to have regular, predictable, 'simple' behaviour, and most research effort has been directed to this end rather than to discovery and understanding of exotic dynamical behaviour. For the same reason, the tools of the control theorist, ranging from the most mathematically esoteric to the most pragmatically useful, do not often seem to be what one needs to understand chaotic phenomena.

Yet chaos is certainly possible in feedback systems. Even ignoring standard results about converting between feedback and state formulations [22], we can easily regard the prototype chaotic difference equation as a feedback system. Let x_t be the breeding population of blowflies [20, 21] at time t and let $y_t = g(x_t)$ be the number of eggs produced. Let $z_t = h(y_t)$ be the number of eggs which produce mature adults one time unit later, so $x_{t+1} = z_t$. Then we have a feedback system consisting of the function g followed by the function h fed back via a time delay to the input of the function g:

$$x_{t+1} = h(g(x_t))$$

If the composition of g and h has certain properties, we can, of course, get chaos [16, 20, 21, 26]. For example, when $f(x) = h(g(x))$ describes a humped map, then, if the hump is peaked enough, we may get chaos. As this is one of the most extensively studied problems in dynamical systems, we refer the reader to the literature [16, 20, 21, 26, 31].

There are also examples of continuous-time feedback systems with chaotic behaviour, as we shall see. Indeed, once one turns from engineering to biology, there seem to be many examples of feedback systems which could be chaotic under the right conditions. Most of this chapter is devoted to presenting examples of chaotic feedback systems, with discussion of how they may be analysed, but I shall return at the end to a discussion of the general relationship between feedback and chaos.

5.3 Discrete time feedback systems

It is common to model feedback systems as having a linear dynamical part (a linear difference or differential equation) followed by a nonlinear instantaneous part (a vector function) whose output is the input of the linear part. Thus, in the case of discrete time systems,

(5.4)
$$x_{t+1} = \mathbf{A}x_t + \mathbf{B}u_t$$
$$y_t = \mathbf{C}x_t$$
$$u_t = f(y_t)$$

where $\mathbf{A}$, $\mathbf{B}$ and $\mathbf{C}$ are matrices and f is a nonlinear function. Although this may not be the most natural model, it is generally applicable and we shall use it for definiteness. If **B** has a single column and **C** has a single row, then both y_t and u_t are scalar-valued and the system is called a *single-loop* feedback system, because there is only one 'feedback loop' consisting of the signals u and y which are scalar valued.

Writing (5.4) as

$$(5.5) \qquad x_{t+1} = \mathbf{A}x_t = \mathbf{B}f(\mathbf{C}x_t)$$

we obtain a special difference equation. In the extreme case where x_t is scalar valued, we have a one-dimensional difference equation to which the standard results can be applied [16, 26, 31], and if x is vector-valued, we may use results such as Marotto's snap-back repeller theorem [18] or its generalisation by Kloeden [15]. To illustrate Marotto's approach, let us look at an example given by Baillieul *et al.* [1]. We will then indicate how the same problem may be studied from a more control-theoretic viewpoint, as Baillieul *et al.* originally did.

A standard model of part of a power conversion network is a first-order pulse-width modulated feedback system [1]. This corresponds to (5.5) with x_t scalar, **A** and **B** scalar, **C** = 1, and a piecewise smooth monotone form for f. With the choice of f adopted by Baillieul *et al.* (5.5) becomes

$$(5.6) \qquad x_{t+1} = F(x_t)$$

where

$$(5.7) \qquad F(x) = \begin{cases} mx + b_1 & x \leq \gamma \\ px + b_2 & \gamma < x < \delta \\ mx & x \geq \delta \end{cases}$$

for certain constants $\gamma, \delta, m, p, b_1, b_2$ (see Fig. 5.1). We assume $\gamma > 0$.

For there to be a nontrivial solution, the 45° line in Fig. 5.1 must intersect the graph of F. Except for an obvious degenerate case, when $m = 1$, there can only be one such solution at, say, $\hat{x}$. The interesting case is where $\gamma < \hat{x} < \delta$ and $p < -1$, so that $\hat{x}$ is a repeller. Notice that there are two points z_1 and z_2 in Fig. 5.1, which are pre-images of $\hat{x}$, i.e. $F(z_1) = \hat{x}$ and $F(z_2) = \hat{x}$. If there is some point y close to $\hat{x}$ which iterates out to one of the points z_k, i.e. if $x_0 = y$ and, say, $x_r = z_2$ for some integer r, then the trajectory starting from r works its way out to z_2 after which it snaps back to $\hat{x}$, where it remains. In such a case, $\hat{x}$ is called a *snap-back repeller* and Marotto [18] showed that the existence of a snap-back repeller is sufficient for chaos, in the usual sense of there being infinitely many periodic trajectories and infinitely many trajectories which are not asymptotically periodic.

In the present case we can easily write down explicit conditions for $\hat{x}$ to be a snap-back repeller. They are (besides $\gamma < \hat{x} < \delta$ and $p < -1$) that a

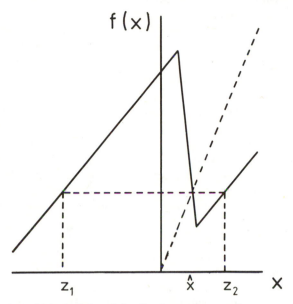

Fig. 5.1 The pulse-width modulated feedback system contains a snap-back repeller, $\hat{x}$, when z_1 or z_2 is in the basin of repulsion of $\hat{x}$. Here z_1 and z_2 are the two pre-images of $\hat{x}$.

pre-image of z_1 or of z_2 should lie in (γ, δ), since any point in $(\gamma, \delta$ has pre-images arbitrarily close to $\hat{x}$. We therefore need to find y such that $py = b_2 = z_2$ where $mz_2 = \hat{x} = b_2/(1-p)$. If $y \in (\gamma, \delta)$ then $\hat{x}$ is a snap-back repeller. Now

$$ y = \frac{b_2}{mp(1-p)} - \frac{b_2}{p} $$

so $\hat{x}$ is a snap-back repeller if

$$ (5.8) \qquad \frac{b_2-b_1}{m-p} < \frac{b_2}{p} \left(\frac{1}{m(1-p)} - 1 \right) < \frac{b_2}{m-p} $$

since $\gamma = (b_2 - b_1)/(m - p)$ and $\delta = b_2/(m - p)$. There is another acceptable inequality corresponding to using z_1 instead of z_2. With a little work, it is possible to see that (5.8) and its alternative are equivalent to the conditions for chaos in [1].

Actually, Marotto's theorem [18] requires that F be smooth and also requires that a certain nondegeneracy condition be satisfied. However, a later paper by Kloeden [15] proved that F need only be continuous, as well as introducing other generalisations. The nondegeneracy conditions are trivially satisfied here.

Although we started with a feedback system, we studied it by reducing it to a difference equation. Even if we end up with a difference equation on $\mathbb{R}^n$, the snap-back repeller theorem and its generalisations still hold, though it may not be easy to check that the conditions are satisfied. Nevertheless, it seems worth asking whether the feedback structure can tell us more about the system or about related systems. Could we have perhaps avoided the transformation to state space and used theorems about feedback systems?

One of feedback system theory's favourite tricks is to show that a given system is, in a suitable functional–analytic sense, close to a system whose behaviour is well understood. The circle criterion for stability [3, 22] is a good example, where conditions are given for a nonlinear system to be stable if a linear comparison system is stable. This is not at all the same thing as conventional small-parameter theory: it is often possible to allow discontinuous and even multivalued systems in the feedback loop. The key is usually to choose appropriate function spaces and show that the input and output belong to them [22]. Baillieul *et al.* [1] examined discrete-time feedback systems in this spirit, and produced some useful results.

To state their results in full here would require too much background, but it is possible to understand the general ideas. Suppose we have a *single-loop* feedback system, so f in (5.5) is scalar though x_t may be vector-valued. Suppose there is a system with scalar x_t of the form (5.6), and with a piecewise linear F, which can be shown to be chaotic. If f is close to F (possibly after origin shift and linear transformation) while the linear system corresponding to $(\mathbf{A}, \mathbf{B}, \mathbf{C})$ after the effects of the transformations is close to a simple time delay, then (5.5) is also chaotic. That is, we verify chaotic behaviour by finding a simpler comparison system which is 'close' to the one being studied, and showing the comparison system is chaotic. Closeness is measured by a condition involving the difference between f and F in the Lipschitz norm, together with the l_∞ norm of the difference between the linear system and a simple delay.

In the context of the pulse-width modulation example, this means that the linear part of the system can be high order (i.e. be more complex than a simple delay) and the nonlinear part can diverge from the form in Fig. 5.1, as long as a suitable condition on these variations is satisfied. (Actually, the theorem given in [1] requires F to have a simpler form than we have allowed, but the principle is unchanged.)

Obviously, one could find many more chaotic discrete time feedback systems with little effort, by transforming known chaotic difference equations to feedback form, but the pulse-width modulated system is a realistic example where the parameter values required for chaos are physically possible.

5.4 Continuous time feedback systems

Once we start to think about continuous time feedback systems, there are many examples available but, ironically, it is much harder to get rigorous results. The first example known to the author of an explicitly recognised chaotic feedback system (in either discrete or continuous time) was Sparrow's [32]. He replaced the delay in the one-hump difference map by a suitable high-order continuous-time linear system and showed that the resulting system still appeared to be chaotic on the evidence of numerical simulation.

In a later paper [33], Sparrow displayed chaos in a much lower order linear system in a feedback loop with a piecewise-linear function. Because his feedback system becomes, in state-space form, a pair of linear vector fields in $\mathbb{R}^3$, joined continuously along a plane, he was able to give a rather detailed analysis to supplement the numerical evidence. Later, Brockett [2] obtained similar results independently. It should also be noted that several of Rössler's chaotic differential equations [29] are equivalent to single-loop feedback systems.

Figure 5.2 shows a trajectory in Sparrow's piecewise-linear system, which is equivalent to a linear system with transfer function $1/(s+1)^3$ in a single negative feedback loop with

$$f(z) = \begin{cases} 8.4z - 3.35 & z \leq 3/7 \\ -8.4rz + 0.25 + 3.6r & z \geq 3/7 \end{cases}$$

and $r = 19$. The parameter values are not particularly critical but were chosen to agree with suggestions by Rössler *et al.* [29]. To analyse this system, Sparrow converts it to a differential equation in $\mathbb{R}^3$, namely

$$\begin{aligned} \dot{x} &= -f(z) - x \\ \dot{y} &= x - y \\ \dot{z} &= y - z \end{aligned}$$

and considers return maps to the plane $z = 3/7$. By approximating certain features by a one-dimensional map of the usual kind, he is able to explain the features seen in numerical simulations. This system has also been studied more recently [6].

Holmes and Moon [9] considered a feedback-controlled mechanical

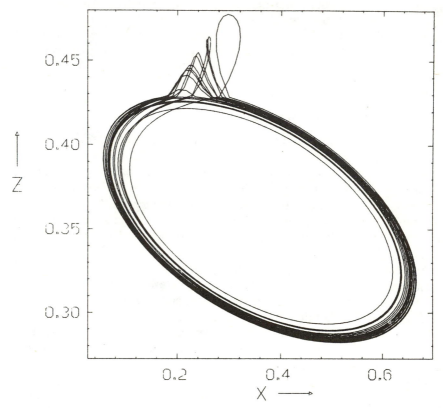

Fig. 5.2 A chaotic trajectory from Sparrow's piecewise-linear-feedback system described in section 5.4. To emphasise chaotic behaviour, the distance $z - (3/7)$ has been multiplied by 10 whenever $z > (3/7)$.

positioning device described by

$$\ddot{x} + \delta\dot{x} + K(x)x = -z + F(t)$$
$$\dot{z} + \alpha z = (x - x_r(t))\alpha G$$

where $F(t)$ is an external force and $x_r(t)$ is the desired position. The nonlinear spring constant is such that there are multiple equilibria; for example,

$$K(x) = \frac{1}{B}(x^2 - 1)(x^2 - B)$$

where $B > 1$, which has (when $F(t) = 0$ and $G = 0$) five equilibria, three of which are stable. In this case, numerical simulation with certain parameter values produces chaotic solutions with a peaked spectrum, indicating an underlying approximate periodicity. The spectrum of the piecewise linear system discussed above would also be peaked: both kinds of system seem to have chaotic solutions that are rather like periodic solutions with quasi-random perturbations.

A somewhat similar system to the above has been modelled by a lagged-feedback Van de Pol oscillator [35] and has been found to produce apparently chaotic output in simulations.

Certain electronic systems may be equally comfortably described by feedback systems or by differential equations; we mention here only Josephson junctions [12, 25], a varactor diode oscillator [19], a synchron-isation system [34] and a certain operational amplifier feedback loop (Y.S. Tang, pers. comm.).

Among the many models of biological feedback systems that may display chaos, we only mention Rapp's epilepsy model [27], Glass and Mackey's two-pacemaker-site heart model [5] and the dopamine dynamics model of King *et al.* [13]. The last of these, which appears to have some relevance to schizophrenia and to Parkinsonism, deals with interactions between neuronal feedback and neurotransmitter kinetics in a two-feed-back loop model. Chaotic solutions of the model are claimed to correlate well with medical observations.

All the continuous time systems mentioned so far have been studied by conversion to differential equations. Certain feedback system oriented methods may be useful, for example in proving that solutions remain bounded [22], but function-space and Fourier-transform approaches have so far been relatively unhelpful. It is interesting to note that the describing function method (also called the method of harmonic balance [22]) may be helpful in cases like the two we have discussed: it seems to be very good at picking out the underlying periodic component of systems with peaked spectrum [23].

Because of the widespread use of transform calculus in control theory, techniques for studying the spectra of feedback systems would seem promising for chaotic solutions. A few years ago, there was a good deal of interest in spectral analysis [10–12] and it turned out that the situation mentioned above, with sharp peaks in a broadband noise-like background, was fairly common. This is discussed in a little more detail in the review paper by Mees and Sparrow [23]. Interest seems now to have moved away from the spectrum on the grounds that it discards too much information and is in any case not well defined unless the system is ergodic. (The Fourier transform of one trajectory, even over a very long time, is not necessarily representative of transforms of other trajectories.) Neverthe-less, more recent attempts to calculate Lyapunov exponents and fractal dimensions [4, 8] seem to be no more useful, and it is possible that spectral methods will become fashionable again. Indeed, a form of correlation integral (though not the familiar cross-correlation with its spectral rep-resentation) has recently been claimed to have great potential in studying chaos and even in distinguishing it from noise [7].

5.5 The relevance of chaos in feedback loops

For the engineer designing a feedback system, it would appear that chaos is always to be avoided: 'noisy' oscillations with little information content, or sudden unpredictable excursions of physical variables, are seldom likely to be desirable. Feedback tends to be used to stabilise systems, not to randomise them. Similarly, natural systems would probably evolve to avoid chaos; as May [20] pointed out, the dramatic population crashes in his ecosystem models would probably lead to extinction. Indeed, successful self-adapting systems of any kind would probably be resistant to chaotic or otherwise irregular inputs from their environments.

The author and colleagues have argued elsewhere [28] that frequency-regulated control systems may be immune to chaos in the same way as they are immune to noise. This will be so in the 'noisy limit cycle' version of chaos where the spectrum is peaked sharply, if control information is transferred via frequency rather than via amplitude. This seems to be the case in part of the regulatory loop of the salivary gland of *Calliphora erythrocephalla*.

Curiously, an (at first sight) almost diametrically opposed argument may be advanced without contradicting the above. Chaos may actually be *helpful* in certain circumstances. Imagine a dynamical system with multiple locally stable equilibria (or limit cycles). If this is perturbed by 'noise', which may in fact be chaotic output from a system interconnected with the first, then the state of the system will tend to migrate between the basins of the different equilibria, spending amounts of time in the various basins which depend on basin size and depth, and on noise amplitude. If one particular equilibrium is desirable, this will ensure that it is actually reached, even though it will also ensure that the system does not remain there all the time.

This type of dynamical system, with noise input replacing chaotic input, has been used successfully to model the slow cooling involved in chemical annealing [24] and has recently been applied to problems of global optimisation of difficult problems with many local optima [14,17]. It seems a good candidate for many problems in neurological computation, in part because it fits naturally with models of parallel processing. Sejonwski and Hinton [30] have applied these ideas to models of visual computation for image recognition. These authors specifically consider the visual cortex as a noisy environment and have had some success, in preliminary studies, in showing that the noise may help to achieve a global optimum correspond-ing to correct recognition. In the context of the above remarks about chaos, we can imagine that some kinds of chaotic input would be just as useful as noise. It is also possible to conceive of a single model in which the chaos or noise is intrinsic rather than externally generated; a fairly realistic model would probably have to allow for intrinsic and extrinsic noise and chaos, and it is not possible *a priori* to say which is the most important.

In all, it seems that feedback systems are ripe for study as chaos generators. We have seen that they can behave chaotically but we have seen that, so far, most work has been solely in terms of state descriptions. Much remains to be done.

References

[1] Baillieul, J., Brockett, R. W. and Washburn, R. B. Chaotic motion in nonlinear feedback systems. *IEEE Trans. CAS-27* 990–7, (1980).

[2] Brockett, R. W. On conditions leading to chaos in feedback systems. Preprint. Division of Applied Sciences, Harvard University (1982).

[3] Desoer, C. A. and Vidyasagar, M. *Feedback Systems: Input–Output Properties*. Academic Press, New York (1979).

[4] Frederickson, P., Kaplan, J. L., Yorke, E. D. and Yorke, J. A. The Lyapunov dimension of strange attractors. *J. Diff. Eqns* **49**, 185–207 (1983).

[5] Glass, L. and Mackey, M. C. Pathological conditions resulting from instabilities in physiological control systems. *Ann. NY Acad. Sci.* **316**, 214–35 (1979).

[6] Glendinning, P. and Sparrow, C. T. Local and global behaviour near homoclinic orbits. *J. Stat. Phys.* **35**, 645–97 (1984).

[7] Grassberger, P. and Procaccia, I. Measuring the strangeness of strange attractors. *Physica 9-D*, 189–208 (1983).

[8] Greenside, H. S., Wolf, A., Swift, J. and Pignataro, T. Impracticability of a box-counting algorithm for calculating the dimensionality of strange attractors. *Phys. Rev.* **A25**, 3453–6 (1982).

[9] Holmes, P. and Moon, F. C. Strange attractors and chaos in nonlinear mechanics. *J. Appl. Mech.* December (1983).

[10] Huberman, B. A. and Rudnick, J. Scaling behaviour of chaotic flows. *Phys. Rev. Lett.* **45**, 154–6 (1980).

[11] Huberman, B. A. and Zisook, A. B. Power spectra of strange attractors. *Phys. Rev. Lett.* **46**, 626–8 (1981).

[12] Huberman, B. A., Crutchfield, J. P. and Packard, N. W. Noise phenomena in Josephson junctions. *Appl. Phys. Lett.* **37**, 750–2 (1980).

[13] King, R., Barchas, J. D. and Huberman, B. A. Chaotic behaviour in dopamine neurodynamics. Preprint, Stanford University School of Medicine (1984).

[14] Kirkpatrick, S., Gelatt, C. D. and Vecchi, M. P. *Optimization by simulated annealing*. Research Report R.C. 9355, IBM, Yorktown Heights (1982).

[15] Kloeden, P. E. Chaotic difference equations in $\mathbb{R}^n$. *J. Austral. Math. Soc. (Series A)* **31**, 217–25 (1981).

[16] Li, T. and Yorke, J. A. Period three implies chaos. *Amer. Math. Monthly,* **82**, 985 (1975).

[17] Lundy, M. and Mees, A. I. Convergence of the annealing algorithm. *Math. Programming* in press (1985).

[18] Marotto, F. R. Snap-back repellers imply chaos in $\mathbb{R}^n$. *J. Math. Anal. Appl.* **63**, 199–223 (1978).

[19] Matsumoto, T., Chua, L. O. and Tanaka, S. *The simplest chaotic non-autonomous circuit.* E.R.L. memorandum M84/28, University of California at Berkeley (1984).

[20] May, R. M. Deterministic models with chaotic dynamics. *Nature, Lond.* **256**, 165–6 (1975).

[21] May, R. M. and Oster, G. Bifurcations and dynamic complexity in simple ecological models. *Amer. Natur.* **110**, 573–99 (1976).

[22] Mees, A. I. *Dynamics of Feedback Systems.* Wiley, New York (1981).

[23] Mees, A. I. and Sparrow, C. T. Chaos. *IEEE Proc.* **128(D)**, 201–5 (1981).

[24] Metropolis, N., Rosenbluth, A. W., Rosenbluth, M. N. and Teller, A. H. Equation of state calculation by fast computing machines. *J. Chem. Phys.* **21**, 1087–92 (1953).

[25] Odyniec, M. and Chua, L. O. Josephson-junction circuit analysis via integral manifolds. *IEEE Trans. CAS-30*, 308–20. (1983).

[26] Preston, C. J. Analysis of the iterates of a one-hump function. Preprint, Dept. Mathematics, Cambridge University (1975).

[27] Rapp, P. E. Preprint, Dept of Physiology, Medical College of Pennsylvania, Philadelphia (1984).

[28] Rapp, P. E., Mees, A. I. and Sparrow, C. T. Frequency dependent biochemical regulation is more accurate than amplitude-dependent control. *J. Theor. Biol.* **90**, 531–44 (1980).

[29] Rössler, R., Gotz, F. and Rössler, O. E. Chaos in endocrinology. *Biophys. J.* **25**, 216 (1979).

[30] Sejonwski, T. J. and Hinton, G. E. Parallel stochastic search in early vision. *Vision, Brain and Cooperative Computation,* eds. M. Arbib and A. R. Hanson (1984).

[31] Sharkovskii, A. N. Coexistence of cycles of a continuous map of a line into itself. *Ukr. Math. Z.* **16**, 61–71 (1964).

[32] Sparrow, C. T. Bifurcation and chaotic behaviour in simple feedback systems. *J. Theor. Biol.* **83**, 93–105 (1980).

[33] Sparrow, C. T. Chaotic behaviour in a 3-dimensional feedback system. *J. Math. Anal. and Applics* **83**, 275–91 (1981).

[34] Tang, Y. S., Mees, A. I. and Chua, L. O. Synchronization and chaos. *IEEE Trans. CAS-30,* 620–6 (1983).

[35] Ueda, Y., Doumoto, H. and Nobumoto, K. An example of random oscillations in three-order self-restoring system. *Proc. EEC Joint Meeting at Kansai, Japan* (1978).

6
The Lorenz equations

C. Sparrow

Department of Pure Mathematics and Mathematical Statistics,
University of Cambridge, 16 Mill Lane, Cambridge CB2 1SB, UK

6.1 Introduction

The Lorenz equations, named after Ed Lorenz who first introduced them as a model of a two-dimensional convection [21], have been important for a number of different reasons at various times in the past 20 years or so. Initially they were remarkable just because they are a simple three-dimensional nonlinear system of autonomous ordinary differential equations showing chaotic behaviour; though many such systems are now known (as described elsewhere in this volume), in 1963 such systems were almost unheard of. So much so, in fact, that, despite the beauty of Lorenz's original paper, and the remarkable progress he made in understanding the behaviour of his system, the paper (and the ideas) were largely ignored for nearly ten years. Also, of course, the equations were of importance because of their connection with the problem from which they were derived: here was the hope that turbulent phenomena could be modelled by simple finite-dimensional systems, but this aspect too was largely ignored.

In the 1970s, with the burgeoning interest in dynamical systems (and the wider acceptance of the notion of chaotic behaviour), the Lorenz equations featured in many papers. In several cases, important types of behaviour were first investigated in the Lorenz equations although these behaviours have since been recognised as typical of many systems. For instance Manneville and Pomeau's work on intermittence [24] and the Hénon map [18] both appeared in papers on the Lorenz system. At the same time, a specific geometric model of the equations in a small range of parameter values was developed. This work was led by Williams [38, 39] and Guckenheimer [14, 17], though the range of parameter values in question was actually that studied by Lorenz in his original paper. This geometric model (a geometrically constructed system which seems to have the same properties as the Lorenz equations though the connection depends on several global properties of the flow which have not yet been proved) has an attracting set, for a range of parameter values, which is a strange attractor, but which is a slightly different strange attractor at every parameter value in the range (see below). This attracting set, then, is not structurally stable, though it only just fails to satisfy Smale's axiom A [34]. This discovery probably had an important effect on the search for extensions to axiom A and the notion of structural stability; it seems to have more or less stopped it. One is actually led to wonder what effect a knowledge of the Lorenz system, at an earlier stage, would have had on Smale's development of these important ideas.

For the present, the equations are still almost the only candidate for a system having a well-understood (topologically and dynamically) strange attractor in a range of parameter values. Their importance as a model of convection has diminished, it being almost universally accepted that they are a suitable model for the original problem only at relatively uninteresting parameter values, and their claim to physical relevance now rests mainly on their connection with the Maxwell–Bloch equations for lasers, and on convection problems in specially shaped (usually toroidal) regions. On the other hand, with the increasing recognition that the study of homoclinic and heteroclinic bifurcations is central to the understanding of chaotic behaviour, the Lorenz equations are again generating a lot of interest; this is partly due to the very great number and variety of such bifurcations occurring in the system, and partly because many of these can be precisely analysed in terms of their effects on the periodic orbits and other invariant trajectories in the system. It is mainly this aspect of the equations which will be discussed below. More detail and many older references can be found in Sparrow [35].

Finally, and even as I write, new and deeper understandings of the Lorenz system are in the offing [12]. Though excessive concentration on these equations can be criticised (since they are not, in many ways, typical of chaotic systems), it remains true that examples of nearly all the types of chaotic behaviour seen in other three-dimensional dissipative systems of differential equations can be found, for some parameter values, in the Lorenz system. The immense amount of detailed experimental and theoretical work done on the equations often allows these new developments to be seen in a fuller context than is possible with newer and less well understood systems.

6.2 The equations

The Lorenz equations are:

$$\frac{dx}{dt} = \sigma(y-x)$$

(6.1)
$$\frac{dy}{dt} = rx - y - xz$$

$$\frac{dz}{dt} = xy - bz$$

where $t, x, y, z, \sigma, r, b, \in \mathbb{R}$, and σ, r, and b are three parameters which are normally taken, because of their physical origins, to be positive. The equations are often studied for different values of r in $0 < r < \infty$ with σ and b held constant; Lorenz's original paper [21] took $\sigma = 10$ and $b = 8/3$.

When $r < 1$, the origin $(0, 0, 0)$ is globally attracting (all trajectories tend towards it), but for $r > 1$ there are three stationary or equilibrium points. These are the origin, which is unstable, and two points $C^{\pm} = \{\pm b\sqrt{r-1}, \pm b\sqrt{r-1}, r-1\}$. There is an r-value, $r_{\mathrm{H}} = [\sigma(\sigma + b + 3)]/(\sigma-b-1)$ at which a Hopf bifurcation occurs at $C^{\pm}$. If $\sigma-b-1 < 0$, the Hopf bifurcation does not occur for any positive r, and $C^{\pm}$ are stable for all $r > 1$; otherwise $C^{\pm}$ are stable in $1 < r < r_{\mathrm{H}}$ and unstable in $r > r_{\mathrm{H}}$.

Notice that C^{+} and C^{-} are images of each other under the natural symmetry $(x, y, z) \rightarrow (-x, -y, z)$ of the equations, and that they therefore undergo bifurcations simultaneously. Also, the z-axis is invariant under the flow, so any periodic orbits in the system can be partially described by an integer n specifying the number of times that they wind around the z-axis. Such orbits will either be symmetric (i.e. be their own image under the symmetry) or will occur in symmetric pairs, each of which is taken to the other by the symmetry.

It can also be shown that there is a bounded set $E \subset \mathbb{R}^3$, which depends on the parameters, which all trajectories eventually enter and thereafter remain within [21, 35]. This, together with the fact that the flow has a negative divergence, $-\sigma-b-1$, implies that all trajectories eventually tend towards some bounded set Λ with zero volume. The set Λ also depends on the parameters, and those parameter values at which the topology of Λ changes are, by definition, the parameter values at which bifurcations occur in the system. If we can understand the topology and stability of the various components of Λ, then we will know the long-term behaviour of all trajectories. This, in the absence of closed-form solutions to the equations, is our aim.

6.3 Homoclinic orbits

The only bifurcations occurring in the equations which are accessible to simple classical analysis have already been mentioned. These are the bifurcation at $r = 1$, when the origin loses stability and the stationary points $C^{\pm}$ appear, and the Hopf bifurcation at $r = r_{\mathrm{H}}$ when $C^{\pm}$ lose stability by absorbing, as r increases, two unstable periodic orbits [25]. In order to discover more about bifurcations in the system, it is necessary to resort to computer simulation of the equations.

It transpires that the most important bifurcations occurring in the equations are associated with the possibility that there may be *homoclinic orbits* to the origin for some parameter values. A homoclinic orbit is a trajectory which tends, in both forwards and backwards time, towards an unstable (saddle) stationary point. Three examples are shown, schematic-ally, in Fig. 6.1. All these three types of homoclinic orbit actually occur in the Lorenz equations, and the symmetry of the equations ensures that in all three cases there are two homoclinic orbits as shown in Fig. 6.1a rather than the one shown, for clarity, in Figs 6.1b and c.

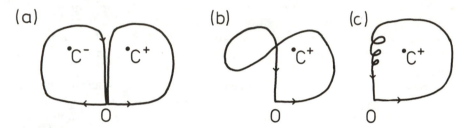

Fig. 6.1 Three homoclinic orbits to the origin which occur in the Lorenz equa-tions. In all three cases the symmetry ensures that both branches of the unstable manifold of O are homoclinic simultaneously (as in (a)); in (b) and (c) only one branch is shown for clarity.

Recent and extensive numerical computations by Alfsen and Frøyland [1] illustrate the amazing wealth of homoclinic orbits in the equations. They produce a picture showing some of the r and b parameter values (with σ fixed at $\sigma = 10$) for which homoclinic orbits to the origin can be seen. A schematic version of their diagram is shown in Fig. 6.2; every point on one of the solid lines on this figure represents a pair of r- and b-values for which homoclinic orbits occur.

The point X marked on Fig. 6.2 (at parameter values $r \approx 30.475$, $b \approx 2.6123$) lies at the centre of a logarithmic spiral of r-, b-values at which there are homoclinic orbits to the origin [11]. At X itself there are heteroclinic orbits (orbits which tend towards one stationary point in forward time and to another in backward time) linking all three stationary points. The situation at X is shown in Fig. 6.3a. The homoclinic orbits to the origin that lie close to X spiral a great number of times around C^+ or C^- before returning to the origin; as you work away from X along the spiral the number of turns around C^+ or C^- decreases. Figure 6.3b shows yet another type of homoclinic orbit; this one involves only the stationary points C^+ and C^-. (Strictly speaking these orbits are heteroclinic but it

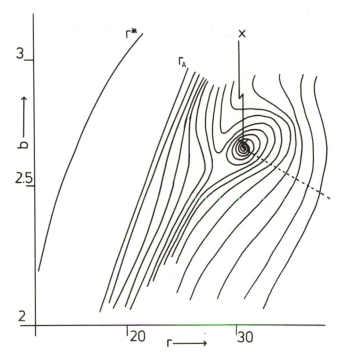

Fig. 6.2 r,b parameter space for $\sigma = 10$. Solid lines indicate homoclinic orbits to the origin. Infinitely many such lines accumulate on the parameter line marked r_A and there is a logarithmic spiral with infinitely many turns spiralling into the point marked X ($r \approx 30.475$, $b \approx 2.623$, cf. Fig. 6.3b). The dashed line marks parameter values for which there is a symmetric homoclinic orbit between C^+ and C^- (cf. Fig. 6.3b). (After Alfsen and Frøyland [1]).

makes sense, because of the symmetry, to identify C^+ with C^- and to think of this situation as a symmetric version of a homoclinic orbit to a single stationary point.) Homoclinic orbits such as this one occur along the dotted line in Fig. 6.2.

Before looking at the implications of this plethora of orbits, it remains to state that only a few of the homoclinic orbits known to occur in the Lorenz equations are displayed in Fig. 6.2. The ones shown are only the simplest; each winds only once around the z-axis when projected on to the x–y plane. There are, in fact, infinitely many other families of homoclinic orbits winding any number of times around the z-axis. Some idea of the complexity involved can be gained by considering that there is an infinite sequence of X-like points accumulating on the point X in Fig. 6.2. These are points in r, b parameter space where the three stationary points are connected together by heteroclinic orbits of a more complicated kind than those shown in Fig. 6.3a. Each such point has its own logarithmic spiral of

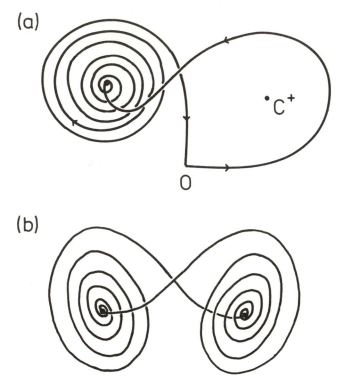

Fig. 6.3 (a) Heteroclinic connections between O and C^- at the point X from Fig. 6.2. Symmetric connections occur between O and C^+ but are not shown. (b) Homoclinic orbits for $C^\pm$.

homoclinic orbits to the origin around it, and a dotted line of homoclinic orbits to the stationary points $C^\pm$ coming out of it. Each of these points then has another sequence of points accumulating on it where yet more complicated heteroclinic orbits connect the three stationary points, and so on, *ad infinitum*. Of course, the various families of homoclinic orbits to the origin cannot cross (at any point in parameter space the two trajectories which tend in backward time to the origin are unique and can be part of at most one homoclinic orbit) and the dotted lines representing homoclinic orbits to the points $C^\pm$ also cannot cross (for similar reasons). None the less, the full picture in just this two-dimensional parameter space is almost beyond imagining. Further details can be found in [11].

6.4 Bifurcations associated with homoclinic orbits

In order to understand why it is that homoclinic orbits are such important features of the Lorenz equations, we will examine the change of behaviour of the system as r passes through a value at which a homoclinic orbit like that shown in Fig. 6.1a occurs. The results depend crucially on the symmetry and on the relative sizes of the eigenvalues of the linearised flow near the origin.

We examine a small region of phase space consisting of a small box, B, around the origin and two thin tubes, S and T, around part of the two

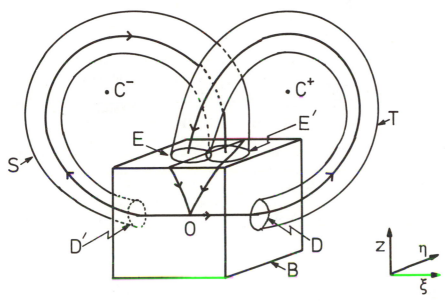

Fig. 6.4 The small region analysed for r-values close to r^* at which a homoclinic orbit occurs. D and D' are discs on the sides of box B where B meets the tubes T and S, respectively. E and E' are similar discs on the top face of B.

branches of the unstable manifold of the origin (see Fig. 6.4). We assume that at some critical r-value, r^*, there is a homoclinic orbit like that shown in Fig. 6.1a, and consider, for r-values close to r^*, whether or not there are any trajectories which remain forever within the regions $B \cup S \cup T$. Providing that we choose B, S and T to be small enough, with r close enough to r^*, and that we are only concerned with the topology of trajectories, it is permissible to regard the flow within B as linear, and to think of the map which takes points in the discs D and D' to points in the discs E and E' as a linear transformation.

In other words, if we choose coordinates ξ, η and z in B so that the linearised flow near the origin takes the form

(6.2)
$$\left. \begin{array}{l} \dot{\xi} = \lambda_1 \xi \\ \dot{\eta} = -\lambda_2 \eta \\ \dot{z} = -\lambda_3 z \end{array} \right\} \quad \lambda_2 > \lambda_1 > \lambda_3 > 0$$

we can use these equations to work out the point $(\pm c, \eta^1, z)$ on the side of B where a trajectory emerges from B if it starts at a point (ξ^0, η^0, c) on the top face of B. (We assume that the box B is a cube with faces which are part of the planes $|\xi| = c$, $|\eta| = c$ and $|z| = c$.) Then, for a trajectory starting at a point $(\pm c, \eta, z)$ on the side of B, providing η and z are small (within D or D', say), we can write

$$(6.3) \qquad \begin{pmatrix} \xi^2 \\ \eta^2 \end{pmatrix} = \begin{pmatrix} a \\ b \end{pmatrix} + \mathbf{A} \begin{pmatrix} \eta^1 \\ z \end{pmatrix}$$

for the point (ξ^2, η^2, c) at which this trajectory next strikes the top face of B. Here a and b are constants (which may depend on r) and $\mathbf{A}$ is a 2×2 matrix. This procedure can be rigorously justified, but roughly speaking the linear flow in B is justified for B small enough, and the affine transformation is justified because the time taken by trajectories to traverse the tubes T or S is small compared with the time they spend within B. It is also permissible, for r close to r^*, to assume that the eigenvalues λ_i and the matrix $\mathbf{A}$ are constants which do not depend on r, whereas a and b depend linearly on $(r - r^*)$.

Let us, then, carry out this program. For a point (ξ^0, η^0, c) on the top of B ($\xi^0 > 0$) we obtain a point (c, η^1, z) on the side $\xi = c$ of B given by

$$(6.4) \qquad (\eta^1, z) = \left[\eta^0 \left(\frac{\xi^0}{c} \right)^{\lambda_2/\lambda_1}, c \left(\frac{\xi^0}{c} \right)^{\lambda_3/\lambda_1} \right]$$

with a similar formula, given by the symmetry, for $\xi^0 < 0$. Since ξ is small compared with c (for thin-enough tubes) the first coordinate is much smaller than the second ($\lambda_2/\lambda_1 > 1$ and $\lambda_3/\lambda_1 < 1$). Figure 6.5a shows the thin pointed region through which trajectories leave B if they start inside the rectangle $ABCD$ on the top face of B. Combining eqn (6.4) with (6.3) we obtain

$$(6.5) \qquad \begin{pmatrix} \xi^2 \\ \eta^2 \end{pmatrix} = \begin{pmatrix} \alpha(r\text{-}r^*) \\ \eta^* + \beta(r\text{-}r^*) \end{pmatrix} + \mathbf{A} \begin{pmatrix} k_1 \eta^0 \xi^{0\,\lambda_2/\lambda_1} \\ k_2 \xi^{0\,\lambda_3/\lambda_1} \end{pmatrix}$$

where $(0, \eta^*)$ is the point on the top face of B where the unstable manifold of the origin returns to the stable manifold of the origin when $r = r^*$. We can determine the sign of α and η^* experimentally; for the case usually studied ($\sigma = 10$, $b = 8/3$, and $r^* \approx 13.926$), both are negative. Also, for the same case, we determine the sign of the elements of $\mathbf{A}$ which multiply the larger of the two coordinates from eqn (6.4). This leads to a series of return maps, for r-values $r_1 < r^*$, $r = r^*$ and $r_2 > r^*$ like

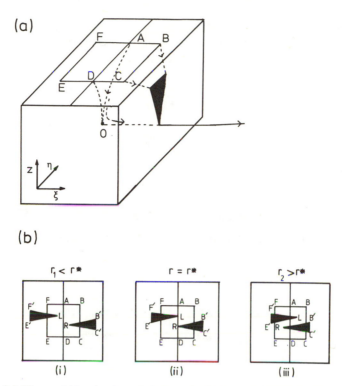

Fig. 6.5 (a) Flow within the box *B* is linear. Trajectories started within ABCD emerge from *B* through the longer and thinner shaded region. (b) Return maps on the top face of *B* for (i) $r_1 < r^*$; (ii) $r = r^*$; (iii) $r_2 > r^*$. The points *B'*, *C'*, *E'*, *F'* are the first returns of points *B*, *C*, *E* and *F* respectively. *R* and *L* are the points where the right- and left-hand branches of the unstable manifold of the origin first strike the top of *B*. Trajectories started at points within *ABCD* or *ADEF* return to points within the shaded regions *RB'C'* and *LE'F'*, respectively.

those shown in Fig. 6.5b. This figure shows the top face of the box *B*; trajectories started within *ABCD* return to the lower shaded region, *RB'C'*, and trajectories started within *ADEF* return to the upper shaded region *LE'F'*. The closer to *AD* (which lies on the stable manifold of the origin) a trajectory starts, the closer to one of the points *R* or *L* it will return; *R* is the point where the right-hand branch of the unstable manifold of the origin first strikes the top face of *B*, and *L* is the equivalent point for the left-hand branch.

The most important feature of these maps is that stretching occurs in the ξ-direction and that there is contraction in the η-direction. The existence of expanding and contracting directions is an essential prerequisite for chaotic behaviour in dissipative systems of this sort. However, looking more closely at Fig. 6.5b, we see that in the cases $r_1 < r$ and

$r = r^*$ there is no chance that there are any trajectories which remain forever within the region of interest except for the homoclinic orbit itself at $r = r^*$; for all other trajectories the modulus of the ξ-coordinate increases on each pass through the top of B and so all trajectories eventually wander out of the region of validity of our analysis (and, in fact, spiral into C_1 or C_2). The case $r_2 > r^*$ is more interesting. The expansion and contraction imply the existence of four regions, shown schematically in Fig. 6.6a, which are mapped by the flow into the four shaded regions of

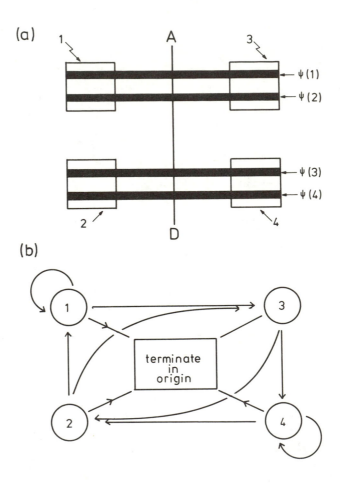

Fig. 6.6 Schematic representation of the behaviour of the return map for $r_2 > r^*$ (Fig. 6.5b). (a) Regions 1, 2, 3, 4 are mapped into longer and thinner regions marked $\psi(1)$, $\psi(2)$, $\psi(3)$ and $\psi(4)$ respectively. (b) Trajectories can be found which follow, in both forwards and backwards time, any route between regions allowed by the arrows.

Fig. 6.6a. Each of the shaded regions is much longer in the ξ-direction and much narrower in the η-direction than the original regions from which they came.

Figure 6.6b shows us the possible routes for trajectories to pass through the four shaded areas. For instance, if we want a trajectory that goes 1–3–4–2–... then we start in region 1; to get to region 3 we must be in the right-hand 'third' of region 1; if we are going to go on from region 3 to region 4, we must be in the right-hand 'third' of that 'third'; and, to go on to region 2, in the left 'third' of that 'third', etc. For any infinite sequence of symbols 1, 2, 3 and 4 allowed by Fig. 6.6b we can find a vertical line of points (arrived at by taking away 'two-thirds' of an interval an infinite number of times) from which trajectories pass through the four shaded areas in the prescribed sequence. In addition, there will be vertical lines of points that generate finite sequences which represent trajectories that eventually strike the top face of B on the line $\xi = 0$ (and thereafter tend towards the origin and never strike the top face of B again). All these vertical lines, from which trajectories remain within $B \cup S \cup T$ for all future time, form a Cantor set, and trajectories that start between lines eventually leave our small region of interest.

If we now consider where trajectories came from, we can divide the four shaded areas into horizontal bands, each of which corresponds to some finite history. For example, if we are in region 1 (Fig. 6.6a) and came from region 2, we know we are in the lower horizontal band, which is the intersection of region 1 and $\psi(2)$. We can repeat this process indefinitely, arriving at a Cantor set of horizontal lines which represent the possible histories (allowed by Fig. 6.6b) of trajectories that remained within our small region of interest for all past time. The points that lie on the intersections of the horizontal and vertical Cantor sets will be points through which trajectories pass that do not leave our small region of interest in either forwards or backwards time. Trajectories passing through any other points leave the region of interest, or have come from outside it, or both.

Each point lying on an intersection will have a unique symbolic representation (which is infinite to the left, but which may terminate on the right) in terms of symbols 1, 2, 3 and 4. Conversely, each sequence of symbols, infinite on the left, will correspond to one point, providing that the sequence can be generated by following the arrows in Fig. 6.6b. This last proviso actually tells us that we only need two symbols, T and S, to describe points. If we write T for 3 or 4, and S for 1 or 2, then any infinite sequence of Ts and Ss has one and only one representation in 1s, 2s, 3s and 4s which obeys the Fig. 6.6 rules. For example, $STTTSST...$ must be 1344213.... Thus, the points that lie in our 'invariant set' (the set of all points lying on intersections of the vertical and horizontal lines), each of which is taken by the return map to another such point, correspond one to one with symbolic sequences, infinite on the left, of Ts and Ss. The way we have chosen the symbols T and S also means, conveniently, that trajectories threading their way through the top face of the box B according to some sequence of Ts and Ss, make the same sequence of passes through the tubes, T and S.

We can tell how trajectories look by examining their corresponding sequence. Periodic trajectories (a countable infinity) will have repeating sequences; trajectories that terminate in the origin (an uncountable infinity) will have terminating sequences; trajectories that are asymptotically periodic (in either forwards or backwards time — an uncountable infinity) will have aperiodic sequences. Notice that there will not be quite so many trajectories as points on the top face of B since trajectories will, in general, intersect this plane in a large (possibly infinite) number of points. When considering points, we have some central mark in the sequence to tell us where we are, and with trajectories we do not. For points, each application of the return map moves the central mark one place to the right. As an example, consider the symmetric 'period 2' periodic orbit represented by the doubly infinite sequence ...*TSTSTSTS*... . There is only one such orbit. This orbit intersects the top face of the box B in two points, one of which lies to the left of the stable manifold of the origin and which is represented by the doubly infinite sequence (with mark) ... *TSTSTSTS*... , the other of which lies to the right of the stable manifold of the origin and is represented by the doubly infinite sequence (with mark) ...*TSTSTSTS*... . Each application of the return map moves the mark one place to the right. Hence, as we follow the orbit around, we oscillate between one point and the other.

We call the collection of trajectories that remains within $B \cup S \cup T$ forever a 'strange invariant set'. It is invariant because the flow takes it into (and on to) itself. The set is created in the bifurcation that occurs as r passes through r^*, and we can now mention some of the interesting properties that lead us to call it 'strange'.

(1) No single trajectory, nor any subset of the strange invariant set, is stable; we can find trajectories, as close as we like to any trajectory in the strange invariant set, which leave $B \cup S \cup T$ in both forwards and backwards time.

(2) Even if we restrict our attention to those trajectories in our strange invariant set, we can see that almost all pairs of trajectories started at points close together on the top face of the box B will not remain close together as we follow them around. This is called 'sensitive dependence on initial conditions', a phenomenon which persists even when the strange invariant set becomes attracting (see below) and which gives the typical 'chaotic', 'turbulent' or 'pseudo-random' behaviour which we associate with a 'strange attractor'. We can see

why this phenomenon should occur. Two points on the top face of *B* are close together if the parts of their symbolic descriptions near the central mark are the same. As we follow trajectories starting at two close-together points, we move to points with the same symbolic descriptions but with the central mark moved further and further to the right. Eventually we can expect the central mark (which tells us which region we are in at the moment) to have moved to parts of the two sequences which are quite different. This argument will only fail to apply to two close-together points on the same vertical line ($\xi \approx$ constant); such points will have converging futures (but distinct pasts).

(3) There is a very real sense in which the strange invariant set is a single object and not just the sum of its parts. It contains an uncountable infinity of dense trajectories; these are trajectories which pass as close as we like to all trajectories in the strange invariant set. (We can construct doubly infinite sequences of *T*s and *S*s which contain every possible finite sequence of *T*s and *S*s; the trajectories corresponding to these sequences pass as close as we like to any other trajectory in the invariant set.)

(4) Periodic orbits are dense in the strange invariant set. Since periodic orbits can contain any finite number of symbols which repeat, it is clear that we can find a periodic orbit which passes arbitrarily close to any other trajectory in the strange invariant set.

The strange invariant set that we have been discussing is very similar to Smale's horseshoe example, introduced in his seminal (1967) paper. Our use of 'symbolic dynamics' to describe the set also follows Smale.

It seems, then, that associated with the existence of a homoclinic orbit in the flow, there is a bifurcation which produces a remarkable collection of periodic orbits and other trajectories. Notice in particular that we have not used, anywhere in our analysis, the 'shape' of the tubes *T* and *S*, so we can expect a similar analysis to hold for all homoclinic orbits to the origin which occur in the Lorenz equations. Returning to Fig. 6.2 for a moment, whenever the parameters are altered so that one of the solid lines (representing homoclinic orbits to the origin) is crossed, there will be a bifurcation similar to that described above.

It is also possible to analyse, in a very similar way, the behaviour close to homoclinic orbits involving the stationary points C^+ and C^- [10], and, indeed, the behaviour close to the special point X on Fig. 6.2, which represents parameter values at which all three stationary points are connected by heteroclinic orbits [11]. In the former case, the fact that the linearised flow near C^+ and C^- has a complex pair of eigenvalues alters the results considerably. On the one hand the behaviour is more complicated very close to the critical parameter value, with infinitely many

periodic orbits in existence on both sides of the bifurcation, some of which
may be stable [10, 15, 31, 32]. On the other hand, it is argued in [10] that,
if we look only at parameter values reasonably far from the critical value,
it will usually be the case that the cumulative effect of the whole sequence
of complicated bifurcations occurring near the critical parameter value is
just to add a single symmetric periodic orbit to the system. The analysis for
parameter values near point X is much harder, involving elements of all
the other analyses mentioned so far, but it is at least possible to confirm
that Fig. 6.2 is qualitatively correct [11].

6.5 Strange attractors

For parameter values $\sigma = 10$, $b = 8/3$, and r increasing from 1, numerical
simulations of the Lorenz equations show the following behaviour:

(1) For $1 < r < 13.926$, all numerically computed trajectories spiral
    into the stable stationary points $C^{\pm}$. The right-hand branch of the
    unstable manifold of the origin spirals into C^{+} and the left-hand
    branch into C^{-}.

(2) At $r \approx 13.926$ there is a homoclinic orbit like that shown in Fig.
    6.1a and for $r > 13.926$ the right-hand branch of the unstable
    manifold of the origin spirals into C^{-} and vice versa.

(3) $13.926 < r < 24.06$. All numerically computed trajectories eventu-
    ally spiral into C^{+} or C^{-} but some spend a considerable time
    wandering near the strange invariant set (which was produced in
    the bifurcation that occurred as r passed through its homoclinic
    value) before doing so. The time spent wandering increases rapidly
    as r approaches 24.06.

(4) $r > 24.06$. Some trajectories wander forever near the strange
    invariant set which has become a strange attractor. For $r < 24.74$ it
    is still possible to find some trajectories which spiral into $C^{\pm}$, but,
    at $r = r_{\mathrm{H}} \approx 24.74$, $C^{\pm}$ lose stability and in $r > 24.74$ all numeric-
    ally computed trajectories wander forever near the strange
    attractor.

It is clear from our analysis in the previous section that we do not
expect the strange invariant set produced at $r \approx 13.926$ to be stable and
we do not expect it to show up in numerical simulations; the behaviour
described in (2) and (3) above ($r > 13.926$) cannot be predicted from the
purely local analysis. However, we can understand it in similar terms if we
numerically (as opposed to analytically) compute return maps on a suitable
plane. A suitable plane at these parameter values turns out to be the plane
$z = r-1$, which includes $C^{\pm}$, and the return map obtained, for r-values
just less than and just greater than $r \approx 24.06$, is shown schematically in

(a)

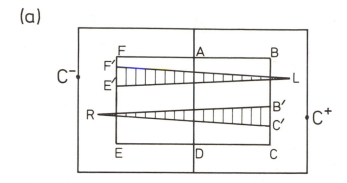

(b)

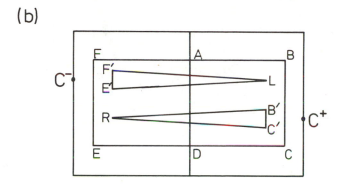

Fig. 6.7 Schematic versions of return maps calculated numerically on the plane $z = r - 1$. The labelling scheme is the same as for Fig. 6.5b. (a) Trajectories spiral into $C^{\pm}$ only after falling into one of the two doubly hatched regions which lie near R and L outside $BCEF$. These decrease in size as r increases. (b) All trajectories started with $BCEF$ remain near the strange attractor forever.

Fig. 6.7. For $r < 24.06$ (Fig. 6.7a) the return map looks similar to Fig. 6.5b for $r_2 > r^*$. Providing we assume that there is still contraction in one direction and expansion in the other (which we can no longer prove but which appears to be true) the same analysis will hold. As r increases towards $r \approx 24.06$ the strange invariant set seems to spread out so that the doubly hatched regions on Fig. 6.7a decrease in size. Trajectories which start within $BCEF$ continue to wander near the strange invariant set until they eventually fall into one of these doubly hatched regions outside $BCEF$ (from whence they spiral into C^{+} or C^{-}); so, as r approaches 24.06, trajectories wander for longer and longer. Of course, trajectories started very near the origin, which first strike the return plane close to R or L, spiral at once into $C^{\pm}$; this is how one determines experimentally the precise parameter value at which the strange set becomes attracting.

For $r > 24.06$, return maps, for suitable choices of initial rectangle

BCEF, look like that shown in Fig. 6.7b. The doubly hatched regions are absent and any trajectory started within *BCEF* returns to a point within *BCEF* and so continues to wander near the strange set forever. Figure 6.8 shows a typical trajectory for $r = 28.0$.

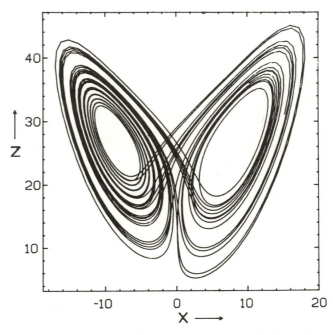

Fig. 6.8 A numerically calculated trajectory for $r = 28.0$, $b = 8/3$, $\sigma = 10$. The trajectory is projected on to the x, z plane and does not close up.

The supposed expansion in directions perpendicular to *AD* still ensures that there can be no stable periodic orbits, but an analysis similar to that in the previous section does not have quite the same result. This is because the images $RB'C'$ and $LE'F'$ of *ABCD* and *ADEF* no longer stretch right across the original rectangles, so we cannot guarantee the existence of a trajectory corresponding to every sequence allowed in Fig. 6.6b. Some trajectories have been lost from our strange invariant set.

The normal method of analysing this strange attractor is to construct a model which has the properties we like to think (and do think) that the actual Lorenz system has, and then to reduce this model to a one-dimensional mapping of an interval by 'factoring out' the contracting direction [17, 27, 35]. Such an analysis leads to the conclusion that in every neighbourhood of every r-value we can find uncountably many different Lorenz attractors, each containing slightly different periodic orbits and trajectories. We will not attempt this analysis here but instead will offer an intuitive (and practical) argument for why this should be so.

Once the invariant set is attracting, there is the possibility that the unstable manifold of the origin, which first strikes the return plane at R or L and which then wanders chaotically around the strange attractor forever, may eventually strike AD and then tend back towards the origin. In other words, homoclinic orbits are possible. The chaotic nature of the motion suggests that in any neighbourhood of any r-value there will be many (countably many) r-values for which this occurs, though the different homoclinic orbits involved will happen only after the unstable manifold has made very different sequences of passes through the return plane. Each such homoclinic orbit can be analysed locally in the same way as outlined in section 6.4. This time some of the constants have a different sign and the analysis suggests that each such homoclinic orbit will remove a strange invariant set of periodic orbits and trajectories from the system. The analysis still only concerns a small region contained within thin tubes around the homoclinic orbit. The fact that trajectories which wander out of the region may later return (after wandering chaotically near the strange attractor) is of no concern; we are concerned only with trajectories that remain forever within the small region, and a strange collection of these is removed.

This argument, though less precise, cannot replace the usual arguments which are needed to justify our assumption that there will be countably many homoclinic orbits in any r-interval. However, it does have one great advantage. Even if it eventually transpires that the Lorenz equations do not satisfy the conditions necessary to justify the rigorous analysis (but see [33]), it is none the less true that a great many (infinitely many) homoclinic orbits do occur in the system though perhaps not distributed densely through all r-intervals. Our intuitive arguments will still apply to these homoclinic orbits and give some understanding of the way in which the structure of the attractor changes.

Before moving on, it is sensible to relate the contents of the last two sections to Fig. 6.2. The homoclinic orbit marked r^* in that figure is the one at which the strange invariant set is first created as described in section 6.4. The parameter value r_A, which is the accumulation point of an infinite sequence of homoclinic orbits occurring as r decreases, is the parameter value at which the strange set becomes an attractor. When $b = 8/3$, the remarks of this section apply for r-values $24.06 < r < 28$ but cease to apply when the spiral of homoclinic orbits about X in Fig. 6.2 is crossed.

6.6 Other parameter values

We have only looked, in this chapter, at a small subset of the parameter values that have been examined by researchers in recent years. As a consequence, it has only been possible to describe a few of the phenomena which have been observed.

From a purely phenomenological point of view, the most obvious omission is a description of the various stable periodic orbits observed, together with the period-doubling windows which normally accompany them [7, 22, 35]. For example, as r is increased above an r-value near 30 ($\sigma = 10$, b, $= 8/3$), the chaotic behaviour is interrupted by parameter intervals in which stable periodic orbits are seen. See, for example, Fig. 6.9. For large-enough r-values ($r > 312$) there is a stable symmetric periodic orbit which winds just once around the Z-axis and which then persists for all larger r-values (Fig. 6.9c). This behaviour (alternating windows of stable and chaotic behaviour) is typical of most chaotic systems of ordinary differential equations, and raises the usual questions. In the context of the Lorenz equations and the previous sections of this chapter, there are two related ways to explain this onset of stability. First, the changeover from the strange-attractor regime to the stable periodic orbit regime is associated with passing through the top of the spiral of homoclinic orbits in Fig. 6.2; the usual parameter value $b = 8/3$ lies just above the point X of Fig. 6.2. Secondly, if we look at return maps, as r passes through the critical region they start to change so that they resemble Fig. 6.10. The folds in the return maps prevent the relatively simple analysis of the strange attractor from remaining true, since points which are separated by the expansion in one direction can, if they are later on the opposite sides of the fold in the map, be forced back together again by the contraction in the other direction. Hence stable orbits are possible. The first point definitely implies the second, but it must be said that other mechanisms exist in the equations for introducing folds into the return maps, so that a similar transition is observed even for b-values which do not mean that increasing r involves passing through the spiral of Fig. 6.2.

We have also not mentioned the many different styles of analysis which have been developed for special parameter ranges. These include:

(1) The study of one-dimensional maps derived from the system. These may be useful as heuristic guides (but not rigorous ones) in some parameter ranges, e.g. larger b ($\sigma = 10$) [29] or the limit $\sigma \approx r$, $r \to \infty$ [5, 6], or as the basis of the rigorous treatment of the strange-attractor parameter range [17, 27, 35].

(2) The method of averaging applied in the limit $r \to \infty$ [4, 28, 35]. This theory enables one to predict that for reasonably large b-values (including $b = 8/3$) there will be a single stable symmetric orbit for large enough r, but that for smaller b ($b < 2$?) there will be, in addition, an infinite sequence of homoclinic orbits to the origin (and other complicated behaviour) which never terminates as $r \to \infty$ [4, 35].

(3) The theory of knottedness of orbits in the equations [3].

(4) Various pieces of analysis designed to approximate, in some

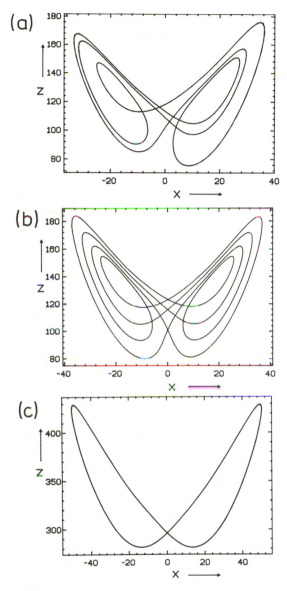

Fig. 6.9 Stable periodic orbits for various values of the parameter r ($\sigma = 10$, $b = 8/3$). (a) $r = 126.52$; (b) $r = 132.5$; (c) $r = 350.0$.

analytic fashion or other, the strange attractor regime [13, 26].

(5) Various statistical studies of the equations in chaotic regimes, including the determination of fractal dimensions [19, 23, 30].

(6) As mentioned above, various studies of special properties of the system for particular parameter intervals in the chaotic/stable

periodic regimes. These include intermittency [24] and noisy periodicity [22], both of which occur widely in other systems.

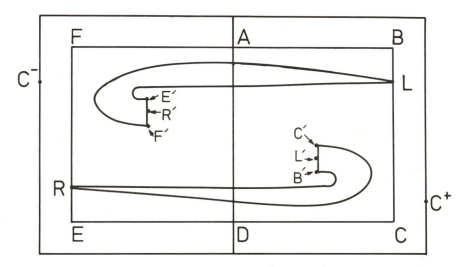

Fig. 6.10 Return maps in the parameter range $r > 30$ ($\sigma = 10$, $b = 8/3$). As r increases, the points R', L' (the first returns of R and L) cross AD ($r \approx 54.6$ and causing a homoclinic orbit) and eventually the fold also crosses AD so that all trajectories oscillate regularly back and forth, from side to side of AD, forever.

What we have done is to concentrate on two particular themes. One of these is the study of the strange attractor regime and the creation of the attractor through a homoclinic bifurcation. This deserves special mention because, on the assumption that various conditions which appear to be satisfied by the return maps are actually satisfied, this attractor is probably the only well understood strange attractor known in a system of 'natural' three-dimensional differential equations; we have strong reasons to suppose that there can be no stable orbits in a relatively large parameter range, as opposed to the normal 'chaotic attractors' where one merely cannot observe them but has no arguments to suggest they cannot exist (they may be of extremely high period or have very complicated basins of attraction — see [8]).

Our other theme, the more general study of homoclinic orbits and bifurcations, is perhaps of even greater importance because it can be applied to other systems of equations as well [2, 9, 10, 20]. We have only really looked closely at the local analysis of one particular kind of homoclinic orbit (in section 6.4); such local analyses are interesting in their own right, but are relatively well known [37]. The real usefulness of these studies comes, though, when the global implications are considered. These global considerations do rely on numerical computation to a large extent, are nonrigorous, and do not provide predictions of actual behaviour at actual parameter values. But consider:

(1) It is possible to follow periodic orbits numerically with changing parameter (even if they are unstable); every orbit so far examined in the Lorenz equations can be traced, at least in one direction, to a homoclinic bifurcation. This seems to be generally true in most other systems of chaotic equations as well.

(2) It is frequently possible to determine how many periodic orbits of what topological type are created or destroyed in each such bifurcation. This may either be done precisely (as with our homoclinic orbits to the origin) or more loosely, as explained in section 6.4 (for other types of homoclinic orbit).

(3) It is frequently possible that if one homoclinic orbit occurs at one parameter value, and another occurs at another parameter value, then certain general properties of an intermediate sequence of homoclinic orbits can be determined. This theme is developed in detail, for the Lorenz equations, in [35], and for a very different system of equations in [10].

(4) It happens that many systems of equations, particularly those of some physical relevance, have simple behaviour for extreme values of a parameter. In the case of the Lorenz equations we know the behaviour in $r < 1$ and for r large.

(5) Putting all of the above four points together, it is possible to develop a type of book-keeping for periodic orbits. Such an exercise can span huge parameter ranges in which quite different behaviours are observed (but all of which are intimately tied up with the existence of periodic orbits: even the strange attractor is densely packed with unstable orbits). If this book-keeping adds up — all orbits which are created must either be destroyed or persist for all parameter values — all is well and good. If not, new bifurcations must be searched for. In the case of the Lorenz equations, several new phenomena have been located in this way [10, 35], and the technique is showing promise in other systems [10].

(6) Finally, and importantly, the study of homoclinic orbits can be undertaken relatively easily and cheaply on a computer.

We have, then, arrived at the beginning rather than at the end of a story. It is probably fair to say that continued detailed study of the Lorenz equations will go on throwing out new ideas of general applicability, despite their rather special properties (e.g. symmetry). Regardless of the physical relevance (or otherwise!) of some of these studies, this process will be worth while as long as the study of chaotic ordinary differential equations remains the Pandora's box it is today. And in the Lorenz equations, it is easier to make connections between apparently unrelated things than in less well known, less well studied and less well understood systems.

It only remains to guide the interested reader to some other, more measured, introductions to the Lorenz equations, not so far referred to in this chapter. Guckenheimer and Holmes develop the theory of the strange attractor in a very thorough and comprehensible way in their book [16], and reference [36] is probably also worth while for a slightly different emphasis.

Acknowledgements and Apologies

I have made few attempts in this chapter to give the first references to ideas or discoveries. Rather, I have concentrated on those I know and on those that I think are easiest to read. More references can be found in [35]. I apologise to any who are offended. It is with more pleasure that I thank all those who have helped my ideas about the Lorenz equations to develop; they are many, but I must mention particularly Peter Swinnerton-Dyer, Bob Williams and Paul Glendinning.

References

[1] Alfsen, K. H. and Frøyland, J. Systematics of the Lorenz model at $\sigma = 10$. Preprint, Department of Physics, University of Oslo, Norway (1984).

[2] Bernoff, A. J. Heteroclinic and homoclinic orbits in a model of magneto convection. Preprint, DAMTP, Cambridge (1984).

[3] Birman, J. S. and Williams, R. F. Knotted periodic orbits in dynamical systems — 1: Lorenz equations. *Topology* **22** (1), 47–82 (1983).

[4] Fowler, A. C. Analysis of the Lorenz equations for large r. Preprint, Mathematics Institute, Oxford (1982).

[5] Fowler, A. C. and McGuinness, M. J. A description of the Lorenz attractor at high Prandtl number. *Physica 5-D*, 149–82 (1982).

[6] Fowler, A. C. and McGuinness, M. J. Hysteresis, period doubling and intermittence at high Prandtl number in the Lorenz equations. *Stud. Appl. Math.* **69**, 99–126 (1983).

[7] Franceschini, V. A Feigenbaum sequence of bifurcations in the Lorenz model. *J. Stat. Phys.* **22** (3), 397–406 (1980).

[8] Gambando, J. M. and Tresser, C. Some difficulties generated by small sinks in the numerical study of dynamical systems: two examples. *Phys. L.* **94A** (9), 412–14.

[9] Gaspard, P., Kapral, R. and Nicolis, G. Bifurcation phenomena near homoclinic systems: a two-parameter analysis. *J. Stat. Phys.* **35**, 697 (1984).

[10] Glendinning, P. and Sparrow, C. Local and global behaviour near homoclinic orbits. *J. Stat. Phys.* **35**, 645–96 (1984).

[11] Glendinning, P. and Sparrow, C. T-points: a codimension two heteroclinic bifurcation. Preprint, University of Cambridge (1985).

[12] Glendinning, P. and Sparrow, C. Heteroclinic loops, homoclinic orbits and periodic orbits in the Lorenz system. In preparation.
[13] Graham, R. and Scholz, H. J. Analytic approximation of the Lorenz attractor by invariant manifolds. *Phys. Rev.* **22A**, 1198–204 (1980).
[14] Guckenheimer, J. A strange strange attractor. In *The Hopf Bifurcation and its Applications*, eds J. Marsden and M. McCracken. Springer, New York (1976).
[15] Guckenheimer, J. (1981). On a co-dimension two bifurcation. In *Dynamical Systems and Turbulence*, Warwick 1980, eds D. Rand and L-S. Young. *Lecture Notes in Mathematics* **898**. Springer, New York (1979).
[16] Guckenheimer, J. and Holmes, P. *Nonlinear Oscillations, Dynamical Systems, and Bifurcations of Vector Fields. Appl. Math. Sci.* **42**. Springer, New York (1983).
[17] Guckenheimer, J. and Williams, R. F. Structural stability of the Lorenz attractor. *Publications Mathématiques IHES* **50**, 73–100 (1980).
[18] Hénon, M. and Pomeau, Y. (1976). Two strange attractors with a simple structure. In *Turbulence and Navier–Stokes Equations*, ed. R. Temam. *Lecture Notes in Mathematics* **565**, 29–68. Springer, New York (1976).
[19] Knobloch, E. On the statistical dynamics of the Lorenz model. *J. Stat. Phys.* **20**, 695–709 (1979).
[20] Knobloch, E. and Weiss, N. O. Bifurcations in a model of magnetoconvection. *Physica 9-D*, 379–407 (1983).
[21] Lorenz, E. N. Deterministic non-periodic flows. *J. Atmos. Sci.* **20**, 130–41 (1963).
[22] Lorenz, E. N. Noisy periodicity and reverse bifurcation. In *Nonlinear Dynamics* ed. R. H. G. Helleman, *Ann. NY Acad. Sci.* **357**, 282–91 (1980).
[23] Lucke, M. Statistical dynamics of the Lorenz model. *J. Stat. Phys.* **15** (6), 455–75 (1981).
[24] Manneville, P. and Pomeau, Y. Different ways to turbulence in dissipative dynamical systems. *Physica 1-D*, 219–26 (1980).
[25] Marsden, J. and McCracken, M. *The Hopf Bifurcation and its Applications. Appl. Math. Sci.* **19**. Springer, New York (1976).
[26] Miari, M. Qualitative understanding of the Lorenz equations through a well known second order dynamical system. Preprint, Dipartimento di Fisica dell'Università di Milano (1984).
[27] Rand, D. The topological classification of Lorenz attractors. *Math. Proc. Camb. Phil. Soc.* **83**, 451–60 (1978).
[28] Robbins, K. Periodic solutions and bifurcation structure at high *r* in the Lorenz system. *SIAM J. Appl. Math.* 457–72 (1979).
[29] Schmutz, M. and Rueff, M. Bifurcation schemes in the Lorenz model. *Physica 11-D*, 167–78 (1984).
[30] Shimada, I. and Nagashima, T. A numerical approach to ergodic problem of dissipative dynamical systems. *Prog. Theor. Phys.* **61**, 1605–16 (1979).
[31] Sil'nikov, L. P. A case of the existence of a denumerable set of periodic motions. *Soviet. Math. Dokl.* **6**, 163–6 (1965).
[32] Sil'nikov, L. P. A contribution to the problem of the structure of an extended neighborhood of a rough equilibrium state of saddle-focus type. *Math. USSR Sbornik* **10**, 91–102 (1970).
[33] Sinai, J. G. and Vul, E. B. Hyperbolicity conditions for the Lorenz model. *Physica 2-D*, 3–7 (1981).
[34] Smale, S. Differentiable dynamical systems. *Bull. Amer. Math. Soc.* **73**, 747–817 (1967).

[35] Sparrow, C. T. *The Lorenz Equations: Bifurcations, Chaos and Strange Attractors. Appl. Math. Sci.* **41**. Springer, New York (1982).
[36] Sparrow, C. T. An introduction to the Lorenz equations. *IEEE Trans. CAS-30* (8), 533–41 (1983).
[37] Tresser, C. On some theorems by Sil'nikov. *Ann. Inst. H. Poincaré* **40**, 441–61 (1983).
[38] Williams, R. F. The structure of Lorenz attractors. In Turbulence Seminar, *Lecture Notes in Mathematics* **615**. Springer, Berlin (1977).
[39] Williams, R. F. Structure of Lorenz attractors. *Publications Mathématiques IHES*, **50**, 59–72 (1980).

7
Instabilities and chaos in lasers and optical resonators

W. J. Firth

Department of Physics, Strathclyde University,
Glasgow G4 0NG, UK

7.1 Introduction

Turbulence in lasers and other optical systems is a newly recognised rather than new phenomenon. What is especially interesting about such systems is the possibility that experiments can be constructed which are sufficiently simple and close to theoretical models that routes to chaos can be studied in detail. That is an as yet unfulfilled hope, but the recent pace of experimental progress has been so great that there is every reason for optimism.

This chapter aims to introduce the reader to the most important concepts and systems in the field of optical chaos. It is necessarily selective and undoubtedly subjective in choice of material, and the author apologises where appropriate. The cited works, in the reference lists, will introduce the interested reader to the wider literature. A broad selection of papers on chaos in passive resonators is available in *Optical Bistability* **2** [8], and the January 1985 issue of the *Journal of the Optical Society of America, Series B*, devoted to instabilities in lasers, will be a very useful source in that field.

The chapter opens with an introduction to the semiclassical theory of light-atom interactions and optical resonators. The remainder of the chapter selects three topics which are, in the author's view, the most important and/or mature in the field: homogeneously broadened single-mode lasers, passive resonators, and inhomogeneously broadened lasers. The treatment tries to concentrate on the physical ideas and interactions involved, rather than on detailed mathematical analysis and computer results, which are available in the original sources as well as elsewhere in this book: most calculations lead to one or more of the so-called 'universal' routes to chaos. Where possible, the status of experiments is described:

only for inhomogeneously broadened lasers are the experiments, thanks largely to the efforts of Lee Casperson and Neal Abraham, as well developed as the theory.

7.2 Semi-classical two-level system theory [4]

All the results in this section can be derived in the *semi-classical approximation*, in which the atoms are treated quantum mechanically, but the optical fields as entirely classical. There has been considerable interest, and some work, in fully quantum-optical treatments of these problems, but 'quantum-optical chaos' is essentially a topic for future reference.

We therefore write the system Hamiltonian as

$$H = H_0 + H'$$

where

$$H_0 = \Sigma \; \hbar\omega_n \mid n\rangle\langle n \mid$$

is the unperturbed (dark) atomic Hamiltonian, expressed in its own basis set $\mid n\rangle$, with corresponding energy $E_n = \hbar\omega_n$. The radiation field is treated as a perturbation, with the form (in the dipole approximation)

$$H' = V(t) = e\mathbf{r} \cdot \mathbf{E}(t)$$

where $\mathbf{E}(t)$ is the radiation field and $e\mathbf{r}$ the electron dipole operator. In the basis set $\mid n\rangle$ H' has matrix elements

$$\begin{aligned} H'_{mn} &= \langle m \mid e\mathbf{r} \mid n\rangle \cdot \mathbf{E}(t) \\ &= \mu_{mn} E(t) \end{aligned}$$

where we have neglected as an inessential complication effects due to the vector nature of $\mathbf{E}$.

Next we restrict ourselves to harmonic (optical) fields $E(t)$, with angular frequency ω:

$$E(t) = \mathrm{E}e^{-i\omega t} + \mathrm{E}^*e^{i\omega t}$$

Assuming, for definiteness, that the atom is initially in state $\mid 1\rangle$, and expressing its subsequent evolution as

$$\psi(t) = \sum_n c_n(t) \mid n\rangle \; : \; c_n(0) = \delta_{n1}$$

then standard application of time-dependent perturbation theory gives, to lowest order

$$c_1(t) = 1$$

$$c_n(t) = \frac{i\mu_{n1}\mathrm{E}}{\hbar(\omega_n-\omega_1-\omega)} - \frac{i\mu_{n1}^*\mathrm{E}^*}{\hbar(\omega_n-\omega_1+\omega)}, n \neq 1$$

Clearly the largest excitation will occur in cases where $\hbar\omega$ is resonant with an atomic energy spacing ($\hbar\omega_n - \hbar\omega_1$). We assume, for simplicity, that only for n equal to 2 is there such a resonance (i.e. the atom is free from degeneracies). This allows us to drop all other energy levels from the analysis, and consider only a *two-level atom*. For consistency, we must also neglect the second term above, which is anti-resonant: this is termed the rotating-wave approximation. We can thus, finally, set $\mu_{21}\varepsilon/\hbar$ equal to Ω (the *Rabi frequency*), and obtain the simplified Hamiltonian

$$H = \hbar\omega_0 \, | \, 2 \rangle\langle 2 \, | + \hbar\Omega e^{-i\omega t} \, | \, 2 \rangle\langle 1 \, | + \hbar\Omega^* e^{i\omega t} \, | \, 1 \rangle\langle 2 \, |$$

where $\omega_0 = (\omega_2 - \omega_1)$ and we have arbitrarily set $\omega_1 = 0$.

This simplified single-atom Hamiltonian can be explicitly solved, of course: the solution consists of an oscillation of $| \, c_1 \, |$ and $| \, c_2 \, |$ at frequency Ω — so-called *Rabi-flopping*. We are interested, rather, in ensembles of atoms, at different space-points and, in many cases, with different thermal velocities. Further, these atoms interact with each other and their environment in unknown ways. These complexities are best handled by the density-matrix formalism. The density matrix is defined as

$$(7.1) \qquad \rho_{mn} = \sum_{i \atop \text{atoms}} W(i) \, c_{mi} c_{ni}^* \, | m \rangle\langle n |$$

The density matrix is a weighted average over atomic variables over which we have no knowledge or control. For our purposes, the averaging leaves ρ as a function of position (with some coarse-graining on the optical wavelength scale) and velocity component, v, parallel to the optical propagation direction (z-direction). The definition of ρ is such that the averaged expectation value of any quantum operator $\hat{O}$ is given by

$$\langle \hat{O} \rangle = Tr(\hat{O}\rho)$$

As a quantum-mechanical operator, ρ obeys the Heisenberg equation of motion

$$\dot{\rho} = \frac{i}{\hbar}[H, \rho]$$

A major advantage of the density-matrix formulation is that it is easy to augment this equation by terms representing the interactions not accounted for in H. In the cases of interest, these can be represented by simple excitation and decay terms.

The diagonal elements of ρ (here simply ρ_{11} and ρ_{22}) are, by definition, the mean populations of states $| \, 1 \rangle$ and $| \, 2 \rangle$. Their dynamics can be described by

$$\dot{\rho}_{ii} = \frac{i}{\hbar}[H', \rho]_{ii} + P_i - \gamma_i\rho_{ii} \, , \, i = 1, 2$$

P_i describes incoherent pumping processes and γ_i decay processes: if Ω

is zero, then the commutator vanishes, and we obtain the steady-state solution for the 'dark' populations:

$$\rho_{ii}^0 = P_i/\gamma_i \, , i = 1, 2$$

In a laser, the pumping processes (electrical, chemical, etc., as well as collisional) must be such as to maintain a population inversion, i.e. $\rho_{22} > \rho_{11}$.

The decay processes contributing to γ_i represent all collisional and other processes which destroy c_i, including spontaneous emission.

The off-diagonal element ρ_{21} (= ρ_{12}^*) is described by

$$\dot{\rho}_{21} = \frac{i}{\hbar}[H, \rho]_{21} - \Gamma\rho_{21}$$
$$= i\omega_0\rho_{21} + \frac{i}{\hbar}[H', \rho]_{21} - \Gamma\rho_{21}$$

Physically, ρ_{21} represents a coherence between $|1\rangle$ and $|2\rangle$, since it vanishes in the sum (7.1) unless there is correlation between the amplitudes c_1 and c_2. By hypothesis, such a correlation is induced by H' (Rabi-flopping), but by no other means. The decay constant Γ includes any process that destroys c_1, c_2 *or their correlation*. The latter class include *phase-changing collisions*, in which the interatomic forces in a collision temporarily change ω_0 without inducing a transition.

Physically, a non-zero value of ρ_{21} implies a *polarisation* of the atomic medium, since

$$\langle \mathbf{P} \rangle = Tr(\boldsymbol{\mu}\rho) = \mu_{12}\rho_{21} + \mu_{21}\rho_{12} = 2Re\mu_{12}\rho_{21}$$

The atomic polarisation acts as a source term in the Maxwell equation:

$$\square\, \mathbf{E} = \ddot{\mathbf{P}}$$

We thus have, within the limits of the approximations made, a complete, self-consistent set of equations for the evolution of the field-atom system, subject to appropriate boundary conditions. One further simplifying approximation is often made: recognising that all rates of change are commonly slow compared with the optical cycle time $2\pi/\omega$ (and the wavelength $\lambda = 2\pi/k = 2\pi c/\omega$), one can reduce the Maxwell equation to first order (slowly varying amplitude approximation), and also take out the rapidly varying part of ρ_{21}, setting $\rho_{21} = r_{21}\exp(-i\omega t)$.

This leads to the Maxwell–Bloch equations, which are the basis for the subsequent treatment:

(7.2)
(a) $\left(\dfrac{\partial}{\partial z} + \dfrac{1}{c}\dfrac{\partial}{\partial t}\right)\Omega e^{i\varphi} = G\langle r_{21}(v)\rangle$
(b) $\dot{\rho}_{22} = P_2 - \gamma_2\,\rho_{22} - \frac{1}{2}\,Im\,\Omega\,r_{21}$
(c) $\dot{\rho}_{11} = P_1 - \gamma_1\,\rho_{11} + \frac{1}{2}\,Im\,\Omega\,r_{21}$
(d) $\dot{r}_{21} = i\,(\Delta - \dot{\varphi} - kv)\,r_{21} - \Gamma r_{21} + (\rho_{11} - \rho_{22})\,\Omega e^{i\varphi}$

where the Rabi frequency has been written explicitly in polar form, G is a coupling constant and $\langle \rangle$ allows for a velocity average over the Maxwellian distribution of longitudinal velocities of the atoms. In (7.2d) Δ equals $(\omega - \omega_0)$, the $\dot{\varphi}$ term recognises that a changing phase of the field is equivalent to a frequency shift, and kv expresses the Doppler shift of the field frequency on transformation into the atomic rest frame.

The presence of the continuous parameter v and the space derivative $\partial/\partial z$ clearly endows the Maxwell–Bloch system with a phase space amply large enough for chaotic behaviour: our first priority is to constrain the phase space by examining various idealised limits. In particular, we can eliminate the space derivative by placing the system in an optical resonator: indeed most experimental evidence of optical chaos involves resonator systems. We proceed, therefore, to a brief survey of the relevant aspects of resonator theory [24, 34].

The simplest optical resonator — at least from a theoretical standpoint — is the unidirectional ring cavity (Fig. 7.1), comprising four 45° tilted

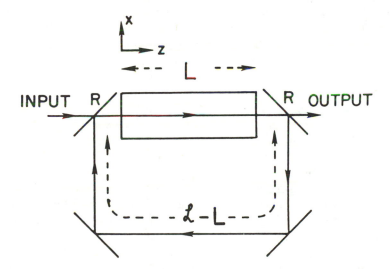

Fig. 7.1 Schematic of ring resonator containing a medium with nonlinear optical properties.

mirrors, two perfectly reflecting, and two of transmittivity T (amplitude transmittivity $\sqrt{T}$), and reflectivity R (respectively $\sqrt{R}$). The cavity contains a two-level medium of length L, and the total optical path is $\mathcal{L}$. The boundary conditions at the entrance and exit from the medium are, respectively:

$$E(0,t) = \sqrt{T}\, E_{\mathrm{I}} + R e^{-i\theta}\, E(L, t - \Delta t)$$

and

$$E_T(t) = \sqrt{T} E(L, t)$$

The latter relation simply implies a unique relationship between the transmitted field amplitude E_T and the internal field. The former enshrines the major cavity properties: E_I is the incident field amplitude (normally zero in a laser); Δt is the time for light to propagate from the cell output around to the input, equal to $(\mathscr{L} - L)/c$; θ is the cavity mistuning, whose physical significance is best examined for the empty cavity, in steady state, whence

$$E(L, t - \Delta t) = E(0, t) = E$$

and thus

$$E = \sqrt{T} E_I / (1 - Re^{-i\theta})$$

The cavity field thus has resonances spaced at intervals of 2π in θ, which are very sharp if R is close to unity (high-finesse resonator). These resonances correspond to constructive interference of the circulating field with the incident field, and θ thus sweeps through 2π as the total cavity length $\mathscr{L}$ is fine-tuned over an optical wavelength λ. A laser (without injected signal) necessarily operates at a frequency such that the *total* mistuning (θ plus any phase shift due to the medium) is equal to zero (modulo 2π). Each such frequency is termed a longitudinal mode frequency of the cavity, and their frequency separation $\sim c/\mathscr{L}$ is termed the *free spectral range*. Instabilities, both in lasers and passive resonators ($E_I \neq 0$, lossy cavity), are termed *single mode* or *multimode* according to whether the resulting frequency spectrum for $E(0, t)$ is or is not narrow compared to the free spectral range.

Single-mode instabilities in high-finesse resonators are particularly simple because the propagation and time-delay effects can be reduced, using a first-order Taylor expansion, to a simple first-order differential equation for the mode amplitude, as will be described in the next section.

Note that the cavity feedback involves a finite *time delay*, of order $t_R = \mathscr{L}/c$: in certain circumstances this enables the Maxwell–Bloch equations (required to relate $E(L, t + L/c)$ to $E(0, t)$) plus cavity boundary conditions, to be expressed as a discrete *mapping* in steps of t_R — this has been extensively studied in the context of passive resonators, and will be elaborated below.

Practical resonators share most of the properties of this ideal unidirectional ring resonator, but with complicating features. One of these is the *transverse mode* structure. The ideal resonator, with plane, and necessarily finite, mirrors, actually supports an infinity of transverse mode structures, of varying complexity in amplitude and phase, transverse to the resonator axis. This introduces undesirable extra degrees of freedom into the system, which can, however, often be suppressed by an appropriate choice of mirror curvature and aperture, assisted in some cases by wave-guiding properties of the medium itself.

A ring laser has a further degeneracy: each longitudinal mode frequency supports a *pair* of counterpropagating modes, which will ordinarily have the same threshold, and thus both be excited. To suppress one of these, and produce a *unidirectional* laser, it is necessary to employ a non-reciprocal element — usually based on the Faraday effect — which makes the cavity more lossy for one of these modes than the other.

This last degeneracy does not occur in a *Fabry–Perot* resonator, in which the light bounces between parallel mirrors. This is by far the commonest cavity for lasers in general, but has major drawbacks from a theoretical standpoint: the standing-wave pattern greatly complicates the atomic response and can also lead to multimode operation, and time and space are much more intimately mixed in the feedback process. None the less, Fabry–Perot experiments seem to give rise to broadly the same sorts of instability as those predicted from ring cavity analyses.

7.3 Single-mode homogeneously broadened laser

The first laser, operated in 1960 by Maiman at Hughes Research Laboratory, was based on flashlamp-excited ruby. This laser provided an early impetus for studies of instabilities by tending to produce noisy, spiked output even under quasi-steady excitation. In these early years, equations similar to the Lorenz system (Chapter 6) [35] were written down, both for lasers and masers, but failure to find a practical system in either spectral region which was adequately described by these equations, plus of course the abundance of other avenues opened up by the development of lasers, led to a rather quiescent phase lasting till the resurgence of interest in the late 1970s.

The basic equations for a single-mode (unidirectional) homogeneously broadened laser in a high-finesse cavity, tuned to resonance so that $\Delta = \theta = 0$, may be written as the real system

$$
\begin{aligned}
\text{(a)} \quad & \dot{x} = -\gamma_c (x + 2Cp) \\
\text{(7.3)} \quad \text{(b)} \quad & \dot{p} = -\Gamma (p - xD) \\
\text{(c)} \quad & \dot{D} = -\gamma (D + xp - 1)
\end{aligned}
$$

Here x is a scaled electric field (or Rabi frequency), and (7.3a) arises in the single-mode limit of the Maxwell equation (7.2a): γ_c, equal to $c(1 - \sqrt{R})/\mathcal{L}$, describes the decay of the cavity field due to mirror transmission, and C, the cooperativity parameter, is defined as $\alpha_0 L\sigma/2T$, where α_0 is the small-signal absorption coefficient and $-\sigma$ the population inversion per atom due to the pumping processes (in the dark). C is thus negative for a laser: its definition arises from the field of optical bistability (see below) where the medium is lossy and thus C is positive. D and p are scaled atomic population differences ($\rho_{22} - \rho_{11}$) and polarisation (ρ_{21}) respectively: scaling is such that in the small signal limit $D = 1$, $p = x$, and then, for consistency in (7.3a), $2C = -1$: this last condition thus defines the laser threshold; for $|2C| < 1$, the only solution of the laser equations is the trivial one $x = p = 0$, $D = 1$, which is stable. For $|2C| > 1$, this solution still exists, but becomes unstable, whereas the nontrivial solution

$$x^2 = -2C - 1 \; ; \; D = -1/2C \; ; \; p = xD$$

is stable — at least for small enough $|C|$, since the system (7.3) is formally equivalent to the Lorenz system [35], though this fact was not recognised until 1975 [17]. Motivated by practical considerations, analysis had tended to concentrate on the rate-equation limit of the above system (and more complicated systems describing real lasers). This involves adiabatic elimination of the polarisation ($\Gamma \to \infty$), and replacing x by the 'intensity' I, equal to x^2, yielding the system

(7.4) (a) $\dot{D} = -\gamma (D + ID - 1)$
       (b) $\dot{I} = -2\gamma_c I (1 + 2CD)$

which has only damped oscillatory solutions (relaxation oscillations).

Turning to the full three equations, the instability condition becomes

$$\gamma_c > \gamma + \Gamma$$

This condition, derived in 1964 by Korobkin and others [16, 25] is termed the 'bad cavity' requirement, since it demands that cavity loss damp the field more strongly than the damping of either the polarisation or the population. These same authors also derived the instability threshold condition, here expressible as:

$$-2C > 1 + \frac{(\Gamma + \gamma + \gamma_c)(\Gamma + \gamma_c)}{\Gamma (\gamma_c - \Gamma - \gamma)}$$

Consideration of the effect on the above expression of varying the various decay rates shows that the right side of the above inequality will always be greater than nine, which implies a pumping rate at least nine times above the laser threshold. This is the major stumbling block to experimental realisation of the Lorenz system in lasers. One can get high gain by lengthening the laser, but then the free spectral range falls, and the laser will go multimode except close to threshold. To prevent this, a short laser is necessary, which means a high density of atoms, which increases Γ and γ, which demands an increase in γ_c, which raises the threshold (actual pump rate at threshold is proportional to γ_c) which demands still higher density, etc. — a vicious circle. Interestingly, in masers, the first electromagnetic wave devices for which equations of the Lorenz type were developed, the problem was that γ_c in the then best-available cavities was so large compared to γ and Γ that $|2C|$ (which is $\sim \gamma_c/\Gamma$ in this limit) was again impossibly high. Subsequent development of superconducting microwave cavities may have changed the situation, but perhaps of more interest is the very recent suggestion [36] to operate lasers in the intermediate regime — the far infrared. In far-infrared lasers, γ and Γ are in the megaHertz range, which readily allows optimisation of γ_c so as to minimise the threshold value of D_0. Undamped relaxation oscillations had already been seen in such lasers, and at the time of writing there is the exciting prospect that the Lorenz system of equations may soon develop an experimental significance in laser physics to match their theoretical impact. In the interim, however, a number of schemes have been devised in which instabilities and chaos have been predicted in experimentally realisable systems. Undoubtedly the most significant of these in the context of lasers has been that of inhomogeneous broadening, to which we will return.

First, we discuss the laser with injected signal, which has been predicted [27] to exhibit a number of instabilities and chaos. Physically, we then have two types of laser response. There is the 'spontaneous' laser action, cavity resonant, and, we will assume, resonant also with the atomic transition on which the population inversion exists. The second response is a regenerative amplification of the injected signal, which is detuned from resonance. For a linear driven oscillator, these responses correspond to the complementary function (here a growing exponential because of the gain) and the particular solution, respectively. In the laser the nonlinearity leads to competition between these two types of behaviour, and gives rise to unstable behaviour as the injected signal is increased from zero (stable self-excited laser, as discussed above) to a value high enough to slave the laser to the injected signal frequency, where the response is again stable. We will deal only with self-excited instabilities, but predictions of chaos also exist for the more complex case where the injected signal, or the pumping rate, are modulated. Generally speaking, instability thresholds are lower than in the quasi-Lorenz system, and the restrictions on decay rates less severe. There is not yet, however, any experimental evidence for these phenomena: injection of a low-power continuous signal is routinely used to induce single-longitudinal mode operation in, for example, high-power CO_2 lasers, but this application corresponds only to the small-injected signal limit of the system under discussion.

We generalise the quasi-Lorenz system by introducing an injected amplitude y with the same scaling as x, but assumed real and positive, while x and p are now complex, in general, because of the detuning Δ and the mistuning θ (which we scale to Γ). The resulting set of equations is

$$
\begin{array}{ll}
\text{(a)} & \dot{x} = -\gamma_c((1 + i\theta)x - y + 2Cp) \\
(7.5) \quad \text{(b)} & \dot{p} = -\Gamma((1 + i\Delta)p - xD) \\
\text{(c)} & \dot{D} = -\gamma(D + \mathrm{Re}\, xp^* - 1)
\end{array}
$$

For finite y, the above system has the steady-state solution

$$
(7.6) \qquad y^2 = |x|^2 \left[\left(1 + \frac{2C}{1 + \Delta^2 + |x|^2} \right)^2 + \left(\theta - \frac{2C\Delta}{1 + \Delta^2 + |x|^2} \right)^2 \right]
$$

Examination of this solution reveals that it is stable if y exceeds a certain minimum value, y_{th}, sufficient to quench the spontaneous lasing. En route from $y = 0$ to this stable 'injection locking', Lugiato *et al.* find, for small y, a pulsation of frequency corresponding to the detuning and of amplitude $\sim y^{\frac{1}{2}}$, imposed on the $y = 0$ value $|2C + 1|^{\frac{1}{2}}$ of x. For the parameters $C = -500$; $\Delta = \theta = 1$; $\gamma = \gamma_c = \Gamma$, this pulsation becomes irregular for $y \sim 120$, is highly chaotic at $y \approx 250$, then shows an inverse period-doubling cascade to regular oscillation at $y \approx 300$. Further increase of y leads to a 'breathing' behaviour: a slow modulation of the oscillation, which develops into an output consisting of 'spikes' followed by quiescent intervals. The latter increase in duration until they evolve into the stable solution at y_{th}, which is close to 312 for these parameters.

For other parameters, including the experimentally accessible $C = -20$, similar behaviour is seen, except that the breathing and spiking seem to be absent.

It should be emphasised that these phenomena are distinct from the Lorenz instability: in fact this domain ($\gamma_c = \Gamma = \gamma$) is absolutely stable for $y = 0$, the Lorenz limit. Injection into an already unstable laser stabilises it at high enough y-values: at low values of y the behaviour is, unsurprisingly, irregular. In fact Lugiato *et al.* find empirically that $\gamma_c \approx \Gamma$ is necessary for chaos: γ can be much smaller than Γ without hindrance.

To conclude this section, we can, at last, refer to an experiment. As noted, the rate equations (7.4) are stable, but the standard technique of modulating one of the parameters, for example γ_c, extends the phase space sufficiently to allow chaotic behaviour. This has been achieved in a CO_2 laser [5]. Bistability — the coexistence of stable oscillations at either one-third or one-quarter of the modulation frequency — was observed. Increasing the modulation depth, the two attractors become strange. A particularly interesting feature of the results was the appearance of a low-frequency divergence of the spectrum when the two attractors become strange. The authors attributed this to low-frequency jumping between the two attractors, and suggested that such behaviour may lie at the root of the puzzling low-frequency noise spectra displayed in many nonlinear physical systems — systems as diverse as electrical resistors, biological membranes and automobile traffic flow.

With the honourable exception of this last case, the homogeneously broadened laser has so far proved a fertile field for instability and chaos only for the theorist. In contrast, when the laser is inhomogeneously broadened, the experiments were for long in advance of the theory, as we shall see below.

First, however, we examine *passive* resonators, which are perhaps the simplest of all optical systems, and indeed, in one limit, reduce to a one-dimensional noninvertible map, and thus possess a period-doubling route to chaos.

7.4 Instabilities in passive nonlinear resonators

Passive nonlinear resonator theory, described by equations (7.5) with $C > 0$ and thus $y \neq 0$, was originally developed in the context of *optical bistability* (OB): $|x|$ in the state equation (7.6) can clearly be a multivalued function of y, which corresponds to an optical device that can have two or more transmission states for a single-input field, and can thus act as an optical memory device (Fig. 7.2).

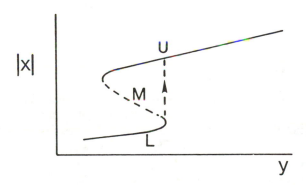

Fig. 7.2 Optical bistability: output $|x|$ from nonlinear resonator may be multiple-valued function of the input y.

Equations such as (7.5) are built on the seminal work of Bonifacio and Lugiato [6] in 1978. For simplicity, these authors treated the high-finesse case in which the dominant dynamic instabilities arise due to excitation of adjacent longitudinal modes, which are also cavity-resonant. When Ikeda in 1979 performed a stability analysis starting from the full Maxwell–Bloch equations, with propagation effects included [22], he discovered an apparently distinct family of instabilities, and indeed chaotic behaviour, which will be the dominant topic of the remainder of this section (as Lugiato

[26] has pointed out, these phenomena are not actually distinct from the Bonifacio–Lugiato instabilities, and can be analysed by similar methods).

The simplest approach to the Ikeda instability makes the limit $\Gamma \to \infty$ in the Maxwell–Bloch equations (7.2). It is then possible (at least in the ring resonator) to integrate the full equation along the characteristic, $z = ct +$ constant, to obtain an explicit relationship between $E(0, t)$ and $E(L, t + L/c)$. Further simplification ensues if we now let $\gamma \to \infty$ and take the *dispersive* or *Kerr limit* $\Delta \to \infty$, $\alpha_0 L \to \infty$, $\Delta/\alpha_0 L$ finite. This reduces the system (7.2) and the cavity boundary condition to a *mapping*, of the form

(7.7) $$E(t + t_{\mathrm{R}}) = A + B \, e^{i(\,|\,E(t)\,|^2 \,-\, \theta)} \cdot E(t)$$

Note that the amplitude $E(t)$ is complex, so that we have a two-dimensional mapping, which is, in fact, *invertible*:

$$E(t) = E' \, e^{-i(\,|\,E'\,|^2 \,-\, \theta)}; \quad E' = (E(t + t_{\mathrm{R}}) - A) \div B$$

We thus have a minimal system for chaos.

The scaled parameters A (transmitted input field), B (amplitude feedback fraction) and θ (cavity mistuning) can all be assumed real.

The fixed points S of (7.7) give rise to an $|S|^2$ vs A^2 typical of optical bistability (Fig. 7.2). Their stability is governed by the Jacobian matrix

$$\mathbf{M} = \begin{bmatrix} Z(1 + i|S|^2) & iZS^2 \\ -iZ^*S^{*2} & Z(1 - i|S|^2) \end{bmatrix}; \quad \begin{aligned} Z &= Be^{i(|S|^2 - \theta)} \\ S &= A/(1 - Z) \end{aligned}$$

which gives rise to the characteristic equation:

(7.8) $$\lambda^2 - 2\lambda(\mathrm{Re}Z - |S|^2\mathrm{Im}Z) + B^2 = 0$$

Since $B < 1$ in a passive resonator, the only possible instabilities are $\lambda = \pm 1$. The plus sign can be seen to correspond to the vertical-slope points of the characteristic of Fig. 7.2 (further analysis shows that the intervening negative slope branches are unstable). The negative sign involves a perturbation to S which reverses each t_{R}, and thus gives rise to an oscillation with period $2t_{\mathrm{R}}$. This oscillation is, in many cases, the beginning of a period-doubling cascade to chaos obeying the Feigenbaum scaling relations [13], but the author is not aware of any proof that the map has this property.

Any of A, B, and θ can be considered as control parameters, but from a physical point of view it is interesting to use θ (which can be varied through its full range by translation of one mirror over one wavelength) to minimise the threshold value of A. This can be done analytically, based on (7.8), and Fig. 7.3 shows A^2B vs B for $\lambda = 1$ (bistability) and $\lambda = -1$ ($2t_{\mathrm{R}}$ or $P2$) instabilities. The choice of ordinate arises from the empirical observation that $A^2B \sim 1$ is required for chaos. It will be observed that, whereas OB and $P2$ instabilities have comparable thresholds in low-finesse cavities (small B), in high-finesse cavities ($B \to 1$) optical bistability is much easier

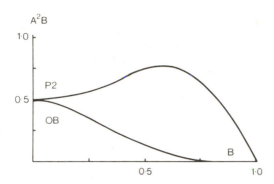

Fig. 7.3 Instability thresholds for the mapping (7.7): lower curve, optical bistability (OB); upper curve Ikeda instability (*P2*).

to achieve than Ikeda instability. This fact, and the decay-rate restrictions discussed below, lie behind the paucity of experimental evidence on Ikeda instabilities: only in two all-optical systems (both pulsed rather than continuous-wave) has $2t_R$ oscillation and chaos been observed to date.

The basic *P2* instability can be given an attractive physical interpretation in terms of standard nonlinear optics [14]. The material nonlinearity responsible for nonlinear refraction, and thus bistability, can also generate sidebands on the pump frequency. Physically, two photons with frequency ω are scattered to form a pair of photons at $\omega \pm \Delta\omega$, so the process is termed four-wave mixing. An additional input signal, the probe, detuned from the pump by $\Delta\omega$, will thus experience gain or loss according to the combined effect of the nonlinearity *and* the mode structure of the cavity. Clearly, it is most advantageous if *both* $\omega + \Delta\omega$ and $\omega - \Delta\omega$ are cavity resonant, i.e. if $2\Delta\omega$ is a multiple of the cavity's free spectral range. An *even* multiple (including zero) means that the pump is itself resonant, and this double resonance lies behind optical bistability and the sideband instabilities analysed by Lugiato and co-workers. An odd multiple, on the other hand, means that the pump is off-resonance, lying exactly halfway between two modes: the beat note between the pump and the resonant sidebands, which will be self-excited if the nonlinearity is strong enough, has period $2t_R$, identifying this as the Ikeda instability.

The nice thing about this system is that the above double resonance can be guaranteed: the refractive-index change induced by the strong pump beam actually moves the comb of longitudinal modes with respect to the pump frequency (it is strictly the product of length and refractive index that determines the mode frequencies). This 'transphasing' of the mode spectrum will alternately give rise to bistable and Ikeda-type double resonances as the pump parameter A is increased: θ determines the starting position of the comb relative to the pump frequency.

The entire period-doubling cascade can be given a similar interpretation. As each new period $2^n t_R$ bursts into oscillation, the effective free spectral range of the cavity is halved, because the generalised condition for resonance must be constructive interference after 2^n round trips, since only after $2^n t_R$ do the cavity's optical properties repeat themselves. At the $2^n t_R$ threshold, these small signal modes are degenerate with the oscillating frequency spectrum, but as A is increased, they transphase in frequency until halfway between the oscillating frequencies. If the gain there is large enough, these new modes burst into oscillation: if not, transphasing continues to a renewed coincidence with the oscillating frequencies, at which point an inverse period-doubling cascade ensues, leading eventually to a steady-state response. Figure 7.4 illustrates this process, showing the small-signal gain spectrum as A is increased through the $P2$ threshold, and the ensuing doubling of the spectrum, followed by transphasing to the verge of $4t_R$ oscillation. It should be noted that the noise spectrum of the system should be very similar to Fig. 7.4, because of the filtering action of the cavity on any broadband noise source.

These considerations constitute a useful base against which to consider

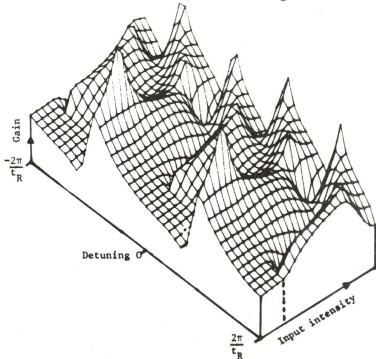

Fig. 7.4 Probe gain spectrum for the mapping (7.7) as the input intensity A^2 is increased. Gain peaks (dressed modes) double their multiplicity at the $P2$ bifurcation (dotted line), then transphase through half their separation and grow as the $P4$ bifurcation is approached.

the complications caused by transverse spatial effects and by finite-medium response time γ^{-1}: there is clearly no reason why the basic gain and trans-phasing mechanism should not survive the introduction of these complications.

Dealing first with finite time constant, perturbation theory readily leads to the conclusion that the gain spectrum of four-wave mixing becomes Lorenzian, with halfwidth $\sim \gamma$, and the associated dispersion means that the free spectral range becomes a function of frequency. The former is usually the more important: clearly if $\gamma t_R \ll 1$, then there will be no significant gain in the Ikeda situation, where the modes straddle the pump frequency: this is the physical origin of the requirement $\gamma^{-1} > t_R$ usually quoted for Ikeda instability. There is no corresponding requirement for bistability, which occurs for $\Delta\omega = 0$, and thus actually benefits from a long time constant, since the gain is $\sim \gamma^{-1}$ for $\Delta\omega = 0$: hence materials such as semiconductors and liquid crystals with small γt_R are ideal for bistability but useless for Ikeda instability.

Another effect of finite γ is to raise the degeneracy by which all symmetrically placed pairs of sideband modes reach threshold simultaneously. On the one hand, the raising of the $\lambda = +1$ degeneracy permits self-pulsing at period t_R, as described by Bonifacio and Lugiato [7], whereas the Ikeda degeneracy splits to yield pulsing at $2t_R/3$, $2t_R/5$, etc. The latter phenomena have been extensively analysed in the limit $B \to 0$, in which the Ikeda mapping becomes one-dimensional and can be realised in a so-called *hybrid* system in which the feedback is electronic rather than optical. In the limit $B \to 0$, the complex map can be approximated by the real noninvertible map

$$I_{n+1} = A^2 + 2AB \cos (I_n + \theta)$$

where $I = E^2$. Introduction of a finite value for γ augments the left side of the above by a term $\gamma^{-1} \dot{I}$, and the index n gives over to continuous time t. As stated, this raises the degeneracy caused by the cosine function, and this system is observed [21] to give a variety of waveforms of period $2T/(2m + 1)$, coexisting in complex ways with each other and with subharmonics and chaos.

The $2t_R/3$ oscillation has also been seen, however, in an all-optical system based on ammonia gas [18].

There is one interesting situation in which the $\gamma t_R > 1$ rule is broken, which can also be interpreted along the above lines. As γ decreases, the four-wave gain increases as its bandwidth decreases. This gain is accompanied by dispersion, just as in a laser, which is such as to *reduce* the frequency interval between two modes lying either side of the pump frequency (mode-pulling). These modes can thus give rise to an Ikeda instability (which will now have a period $\sim\gamma^{-1}$, rather than $2t_R$) *provided* these modes are resolved into two gain peaks: a high-finesse resonator

($\gamma_c \approx \gamma$) is thus required for this version of the Ikeda instability, which gives rise to chaos via a period-doubling cascade in parameter regions corresponding to the upper branch of optical bistability [23]. Though not so far observed, this version of the Ikeda instability is of interest because it can be derived by mean field methods, which otherwise preclude chaos.

Abraham and Firth [1, 2] have undertaken an analysis of folded (Fabry–Perot) resonators for the case $\gamma t_R \sim 1$. They find, in particular, a Feigenbaum period-doubling route to chaos, and are able to show, by an extrapolation technique, that the scaling parameter δ lies within 1% of the Feigenbaum value 4.669... for any of three control parameters A, γt_R, θ (Table 7.1).

Table 7.1 Universal sequence as the control parameter λ approaches the limit point λ_∞ $(\delta_n - \lambda_n - \lambda_{n+1})/(\lambda_{n+1} - \lambda_{n2})$).

Control parameter λ	$\lambda_8 : P4 \rightarrow P8$	λ_∞	δ_8	δ_{16}	δ_{32}
A	4.476	4.463	3.902	4.768	4.612
θ_0	0.09046	0	4.411	4.601	4.658
$(\gamma t_R)^{-1}$	1.826	1.78636	4.506	4.643	4.618

Transverse effects have been examined in a few instances. The most complete of these is due to Moloney *et al.* [31], who find (Fig. 7.5) that the transverse degrees of freedom lead to a Ruelle–Takens route to chaos, via frequency locking of two incommensurate frequencies. A significant recent advance was the demonstration [28] that the basic 'plane-wave' Ikeda map is unphysical, because the $2t_R$ instability has a lower threshold for finite wave vector perturbations than for zero wave vector (plane-wave), with the result that a uniformly excited medium develops an Ikeda oscillation in which neighbouring regions oscillate out of phase. Thus only in cases where the cavity forces a specific transverse profile are transverse effects trivial.

Very considerable theoretical effort has been directed at instabilities in passive resonator systems, of which the above is only a partial, necessarily distorted, summary [9]. Sadly, the experimental situation is much more patchy. The hybrid experiments have been mentioned. Next came an all-optical experiment using an optical fibre as nonlinear medium [32], demonstrating $P2$ and chaos. Unfortunately, the nonresonant nature of the nonlinearity in this experiment meant that it had to compete with other nonlinear processes. In particular, stimulated Brillouin scattering actually had a lower threshold, but was ingeniously eliminated by use of a pulse train, rather than continuous pumping. This complicates the analysis and

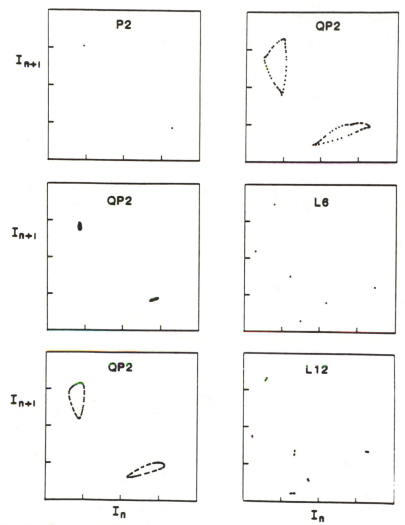

Fig. 7.5 Ikeda instability with transverse effects; the original *P2* is destabilised as the input intensity is increased: The 'quasi-period-2' orbits show points of accumulation and then lock to a *P6* response, which then bifurcates to *P12* en route to chaos. (Courtesy J. V. Moloney.)

reduces the utility of the experiment. Most recently to date have been experiments on ammonia gas resonantly pumped by CO_2 lasers, in which good $2t_R$, $4t_R$ and $2t_R/3$ modulation, and chaos, have been observed in a system which comes close to Ikeda's original proposal [18, 19]. These experiments are, however, still pulsed, and it would be extremely valuable if a true continuous wave all-optical passive resonator system could be developed, to give the sort of experimental impetus that the xenon and helium–xenon lasers have given to laser instabilities.

7.5 Chaos in inhomogeneously broadened lasers

In contrast to the Lorenz model of a homogeneously broadened laser discussed earlier, the situation in high-gain gas lasers with *inhomogeneous* broadening has been one where the experiments have actually been leading the theoretical development. As far back as 1969, Casperson discovered that the high-gain xenon laser (wavelength 3.51 μm) could, even with steady excitation, produce its output as an infinite train of pulses [11]. Moreover, these pulses could, depending on the conditions, repeat regularly, alternate in height or be aperiodic. Casperson pursued two main avenues of approach to a theoretical understanding of these phenomena, viz. coherence effects (finite Γ) and inhomogeneous broadening. He did not, at that stage, incorporate both effects, because 'the mathematics would be more complicated'. Only in 1978 did it become apparent that both effects are necessary, but also sufficient, to produce the observed phenomena. More recently Abraham and co-workers have extended this work to the helium–xenon system, where the helium partial pressure allows the degree of inhomogeneous broadening to be varied as a control parameter, with beautiful results [3], and Mandel [29] had demonstrated theoretically that the Lorenz-model instability evolves continuously into the Casperson instability as the degree of inhomogeneous broadening is varied.

Why is inhomogeneous broadening so crucial? The answer lies in the saturation behaviour. A homogeneously broadened system saturates uniformly in frequency, so that the oscillation frequency, which clearly has maximum gain at threshold, retains its primacy as the laser saturates, preventing other frequencies from reaching threshold (providing it is spatially uniform: in actual lasers higher-order transverse and adjacent longitudinal modes appear because they have different spatial structures from the dominant mode).

The situation in an inhomogeneously broadened medium is quite different. The dominant mode saturates effectively only those atoms whose (Doppler-shifted) resonance frequency lies within about Γ of the mode frequency: other atoms are unsaturated, so that the gain actually *increases* with frequency detuning from the dominant mode: the excited mode burns a *hole* in the gain spectrum. Careful design is thus necessary to restrict laser action to a single mode. Even then, however, an instability can arise due to 'mode-splitting'. As well as the spectral hole burned in the gain, the refractive index contribution of the resonant atoms is also saturated, again over a spectral width $\sim\Gamma$. As a result, the dispersion relation close to the operating frequency develops a 'wiggle', which, for strong enough saturation, may turn into an S-shape. When this happens, we have *three* frequencies which all share the same wave vector, and are thus resonant with the cavity. Since the spectral hole ensures that the two new sideband frequencies actually have higher gain than the central (originating) frequency, they will grow: self-pulsing develops. This phenomenon was termed 'mode-splitting' by Casperson and Yariv [12]. The pulsation period will be of order Γ, but clearly parameter dependent.

Of course, as the sidebands build up, they burn their own holes in the gain and refractive-index spectra, and it is not difficult to envisage the development of pulse shapes as complex as those of Fig. 7.6, or, indeed, period-doubling and chaos in the pulse train: compare the discussion on passive resonators. Deep modulation at frequencies $\sim\Gamma$ is clearly only possible if $\gamma_c > \Gamma$, so we still require a 'bad cavity', and we may note that the spectral holes form and refill on a time scale γ^{-1}, which thus plays an important role in the phenomenon: it has been claimed [20] that 'population pulsations' are the primary source of the instability.

Experimental evidence for chaos based on the mode-splitting instability was first obtained by Casperson in 1978 [10]: the route to chaos was not studied in detail, except that period-two was observed. In recent years Abraham and co-workers have performed detailed studies of routes to chaos in helium–xenon lasers [15] and have observed, as the laser cavity length is fine-tuned, interspersed bands of chaotic output reached, in certain conditions, by period doubling. By varying the discharge current (and hence gain) at fixed cavity length, the Ruelle–Takens route to chaos was observed: two initially incommensurate frequencies (at 0.67 mA) lock to a 4 : 3 ratio (at 0.89 mA) leading to chaos (at 1.6 mA). The intermittency route to chaos was also inferred, with pure xenon in the laser tube, from a sequence in which a single frequency (with its harmonics) progressively broadened on increasing the discharge current, just as one would expect for increasingly frequent bursts of noise against a background of steady oscillation. The chaos itself displays a variety of signatures: Fig. 7.6 displays a 'zoo' of time sequences with corresponding spectra, this time in a ring laser.

From a theoretical standpoint, Minden and Casperson [30] have been able to model the xenon-laser data very well, but found it necessary to include such effects as velocity-changing collisions, and population transfer from upper to lower laser level by spontaneous emission, as well as standing-wave effects. In contrast, most theoretical work has been based, like eqns (7.3), on a single-mode, unidirectional ring laser, with only the simplest decay terms. Mandel [29] has been able to obtain a number of useful analytic results for such a system, usually assuming $\gamma = \Gamma$ and a Lorenzian, rather than Gaussian, velocity distribution function. Recent numerical work by Shih *et al.* [33] has demonstrated that when the known parameters for the helium–xenon system are inserted into eqns (7.2), then all three classic routes to chaos can be obtained by careful scans over realistic ranges of the control parameters.

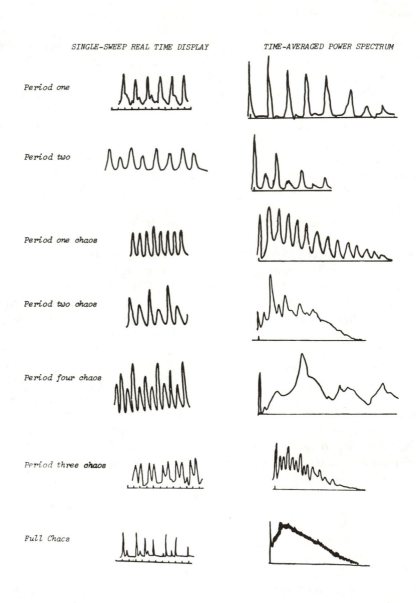

Fig. 7.6 Real-time single-sweep oscilloscope traces and associated power spectra from the output of a compact ring xenon laser ([3], Fig. 14a—courtesy Springer, Berlin).

In summary, single-mode inhomogeneously-broadened lasers have provided a rich corpus of both experimental and theoretical work: detailed quantitative comparison of the two is becoming possible with the development of unidirectional single-mode helium–xenon lasers by the Abraham group.

Historically, one of the earliest frequency-locking phenomena to be studied in laser physics was the locking of three or more longitudinal modes. When many modes are involved, this phenomenon results in very short pulses—typically picoseconds—which have grabbed most of the attention because of their many applications. Recently, however, there have been careful studies of the transition from free-running to mode-locked operation of three-mode lasers. First phase-locking occurs, which causes the appearance of low-frequency 'beat–beat' notes in the spectrum (recall that the cavity-mode spectrum is distorted by the atomic dispersion, so that at threshold the modes will not be quite equally spaced). This 'beat–beat' spectrum may show bands of periodic and chaotic response before the modes lock, giving a quiet low-frequency spectrum [3].

An interesting set of experiments [37], probably related to the above, have shown the three classic routes to chaos in the 'beat–beat' spectrum of a high-gain 3.39 μm helium–neon laser, the control parameter in this case being a slight tilt of one laser mirror, which itself causes the initially single-mode laser to oscillate on three modes.

Three-mode lasers, especially in folded resonators, are probably more complicated than theorists would wish, but the class of lasers in which chaos of this type may be observed is probably very large, and such experiments may well prove useful in testing the universality of the 'universal' routes to chaos.

7.6 Summary

This brief survey has been intended to convey the flavour of the work on chaos in lasers and other all-optical systems which has exploded into the literature in the last five years or so. The theme underlying much of this work is that broadly satisfactory, and reasonably simple, models exist for the huge variety of nonlinear optical processes which have been discovered since the advent of the laser: there is optimism that nonlinear optics can similarly provide simple systems showing the full range of chaotic behaviour. As we have seen, this optimism is already justified on the theoretical side, and this is providing a fruitful stimulus to the experimentalists.

Nonlinear optics is perhaps unique among systems showing chaos in that a fully quantum mechanical description of the phenomena is conceivable: there is the interesting and exciting prospect that the next few years may reveal what, if anything, is meant by 'quantum turbulence', and all-optical systems such as those described or proposed above are likely to be in the forefront of experimental tests of such concepts.

References

[1] Abraham, E. and Firth, W.J. Periodic oscillations and chaos in a Fabry–Perot cavity containing a nonlinearity of finite response time. *Opt. Acta* **30**, 1541–60 (1983).

[2] Abraham, E. and Firth, W.J. Multiparameter universal route to chaos in a Fabry–Perot resonator. *Optical Bistability* **2**, 119–26 (1984).

[3] Abraham, N.B., Chyba, T., Coleman, M., Gioggia, R.S., Halas, N.J., Hoffer, L.M., Liu, S.N., Maeda, M. and Wesson, J.C. Experimental evidence for self-pulsing and chaos in cw-excited lasers. *Springer Lecture Notes in Physics* **182**: *Laser Physics* 107–31 (1983).

[4] Allen, L. and Eberley, J.H. *Optical Resonance and Two-level Atoms*, Wiley-Interscience, New York (1975).

[5] Arecchi, F.T., Meucci, R., Puccioni, G. and Tredicce, J. Experimental evidence of sunharmonic bifurcations, multistability and turbulence in a Q-switched gas laser. *Phys. Rev. Lett.* **49**, 1217–20 (1982).

[6] Bonifacio, R. and Lugiato, L.A. Bistable absorption in a ring cavity. *Lett. Nuovo Cimento* **21**, 505–9 (1978).

[7] Bonifacio, R. and Lugiato, L.A. Instabilities for a coherently driven absorber in a ring cavity. *Lett. Nuovo Cimento* **21**, 510–16 (1978).

[8] Bowden, C.M., Gibbs, H.M. and McCall, S.L. (eds) *Optical Bistability* **2** (Plenum, New York, 1984) contains numerous articles on instabilities in passive resonators.

[9] Carmichael, H.J. Chaos in nonlinear optical systems. *Springer Lecture Notes in Physics* **182**: *Laser Physics* 64–87 (1983).

[10] Casperson, L.W. Spontaneous coherent pulsations in laser oscillators. *IEEE J. Quant. Elec.* **QE-14**, 756–61 (1978).

[11] Casperson, L.W. Spontaneous pulsations in lasers. *Springer Lecture Notes in Physics* **182**: *Laser Physics* 88–106 (1983).

[12] Casperson, L.W. and Yariv, A. Longitudinal modes in a high-gain laser. *Appl. Phys. Lett.* **17**, 259–61 (1970).

[13] Feigenbaum, M.J. Quantitative universality for a class of nonlinear transformations. *J. Stat. Phys.* **19**, 25–52 (1978).

[14] Firth, W.J., Wright, E.M. and Cummins, E.J.D. Connection between Ikeda instability and phase-conjugation. *Optical Bistability* **2**, 111–18 (1984).

[15] Gioggia, R.S. and Abraham, N.B. Routes to chaotic output from a single-mode, dc-excited laser. *Phys. Rev. Lett.* **51**, 650–3 (1983).

[16] Grasyuk, A.Z. and Oragevskiy, A.N. Transient processes in molecular oscillators. *Radio Eng. Electron. Phys. (USSR)* **9**, 424–30 (1964).

[17] Haken, H. Analogy between higher instabilities in fluids and lasers. *Phys. Lett.* **53A**, 77–8 (1975).

[18] Harrison, R.G., Firth, W.J., Emshary, C.A. and Al-Saidi, I.A. Observation of period-doubling in an all-optical resonator containing NH_3 gas. *Phys. Rev. Lett.* **51**, 562–5 (1983).

[19] Harrison, R.G., Firth, W.J. and Al-Saidi, I.A. Observation of bifurcation to chaos in a passive all-optical Fabry–Perot resonator. *Phys. Rev. Lett.* **53**, 258–61 (1984).

[20] Hendow, S.T. and Sargent, M. III The role of population pulsation in single-mode laser instabilities. *Opt. Commun.* **40**, 385–90 (1982).

[21] Hopf, F.A., Derstine, M.W., Gibbs, H.M. and Rushford, M.C. Chaos in optics. *Optical Bistability* **2**, 67–80 (1984).

[22] Ikeda, K. Multiple-valued stationary state and its instability of the transmitted light by a ring cavity system. *Opt. Commun.* **30**, 257–61 (1979).

[23] Ikeda, K. and Akimoto, O. Instability leading to periodic and chaotic self-pulsations in a bistable optical cavity. *Phys. Rev. Lett.* **48**, 617–20 (1982).

[24] Kogelnik, H. and Li, T. Laser beams and resonators. *Appl. Opt.* **5**, 1550–66 (1966).

[25] Korobkin, V.V. and Uspenskiy, A.V. On the theory of pulsations in the output of the ruby laser. *Sov. Phys. JETP,* **18**, 693–7 (1964).

[26] Lugiato, L.A. Theory of optical bistability. *Progress in Optics* **21**, 69–216, Elsevier, Amsterdam (1984).

[27] Lugiato, L.A., Narducci, L.M., Bandy, D.K. and Pennise, C.A. Breathing, spiking and chaos in a laser with injected signal. *Opt. Commun.* **46**, 64–68 (1983).

[28] McLaughlin, D.W., Moloney, J.W. and Newell, A.C. A new class of instabilities in passive optical cavities, to be published.

[29] Mandel, P. Casperson's instability: analytic results, *Proc. 5th Conf. Coherence and Quantum Optics,* eds L. Mandel and E. Wolf, Plenum, New York (1984).

[30] Minden, M.L. and Casperson, L.W. Dispersion-induced instability in cw laser oscillators. *IEEE J. Quant. Elec.* **QE-18**, 1952–7 (1982).

[31] Moloney, J.V., Hopf, F.A. and Gibbs, H.M. Effect of transverse beam variation on bifurcations in an intrinsic bistable ring cavity. *Phys. Rev.* **A25**, 3442–5 (1982).

[32] Nakatsuka, H., Asaka, S., Itoh, M., Ikeda, K. and Matsuoka, M. Observation of bifurcation to chaos in an all-optical bistable system. *Phys. Rev. Lett.* **50**, 109–12 (1983).

[33] Shih, M.L., Milonni, P.W. and Ackerhalt, J.R. Modelling laser instabilities and chaos. *J. Opt. Soc. Am.* **28**, 130–6 (1985).

[34] Siegman, A.E. Introduction to optical resonators. In *Lasers: Physics, Systems and Techniques,* eds W.J. Firth and R.G. Harrison, pp. 45-101, Edinburgh University Press, Edinburgh (1983).

[35] Sparrow, C. The Lorenz system, Chapter 6, this volume.

[36] Weiss, C.O. and Klische, W. On observability of Lorenz instabilities in lasers. *Opt. Commun.* **51**, 47–8 (1984).

[37] Weiss, C.O., Godone, A. and Olafsson, A. Routes to chaotic emission in a cw He–Ne laser, *Phys. Rev.* **A28**, 892–5 (1983).

8
Differential systems in ecology and epidemiology

W. M. Schaffer and M. Kot

*Department of Ecology and Evolutionary Biology
and Committee on Applied Mathematics,
The University of Arizona, Tucson, Arizona 85721, USA*

8.1 Introduction

This chapter is concerned with the motion of populations, including human diseases, which is to say their ups and downs. Motion, whether in biology or physics, can take many forms. In recent years, the beginnings of a taxonomy have emerged. We begin by listing some of the species in this dynamical bestiary:

(1) *Point attractors*. In the absence of perturbations, the system approaches a stable point. An ecological example would be competition between two species for a common set of limiting resources (e.g. [19]).
(2) *Limit cycles*. The attractor is a closed curve, topologically equivalent to a circle. Limit cycles, including relaxation oscillations, arise naturally in two-species predator–prey models (e.g. [39]).
(3) *Toroidal flow*. The orbit is on the surface of a torus. In this case, there are two possibilities.
    (a) The motion is periodic—after an integer number of axial rotations the orbit comes back on itself exactly.
    (b) The motion is quasi-periodic—i.e. the winding number is irrational. Hence the orbit never repeats itself and, in fact, is dense on the torus. One way of generating quasi-periodicity is to force a system which, in the absence of forcing, exhibits limit cycles (e.g.[32]). An ecological example would be a predator–prey system in a seasonal environment [30].
(4) *Strange attractors*. Recently, there has been intense interest (e.g. [10, 29, 48]) in attractors which are neither periodic nor quasi-periodic. Mathematically, strange attractors can be characterised by the

presence of horseshoes (e.g. see [27]). As a result, there will be at least one dimension in which, on the average, nearby trajectories diverge exponentially. Small differences in initial conditions are thus amplified with the consequence that the long-term behaviour of the system may be indistinguishable from a random process. For some strange attractors, the orbit is on an almost two-dimensional surface, i.e. the fractal dimension [17, 36] is close to two. Then the short-term behaviour of the system can be predicted by taking a Poincaré section (i.e. slicing the orbit with a plane) and assembling a one-dimensional return map (e.g.) [55]. Strange attractors have been found in models of various physical phenomena, for example the onset of turbulence in convective flow [34]. In what follows, we describe an ecological model—one predator, two prey species—with such an attractor.

(5) *Turbulence.* Beyond low-dimensional strange attractors is fully developed turbulence. Here the motion is highly erratic: imagine, for example, water brought to a rolling boil. In this case, predicting the motion of the individual particles is impossible. Instead, one may speak of a distribution of positions and velocities to which the molecules in aggregate converge.

The foregoing classification is incomplete. For example, not all dynamical systems have attractors. None the less, we hope that it will put the reader in the proper frame of mind: motion is all about us and, for the most part, poorly understood. What the dynamicist can provide is a set of possibilities. Whether or not these mathematical constructs are realised in nature is an altogether different question subject to empirical determination. Recently, convincing evidence has been reported for the reality of strange attractors in physical situations, most notably the Belousov–Zhabotinskii reaction (e.g. [46,57]. In the present chapter, we focus on ecological systems. Do the often erratic fluctuations of natural populations, including outbreaks of human diseases, reflect the workings of strange attractors, or is noise the prime mover of ecological processes?

The remainder of the present chapter is organised as follows. Section 8.2 considers the properties of model systems in which chaotic or complex periodic behaviour is known to occur. Section 8.3 gives some real-world examples. We conclude (section 8.4) by contrasting the effects of adding noise, an inevitability in ecology, to simple and chaotic systems.

8.2 Model systems with complex solutions

8.2.1 *One-dimensional maps*

In the ecological literature, the most widely discussed (e.g. [40, 41, 43]) model admitting to chaotic dynamics is the logistic equation

$$(8.1) \qquad\qquad X_{i+1} = sX_i(1 - X_i)$$

Traditionally, eqn (8.1) and its congeners are viewed as literal descriptions of single-species systems with discrete, non-overlapping generations, for example insects with one generation per year. Then s is closely related to the maximum *per capita* rate of increase, i.e. births minus deaths per individual. For ecologists, this raises a serious problem. Most estimates of s for natural populations are substantially smaller than those required for chaos. Attempts to fit real-world data to more realistic models with two or more parameters [28] point to the same conclusion. Accordingly, ecologists have tended to dismiss the potential for complex dynamics in one-dimensional maps as a mathematical curiosity. Note the ensuing difficulty: since most populations in nature are anything but steady (e.g. see [31]), it would appear to follow that chance plays a major role.

An alternative approach is to imagine equations such as (8.1) as arising from chaotic flows that are essentially two-dimensional [55]. In this case, the X_i are points on a Poincaré section, and the parameters reflect the biology of the entire system. The distinction, whether one regards the map as a literal description of a single-species system or what ecologists would call an 'emergent property' of a higher order continuous system, thus turns out to be crucial. When eqn (8.1) is viewed from the latter perspective, there is no reason to exclude values of s in the chaotic region.

8.2.2 *Delay-differential equations*

Related to the logistic are single-species models, in which the current rate of increase depends on the population's density at some time in the past. Of these, the simplest is the delay logistic

$$(8.2) \qquad dX/dt = rX(t)[1 - X(t-T)]$$

The time lag, T, may be interpreted as arising from the fact that the population's resources are not instantaneously renewed.

Equation (8.2) admits only to stable equilibria or stable cycles. However, related models exhibit chaotic solutions. One such model, studied by May [42], was devised to model the population dynamics of baleen whales. In this case, one writes

$$(8.3) \qquad dX/dt = -DX(t) + BX(t-T)[1 - X^Z(t-T)]$$

Here, $X(t)$ is the number of whales that have achieved reproductive maturity, and D is the current death rate. B, the *per capita* birth rate, is lagged to reflect the prolonged period preceding reproduction. The constant z reflects the intensity of density dependence. For $z < 2[(B/D)-1]$, eqn (8.3) always (i.e for all values of T) has a stable equilibrium point. For $D = 1$, $T = 2$ and $B = 2$, increasing z yields a series of period-doubling bifurcations reminiscent of those observed in the case of the Rössler attractor [12]. For larger values of z, May observed an apparently chaotic orbit, which, for still larger z values, collapses back to a (relatively) simple cycle. (May was unable to determine whether or not this collapse entails reverse bifurcations of the type discussed by Lorenz [35] and observed for the Rössler system.) Unfortunately, the parameter values estimated for baleen whales correspond to the region of a stable fixed point. In other words, like the logistic map, May's equation, when viewed as a literal description of a single species system, is unlikely to account for irregular fluctuations in the densities of real populations.

The qualitative behaviour of eqn (8.3) resembles that of the model of blood production due to Mackey and Glass [37]; also see [44]. Using the above notation, we have

(8.4) $$dX/dt = -DX(t) + BX(t-T)/[1 + X^Z(t-T)]$$

The analysis of delay-differential equations in ecology has not, of course, been limited to single-species models. Of relevance to the present chapter is work by Shibata and Saito [56]. These authors consider a two-species generalisation of eqn (8.2) exhibiting period doubling and what Rössler [45] has called 'torus-type' chaos. For some parameter values. Shibata and Saito observed coexisting chaotic solutions; for others, a limit cycle coexists with an apparently chaotic band. As in the case of Gilpin's model (three *ordinary* differential equations) discussed below, one can extract an essentially one-dimensional mapping from the strange attractor by slicing the orbit with a plane and considering the sequence of the resulting intersections.

8.2.3 *Systems of ordinary differential equations*

The delay-differential equations described above correspond, of course, to infinite dimensional systems. Thus there is no *a priori* reason to expect a limit on the dimensions of their attractors, and the observation of low dimensional behaviour is encouraging. By contrast, systems of ordinary differential equations, such as those studied by Lorenz [34] and Rössler [45] are finite dimensional. One such system which has been studied in an ecological context uses (Lotka–Volterra) equations of the form

(8.5) $$[1/X_i(t)] \, dX_i/dt = r_i + \sum_j a_{ij} X_j(t)$$

to model the dynamics of a single predator and two prey species. Gilpin [20] was the first to point out that eqn (8.5) can give rise to chaotic trajectories. Figure 8.1 shows part of the time series generated for the predators and the power spectrum. The latter lacks distinguishable peaks for high frequencies and exhibits and the log linear behaviour observed for other strange attractors [15, 25]. The route to chaos in Gilpin's equations is via period-doubling bifurcations. In Fig. 8.2 a–d we show the results of tuning one of the parameters, in this case the efficiency with which the predator harvests the preferred prey species. Note that the essential dynamics of the strange attractor can be captured by a one-dimensional non-invertible map of the form

(a)

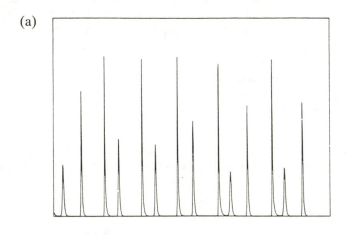

(b)

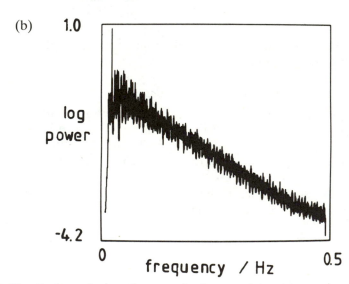

Fig. 8.1 Chaotic dynamics in a three-species (one predator, two prey) model. (a) Numbers of predators vs. time. (b) Power spectrum.

(8.6) $T : X_1 \rightarrow X_{i+1}$

i.e. by an equation similar to the logistic map. Such a map can be constructed either by taking a Poincaré section (Fig. 8.2e) and assembling a return map, or more simply by plotting successive maxima in one of the variables (Fig. 8.2f) in the manner of Lorenz [34]; see also [55, 58]. A similar relation exists between the magnitude of the current maximum and the time until the next outbreak [49]. Thus, were the predator or its

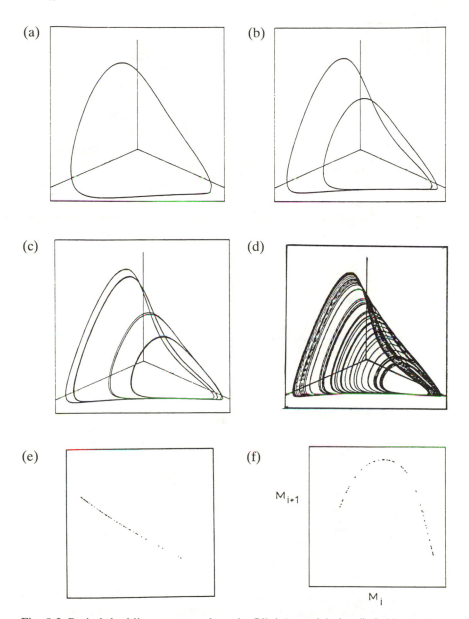

Fig. 8.2 Period doubling route to chaos in Gilpin's model. (a–d) Orbits in phase space. (e) Poincaré section of the chaotic orbit. (f) One-dimensional peak to peak map for the predators.

victims of economic importance, one could predict the timing and severity of the next eruption.

Further tuning of the parameter reveals that the chaotic region contains

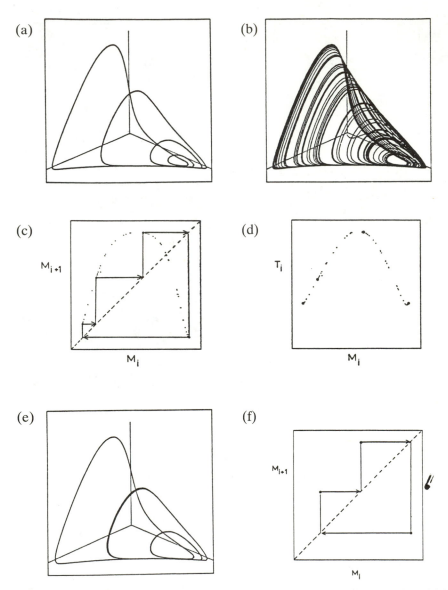

Fig. 8.3 The chaotic region for Gilpin's equations contains periodic orbits which can be identified with those of 1-D maps such as the logistic. (a) A 4-cycle. (b) The 4-cycle plus transient. (c) Peak-to-peak map for predators; circles give the asymptotic dynamics; dots, the transient. (d) Time to next maximum vs. magnitude of current maximum for the 4-cycle. (e) A 3-cycle. (f) Map for the 3-cycle.

periodic windows of the sort observed in 1-D maps (e.g. see [11]). Figure 8.3a shows a 4-cycle. The sequence (RLL) of points on the associated 1-D (Fig. 8.3c) map identifies this cycle with the 4-cycle that occurs in the chaotic region for 1-D maps [11]. Note first that the transient is extremely long and that it fills in both the attractor (Fig. 8.3b) and the 1-D maps (Figs. 8.3c,d). Thus, in the real world, with noise, one would expect to see something resembling the chaotic trajectory for nearby parameter values. (For a discussion of the effects of additive noise on 1-D maps, see [13] and [38]).

The ability to encapsulate the dynamics of an *n*-variable, continuous system with a one-dimensional map indicates that the flow is effectively single-sheeted, even though, as discussed by Lorenz [34] and subsequent authors [55, 60], this 'sheet' in fact consists of an infinite number of closely spaced 'leaves'. As we continue to tune Gilpin's equations, what we shall call single-sheeted chaos gives way to more complex trajectories, at which point a one-dimensional representation becomes untenable (Fig. 8.4). Similar behaviour has been observed in Rössler's equations [15], where it is associated with the reinjection of trajectories into the neighbourhood of an unstable fixed point [18, 21].

8.2.4 *Forced systems*

It is well known that periodic forcing of nonlinear systems can result in periodic, quasi-periodic, and chaotic flows [32]. An interesting example arises in the theory of infectious diseases. Here, there is seasonal variation in transmission rates, especially in childhood diseases, where contact rates among school-age children are higher during the winter [33, 63]. In addition, temperature and humidity may affect dispersal and survival of the infectious agents. The simplest models, called SEIR models [2, 5, 14], for the spread of human diseases consist of coupled differential equations in which the number of individuals in the population are categorised as follows: (1) susceptible; (2) exposed, but not yet infectious; (3) infectious; (4) recovered. Accordingly, one writes

$$
\begin{aligned}
dS(t)/dt &= u[1- S(t)] - bI(t)\,S(t) \\
dE(t)/dt &= bS(t)I(t) - (u + a)E(t) \\
dI(t)/dt &= aE(t) - (u + g)I(t) \\
dR(t)/dt &= gI(t) - uR(t)
\end{aligned}
$$

(8.7)

Here $(1/u)$ is the average life expectancy of individuals in the population; $(1/a)$ the average latency period; and $(1/g)$ the average infectious period. The parameter b is the transmission coefficient.

As given above, eqns (8.7) exhibit weakly damped oscillations about a stable fixed point. This property is at variance with the observation of recurrent epidemics, for example, in measles at intervals of two to five years [7]. One factor which may account for the discrepancy is the aforementioned variation in transmission rates. Thus, several authors [5, 14, 26] have investigated the effects of replacing the parameter b with functions of the sort

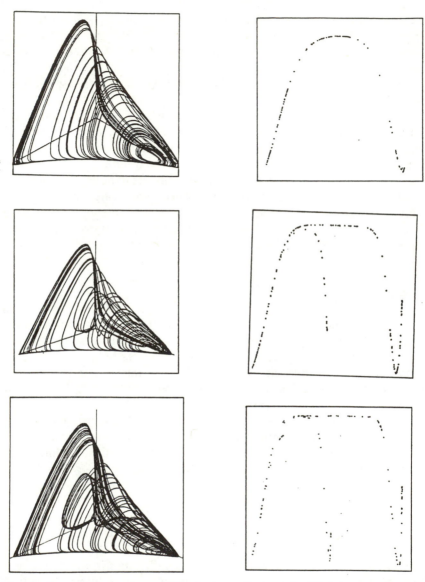

Fig. 8.4 Beyond the region of 'single-sheeted' chaos in Gilpin's equations, one observes more complex behaviour. Each pair of figures gives the orbit (left) and peak-to-peak map for the predators (right).

$$(8.8) \qquad\qquad b(t) = b_0\left[1 + b_1 \cos\left(2\pi t\right)\right]$$

For certain parameter values, this substitution produces resonance, whereby sustained oscillations with periods of an integer number of years result. (Interestingly, the interval between major outbreaks for a number of diseases is in rough accord with the period of the damped oscillations [2].) This observation finds apparent confirmation in the European data where traditional methods of time series analysis suggest prominent period-icities [3].

Numerical studies of eqns (8.7) and (8.8) indicate two additional findings of interest [5, 54]. The first is that the period of solutions arising out of the yearly cycle increases by period doubling. Secondly, for certain parameter values, one observes coexisting solutions of large and small amplitudes. Schwartz's studies [54] further suggest apparently fractal basin boundaries when solutions coexist.

Similar but more explicit results of this sort have been given for a two-species predator–prey system in which the prey's *per capita* rate of increase is subject to seasonal forcing [30]. These authors observed co-existing routes to chaos. Thus, on *increasing* the frequency of forcing, a toroidal scenario is observed, i.e. entrained limit cycle $\rightarrow$ toroidal flow (quasi-periodicity) $\rightarrow$ phase locking (resonance) $\rightarrow$ chaos. Within the regions of phase locking, the sequence of rotation numbers follows a Farey sequence. That is, if, for forcing frequencies ω_1 and ω_2, there exist distinct locking states for which the rotation numbers $r_1 = q_1/p_1$ and $r_2 = q_2/p_2$, then there exists a locking state with $r_3 = (q_1 + q_2)/(p_1 + p_2)$ at ω_3, where $\omega_1 < \omega_3 < \omega_2$. On plotting rotation number vs. ω, this pro-duces a so-called 'devil's staircase'.

However, on *decreasing* the forcing frequency through the same para-meter range, one sees period doubling. Taken together, the findings for both the seasonal SEIR model and the forced predator–prey system suggest that the flows can be viewed as suspensions of a mapping of the plane in which the eigenvalues pass in and out of overlapping resonance structures called Arnold's tongues [4, 6]. An important problem for ecologists is to determine the behaviour of such systems under noise in which the system can shuttle back and forth between basins of attraction.

8.3 Real-world examples

8.3.1 *The embedding problem*

The most convincing way to demonstrate the presence of chaotic motion is to construct a phase portrait, or a slice of it, wherein one plots the values of the several states through time. We grant that there exist quantities such as Lyapunov exponents, entropy, etc., by which one can quantify the degree of chaos. Nevertheless, the fundamental properties of a strange attractor are geometric [24]. As always, a single picture is worth a thousand computations. Unfortunately, ecologists and epidemiologists are rarely able to measure all of the relevant quantities. Hence viewing the motion would seem an impossibility.

How to proceed? The problem turns out to arise in other disciplines, and recently an ingenious solution has been proposed by Takens [59]. The basic idea is as follows. Consider an n-dimensional system for which there are available numerous determinations of the magnitude, $x(t)$, of one of the state variables. Then, for almost every time lag T, the m-dimensional portrait constructed by plotting $x(t)$ vs. $x(t + T)$ vs. $x(t + 2T)$ vs. . . . vs. $x[t + (m - 1)T]$ will have the same dynamical properties as the portrait constructed from the n original variables. By 'same dynamical properties' we mean the same set of Lyapunov exponents [61], fractal dimension [22, 23], etc. Takens showed that m need be no larger than $2n + 1$. However, this is only a sufficiency condition. Often, a smaller number of dimensions will suffice. In particular, an essentially two-dimensional orbit can be embedded in three dimensions, i.e. by setting m equal to three.

As noted above, reconstructing the phase portrait is important since it can help one distinguish chaos, a deterministic phenomenon, from the output of a random process. More precisely, the orbits of chaotic systems have certain fieldmarks, the most important of which is stretching and folding. Application of Takens' reconstruction scheme allows one to determine whether or not an apparently noisy time series possesses these attributes.

8.3.2 *Data for childhood diseases*

Takens' method has been applied with considerable success to physical systems, e.g. the Belousov–Zhabotinskii reaction [46, 57], and Taylor–Couette flow [9] (for a review, see [1]). Is this technique also applicable to ecological systems which are inherently far noisier? We believe that the answer is sometimes yes. In support of this contention, we present data for three childhood diseases: measles, mumps and chickenpox. Two of these exhibit nothing more interesting than a yearly cycle with noise superimposed. In the case of measles epidemics, however, data sets from two North American cities strongly suggest a strange attractor.

Figures 8.5–7 summarise the situation for measles in New York City and Baltimore. Prior to the introduction of the vaccine which led to the disease's effective eradication, major outbreaks occurred every second or third year in New York and at less frequent intervals in Baltimore. For both cities, the data (monthly physicians' reports for the years 1928–63) reveal a strong seasonal component with maximum case rates occurring during the winter. Superimposed on the annual cycle is tremendous between-year variation in incidence. Thus, for New York, the yearly totals ranged from a low of about 2000 cases reported in 1945 to a high of over 79 000 in 1941. For Baltimore, the corresponding figures are about 100 and 18 500. (London and Yorke [33] suggest that the number of cases reported in 1941 may be exaggerated. Specifically, a series of newspaper articles may have stirred unusual public interest and increased the reporting rate beyond the customary one in five to seven cases. The year with the next highest number of cases reported, 41 000 (about which there appears to be no question), was 1954.)

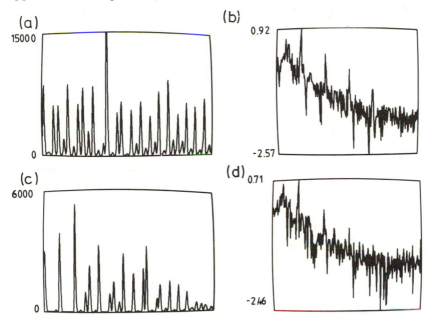

Fig. 8.5 Epidemics of measles in New York City and Baltimore, 1928–63. (a) The numbers of cases reported monthly in New York and smoothed with a three-point running average. (b) Power spectrum computed from the logarithms of the New York data; the principal peak occurs at a frequency corresponding to a period of approximately one year. (c) Smoothed data for Baltimore. (d) Power spectrum for the Baltimore data.

Figure 8.5 gives the two time series—the numbers of cases reported monthly subject to three-point smoothing—and the power spectra (log power plotted against frequency). Spectra were computed for the logarithms of the smoothed data using a Sande–Tukey Radix-2 Fast Fourier Transform [8] with the total power normalised to 1. The power spectra have discernible peaks at a frequency corresponding to a period of 11.9 months, i.e. the yearly cycle. Additionally, for New York, one sees what appear to be higher harmonics. Beyond this, little can be said except that the spectra are quite noisy. In particular, subharmonic peaks at frequencies corresponding to two years are not observed. Thus, there is little evidence for the biennial outbreaks predicted by the models of Dietz [14], Aron and Schwartz [5] and others.

Three-dimensional phase portraits, constructed using Takens' method with the lag set equal to three months, are shown in Figs 8.6 and 8.7. For both cities, most of the trajectory lies on the surface of a cone with its vertex near the origin. It therefore appears that we are dealing with

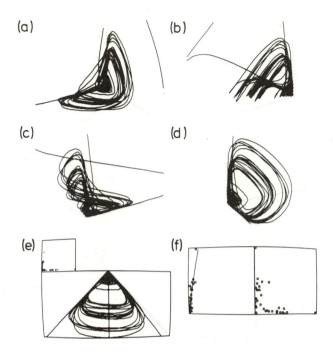

Fig. 8.6 Trajectory reconstructed from the New York data (smoothed and interpolated with cubic splines) using a lag of three months. The motion suggest a Rössler-like attractor in the presence of noise. (a–d) The data embedded in three dimensions and viewed from different perspectives. (e) The orbit viewed from above and sliced with a plane (vertical line) normal to the paper; Poincaré sections are shown in the small box at the upper left. (f) One of the Poincaré sections magnified (left) and the 1-D return map computed therefrom (right). The latter suggests a unimodal curve.

essentially two-dimensional flows that can be embedded in three space. (For the New York data, we calculate a lower bound on the fractal or Hausdorf dimension of 2.55 using the method of Grassberger and Procaccia [22, 23]).

Figures 8.6e and 8.7e confirm the approximately two-dimensional nature of the flows. Here the orbits are viewed from above (main part of the photographs) and sliced with a plane (vertical line) normal to the paper. Poincaré sections are shown in the small boxes at the upper left. Although the sections have noticeable thickness, they are none the less thin in proportion to their length. Note that each box contains two sections which, together, form a rough 'vee'. This is of course what one expects on slicing a hollow cone.

For each city, one of the sections is magnified in Figs 8.6f and 8.7f (left). At the right, we plot the 1-D return map, i.e. successive points on

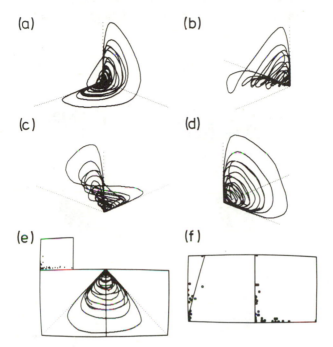

(a)   (b)   (c)   (d)   (e)   (f)

Fig. 8.7 Reconstructed trajectory for the Baltimore data. The 1-D map is highly compressed. Order of photographs as in Fig. 7.

the Poincaré section against each other. The resulting collection of points suggests unimodal, noninvertible maps of the sort associated with the period-doubling route to chaos. For New York, the section and map are especially clean. In passing, we note that the data for 1941 and 1942 (the points at the upper left and lower right of the map) are consistent with unimodality. This is important since the 1941 data may represent an overestimate [33]. For Baltimore, there is more scatter about the section, and the map is extremely compressed at the left. The clustering of points near the origin nevertheless indicates a one-humped curve.

8.4 Effects of noise on systems with simple and complex motion

Examination of the orbits and maps for measles strongly suggests that one is seeing a strange attractor in the presence of noise. In particular, it is possible to demonstrate the presence of the stretching and folding of nearby trajectories that produce the apparent stochasticity inherent in chaotic systems [52]. Other possible examples of low-dimensional chaotic motion in ecology include the lynx cycle of Canada [50], and outbreaks of *Thrips imaginis* [52], an insect pest.

Of course, not all ecological systems admit to such a description. In the case of childhood diseases, for example, the data for chickenpox (Fig. 8.8)

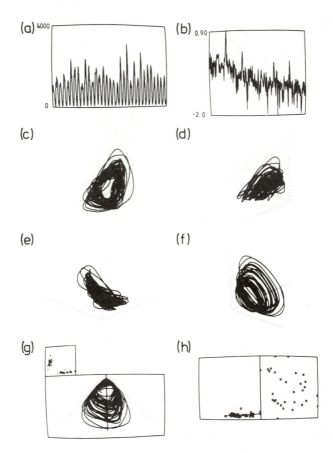

Fig. 8.8 Epidemics of chickenpox in New York City, 1928–63. Despite the fact that the Poincaré section is no thicker than for measles, there is no hint of a unimodal map. (a) The time series. (b) Spectrum. (c–f) Reconstructed orbit. (g, h) Poincaré section and map. The situation for mumps outbreaks is similar [53].

and mumps suggest a simple annual cycle with noise superimposed. In particular, there is no hint of a 1-D map, which is to say that one cannot come up with a one-dimensional rule relating the magnitudes of subsequent outbreaks. This finding raises the question as to the effects of adding noise to systems with simple and complex dynamics. To address this question for 1-D maps, we iterated the mapping

(8.9) $$X \to (1 + z)\, X \exp[r(1 - X)]$$

where z is drawn from a Gaussian distribution with mean 0 and variance s^2.

Note that noise has been added multiplicatively, which is sensible, since the parameters in ecological processes refer to *per capita* rates of birth and mortality. Figure 8.9 shows the results of our investigation. Essentially,

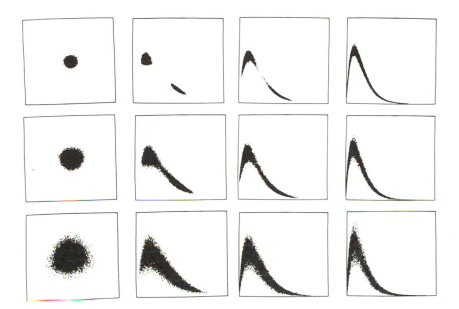

Fig. 8.9 Iterating a 1-D map in the presence of noise. Each of the parts (top to bottom) shows eqn (8.9) iterated for noise levels equal to 0.05, 0.10 and 0.20. From left to right the deterministic dynamics correspond to a stable fixed point, a 2-cycle, 4-cycle and chaos. For the fixed point and cycles, the parameter was chosen to place the map at the superstable point.

one sees that for maps with a stable fixed point, adding noise results in an uninformative cloud of points. However, as one moves (in parameter space) past the bifurcation to a 2-cycle and towards the chaotic region, the addition of noise causes the system to explore the map with increasing fidelity. Thus, an empiricist studying the dynamics of a system with two point cycles or more complex behaviour would probably infer the nature of the underlying dynamics. The same empiricist would probably be at a loss to understand the dynamics of a system with a stable fixed point. In effect, he would see only the noise.

Viewed from the preceding perspective, one might expect (for fixed noise levels) that fractal dimension [17] declines as the deterministic component of the motion becomes more complex. To test this presumption, we iterated eqn (8.9) 5000 times and embedded the resulting time series in successively higher dimensions. For each embedding, we calculated a lower bound, D_c, on the fractal dimension using an algorithm devised by Grassberger and Procaccia [22, 23]. Specifically, we compute the quantity

(8.10) $$C(g) = \lim \{[1/n(n-1)] \sum \theta(g - |X_i - X_j|)\}$$

where $\theta(\cdot)$ is the Heaviside function,

$$\theta(\cdot) = 0, \; g < X_i - X_j$$
$$\theta(\cdot) = 1, \; g > X_i - X_j$$

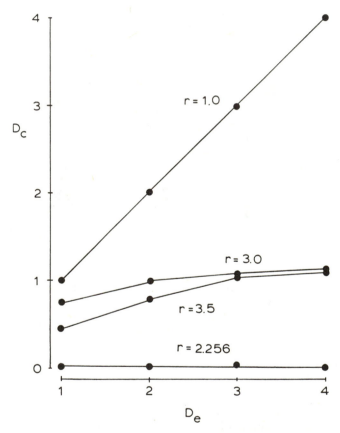

Fig. 8.10 The correlation dimension, D_c, for different embedding dimensions, D_e, for a one-dimensional map, eqn (8.9), iterated in the presence of noise, $s = 0.05$. For maps with a stable fixed point ($r = 1.0$), D_c increases continuously with and is approximately equal to D_e. For maps in the chaotic region ($r = 3.0, 3.5$), D_c levels off at values close to 1. Two point cycles ($r = 2.256$) yield a correlation of zero corresponding to the fact that the time series consists of disjoint subsets.

For attractors that are dense on some subinterval, Grassberger and Procaccia [22, 23] argue that for small g

$$(8.11) \qquad\qquad C(g) \propto g^n$$

and that n is a lower bound on the fractal or Hausdorf dimension, D. The difference arises because Hausdorf dimension is a purely geometric construct, whereas D_c depends on the frequency with which different points on the attractor are actually visited. If this distribution, sometimes called the natural measure of the attractor, is uniform, $D_c = D$. Figure 8.10 summarises our results for $s = 0.05$. Here, we plot $D_c = n$ against D_e, the embedding dimension for several values of r. For $r = 1.0$ (superstable fixed point), the dimension of the time series essentially equals D_e. This is what one expects for a Gaussian process for which the dimension is infinite. For $r = 2.256$ (superstable 2-cycle), a dimension of zero obtains, corresponding to the fact that the points are concentrated into two subsets of the interval (Fig. 8.9). Finally, for maps in the chaotic region ($r = 3.0$ and 3.5), D_c appears to approach an asymptote of about 1.0. Thus, for maps in the chaotic region, it is possible to infer correctly the one-dimensional nature of the process underlying the time series, even in the presence of substantial amounts of noise. For maps with stable fixed points, such an inference does not appear possible. Work in progress [54] suggests analogous results for flows.

Acknowledgements

We thank Bob May for calling our attention to the paper by Inoue and Kamifukumoto. This work was supported by awards from the National Science Foundation and the John Simon Guggenheim Memorial Foundation to the senior author and by the University of Arizona.

References

[1] Abraham, R.H., Gollub, J.P. and Swinney, H.L. Testing nonlinear dynamics. *Physica 11-D* 252–64 (1984).
[2] Anderson, R.M. and May, R.M. Directly transmitted infectious diseases: control by vaccination. *Science* **215**, 1053–60 (1982).
[3] Anderson, R.M., Grenfell, B.T. and May, R.M. Oscillatory phenomena in the incidence of infectious disease and the impact of vaccination: time series analysis. *J. Hygiene, Cambridge* **93**, 587–608.
[4] Arnold, V.I. Loss of stability of self-oscillations close to resonance and versal deformations of equivalent vector fields. *Functional Analysis and its Applications* **11**, 1–10 (1977).
[5] Aron, J.L. and Schwartz, I.B. Seasonality and period-doubling bifurcations in an epidemic model. *J. Theor. Biol.* **110**, 665–80 (1984).
[6] Aronson, D.G., Chory, M.A., Hall, G.R. and McGehee, R.P. A discrete dynamical system with subtly wild behaviour. In *New Approaches to Nonlinear Problems in Dynamics*, ed. P. Holmes. SIAM, Pennsylvania (1982).
[7] Bartlett, M.S. The critical community size for measles in the United States. *J. R. Stat. Soc. Ser. A.* **123**, 36–44 (1960).

[8] Bloomfield, P. *Fourier Analysis of Time Series: an Introduction*. Wiley, New York (1976).

[9] Brandstäter, A., Swift, J., Swinney, H.L., Wolf, A., Farmer, J.D. and Jen, E. Low dimensional chaos in a system with Avogadro's number of degrees of freedom. *Phys. Rev. Lett.* (In press) (1985).

[10] Campbell, D.K., Farmer, J.D. and Rose, H. (eds) *Order in Chaos, Los Alamos Conference*, May 1982, *Physica 7-D* (1983).

[11] Collet, P. and Eckmann, J-P. *Iterated Maps on the Interval as Dynamical Systems*. Birkhäuser, Boston (1980).

[12] Crutchfield, J.P., Donnelly, D., Farmer, D., Jones, G., Packard, N. and Shaw, R. Power spectral analysis of a dynamical system. *Phys. Lett.* **76a**, 1–4 (1980).

[13] Crutchfield, J.P., Farmer, J.D. and Huberman, B.A. Fluctuations and simple chaotic dynamics. *Phys. Rep.* **92**, 45–82 (1982).

[14] Dietz, K. The incidence of infectious diseases under the influence of seasonal fluctuations. *Lecture Notes in Biomathematics* **11**, 1–15 (1976).

[15] Farmer, D., Crutchfield, J., Froehling, H., Packard, N. and Shaw, R. Power spectra and mixing properties of strange attractors. *Ann. NY Acad. Sci.* **357**, 453–72 (1980).

[16] Farmer, J.D. Chaotic attractors of an infinite dimensional dynamical system. *Physica 4-D*, 366–93 (1982).

[17] Farmer, J.D., Ott, E. and Yorke, J.A. The dimension of chaotic attractors. *Physica 7-D*, 153–80 (1983).

[18] Gaspard, P., Kapral, R. and Nicolis, G. Bifurcation phenomena near homoclinic systems: a two parameters analysis. *J. Stat. Phys.* **35**, 697–727 (1984).

[19] Gause, G.F. *The Struggle for Existence*. Haffner, New York (1964).

[20] Gilpin, M.E. Spiral chaos in a predator-prey model. *Amer. Natur.* **113**, 306–8 (1979).

[21] Glendinning, P. and Sparrow, C. Local and global behavior near homoclinic orbits. *J. Stat. Phys.* **35**, 645–95 (1984).

[22] Grassberger, P. and Procaccia, I. Characterization of strange attractors. *Phys. Rev. Lett.* **50**, 346–9 (1983).

[23] Grassberger, P. and Procaccia, L. Measuring the strangeness of strange attractors. *Physica 9-D*, 189–208 (1983).

[24] Grebogi, C., Ott, E., Pelikan, S. and Yorke, J.A. Strange attractors that are not chaotic. *Physica 13-D*, 261–8 (1984).

[25] Greenside, H.S., Ahlers, G., Hohenberg, P.C. and Walden, R.W. A simple stochastic model for the onset of turbulence in Rayleigh–Benard convection. *Physica 5-D*, 322–34 (1984).

[26] Grossman, Z. Oscillatory phenomena in a model for infectious disease. *Theor. Popul. Biol.* **18**, 204–43 (1980).

[27] Guckenheimer, J. and Holmes, P. *Nonlinear Oscillations, Dynamical Systems and Bifurcations of Vector Fields*. Springer, New York (1983).

[28] Hassell, M.P., Lawton, J.H. and May, R.M. Patterns of dynamical behaviour in single-species populations. *J. Animal Ecol.* **45**, 471–86 (1976).

[29] Helleman, R.H.G. Self-generated chaotic behavior in non-linear systems. *Fundamental Problems in Statistical Mechanics. Vol. 5*, ed. E.G.D. Cohen, pp. 165–223, Elsevier North-Holland, Amsterdam (1980).

[30] Inoue, M. and Kamifukumoto, H. Scenarios leading to chaos in a forced Lotka–Volterra model. *Prog. Theor. Phys.* **71**, 930–7 (1984).

[31] Ito, Y. *Comparative Ecology*. Cambridge University Press, Cambridge (1980).

[32] Levi, M. Qualitative analysis of the periodically forced relaxation oscillation. *Memoirs of the American Mathematics Society*, Providence, Rhode Island (1981).

[33] London, W.P. and Yorke, J.A. Recurrent outbreaks of measles, chickenpox and mumps I and II. *Amer. J. Epidemiol.* **98**, 453–82 (1973).

[34] Lorenz, E.N. Deterministic nonperiodic flow. *J. Atmos. Sci.* **20**, 130–41 (1963).

[35] Lorenz, E.N. Noisy periodicity and reverse bifurcations. *Ann. NY Acad. Sci.* **357**, 282–91 (1980).

[36] Mandelbrot, B. *The Fractal Geometry of Nature*. W. H. Freeman, San Francisco (1977).

[37] Mackey, M.C. and Glass, L. Oscillations and chaos in physiological control systems. *Science* **197**, 287–9 (1977).

[38] Matsumoto, K. and Tsuda, I. Noise-induced order. *J. Stat. Phys.* **31**, 87–106 (1983).

[39] May, R.M. *Stability and Complexity in Model Ecosystems*. Princeton University Press, Princeton (1973).

[40] May, R.M. Biological populations with nonoverlapping generations: stable points, stable cycles and chaos. *Science* **186**, 645–7 (1974).

[41] May, R.M. Simple mathematical models with very complicated dynamics. *Nature Lond.* **261**, 459–67 (1976).

[42] May, R.M. Nonlinear phenomena in ecology and epidemiology. *Ann. NY Acad. Sci.* **357**, 267–81 (1980).

[43] May, R.M. and Oster, G.F. Bifurcations and dynamic complexity in simple ecological models. *Amer. Natur.* **110**, 573–99 (1976).

[44] Morris, H.C., Ryan, E.E. and Dodd, R.K. Periodic solutions and chaos in a delay-differential equation modelling haematopoesis. *Nonlinear Anal.* **7**, 623–60 (1983).

[45] Rössler, O. An equation for continuous chaos. *Phys. Lett.* **57A**, 397–98 (1976).

[46] Roux, J.-C., Simoyi, R.H. and Swinney, H.L. Observation of a strange attractor. *Physica 8-D*, 257–66 (1983).

[47] Ruelle, D. Sensitive dependence on initial conditions and turbulent behavior of dynamical systems. *Ann. NY Acad. Sci.* **316**, 408–16 (1979).

[48] Ruelle, D. Strange attractors. *Math. Intell.* **2**, 126–37 (1980).

[49] Schaffer, W.M. Stretching and folding in lynx fur returns: evidence for a strange attractor in nature? *Amer. Natur.* **124**, 798–820 (1984).

[50] Schaffer, W.M. Order and chaos in ecological systems. *Ecology* **66**, 93–106 (1985).

[51] Schaffer, W.M. and Kot, M. Do strange attractors govern ecological systems? *BioScience* **35**, 342–50 (1985).

[52] Schaffer, W.M. and Kot, M. Nearly one-dimensional dynamics in an epidemic. *J. Theor. Biol.* **112**, 403–27 (1985).

[53] Schaffer, W.M., Ellner, S. and Kot, M. MS. The effects of noise on dynamical models of ecological systems.

[54] Schwartz, I.B. MS. Multiple recurrent outbreaks and predictability in seasonally forced nonlinear epidemic models.

[55] Shaw, R. Strange attractors, chaotic behavior and information flow. *Zeitschrift für Natürforschung.* **36a**, 80–112 (1985).

[56] Shibata, A. and Saito, N. Time delays and chaos in two competing species. *Math. Biosciences* **51**, 199–211.

[57] Simoyi, R.H., Wolf, A. and Swinney, H.L. One-dimensional dynamics in a

multi-component chemical reaction. *Phys. Rev. Lett.* **49**, 245–8 (1982).
[58] Sparrow, C. *The Lorenz Equations: Bifurcations, Chaos and Strange Attractors.* Springer, New York (1982).
[59] Takens, F. Detecting strange attractors in turbulence. In. *Lecture Notes in Mathematics,* eds., D. A. Rand and L-S. Young, pp. 366–81. Springer, New York (1981).
[60] Williams, R.F. The structure of Lorenz attractors. In *Turbulence Seminar, Berkeley,* eds., P. Bernard and T. Ratiu, pp. 94–112. Springer, New York (1977).
[61] Wolf, A., Swift, J.B., Swinney, H.L. and Vastano, J.A. Determining Lyapunov exponents from a time series. *Physica 16 D,* 285–317 (1985).
[62] Yorke, J.A. and London, W.P. *Amer. J. Epidemiology* **98**, 469 (1973).
[63] Yorke, J.A., Nathanson, N., Pianigiani, G. and Martin, J. *Amer. J. Epidemiology* **109**, 103 (1979).

9
Oscillations and chaos in cellular metabolism and physiological systems

P. E. Rapp

Department of Physiology and Biochemistry,
The Medical College of Pennsylvania, 3300 Henry Ave., Philadelphia, PA
19129, USA

9.1 The dynamical behaviour of biochemical and physiological systems: an overview

It was once common wisdom that biochemical reactions inevitably converged rapidly to a thermodynamic steady state and that this steady state was unique. Similarly, at the systemic level, a restrictive view of the concept of homeostasis dominated physiological thinking, and it was supposed that physiological control functioned exclusively to restore transiently disturbed systems to a steady state. It is now recognised that this is not the case. Complex dynamical behaviour is an aspect of biological regulation. Two such behaviours, sustained oscillations and chaos, are considered here.

Two periodic biochemical systems will be described in section 9.2, the glycolytic oscillator, and oscillations in intracellular calcium–cyclic AMP control systems. They offer contrasts both in the dynamical basis of rhythmicity and in the analytical methods that have been used to examine them. The mathematical models of the glycolytic pathway take the form of ordinary nonlinear differential equations. Because of the low dimension of these models, it has been possible to establish a fairly complete characterisation of the equations' solutions using techniques from the qualitative theory of differential equations. As will be reported in the next section, these investigations have established to a high degree of confidence that oscillations in the glycolytic pathway are the consequence of allosteric activation of an enzyme by its product. The intracellular calcium–cyclic AMP system is the second example considered in section 9.2. It has been less completely characterised than the glycolytic system. In part this is because it is an inherently more complex system. Adenylate cyclase, the central enzyme in the network, is membrane bound, and the system is multicompartmental since extracellular and internally sequestered calcium are important elements. As in the case of glycolysis, the mathematical description is a system of nonlinear ordinary differential equations, but because of the high dimension of the system, analysis by the methods that succeeded with glycolytic models has not been as successful, and the level of detailed information obtained in glycolytic models has not been attained for calcium–cyclic AMP systems. However, control theory methods, which are ultimately derived from the theory of differential equations, have offered some indication of the dynamical structure of the network. In contrast with glycolysis, allosteric inhibition of an enzyme rather than activation appears to be the crucial process. Though analysis by control theory does not always produce detailed conclusions in large dimensional systems, the results are expressed in a form that encourages qualitative generalisations which assist in constructing an intuitive understanding of what types of metabolic system can oscillate.

The engineering analysis of calcium–cyclic AMP control loops suggested that the introduction of nonmonotonic nonlinearities into these systems could result in a network that can enter a domain of chaotic behaviour. In section 9.3 of this chapter, the implications of this result are considered. Calcium-dependent activation of adenylate cyclase by calmodulin can result in a nonmonotonic dependence of adenylate cyclase activity on calcium. This suggests that the intracellular second-messenger system could become chaotic. Since this biphasic form of activation is observed in neural tissue, this analysis supports other theoretical arguments indicating that chaotic neural behaviour is possible at the cellular level and at the level of small neural networks. These theoretical discussions are consistent with recent experimental results that are also summarised in section 9.3. The fractal dimensions of attractors governing the spontaneous activity of neurones in the somatosensory cortex of the squirrel monkey have been calculated and in some cases the results are consistent with low-dimensional chaotic behaviour.

The incomplete nature of these results, both experimental and theoretical, makes construction of a link between complex dynamical behaviour at the cellular level and clinically observed failures of neural regulation a process more of speculation than deduction. For this reason, these possibilities are given only a brief examination at the end of the chapter. Three neurological disorders are considered: tremor, dyskinesias and epilepsy. On the basis of the preceding discussion of dynamical systems theory, it will be suggested that an epileptic seizure could be an automatic corrective mechanism that re-establishes neural co-ordination that may have been lost as the result of an antecedent dynamical transition to disordered behaviour. Thus, in an abstract sense, the seizure is a restorative phenomenon and not a defect.

9.2 Periodic behaviour in biochemical systems

Oscillatory behaviour is commonly encountered in biological systems. Periods range from fractions of a second, in the case of mammalian central nervous system neurones, to the annual rhythms of plants [120]. The analysis reported here has been motivated by investigations of high-frequency cellular oscillators with periods of the order of minutes or less. However, many of the general conclusions from these studies are applicable to systems with longer periods. Even given this restricted frequency domain, the experimental literature is enormous [90]. Some examples are presented in Fig. 9.1.

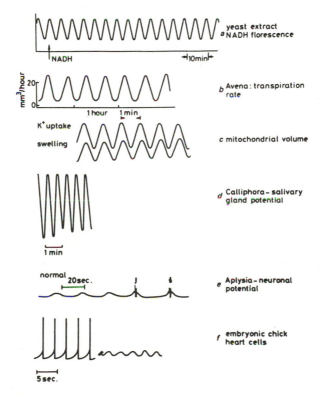

Fig. 9.1 Examples of biological oscillators (reprinted from [89]). (a) Glycolytic oscillations [88]; (b) transpiration rhythms in *Avena* [9]; (c) oscillations in mitochondrial volume and potassium uptake [49]; (d) transepithelial potential oscillations in the blowfly salivary gland [93]; (e) recovery of spike activity in an *Aplysia* neurone on withdrawal of TTX (tetrodotoxin, a Na^+ channel blocker) [56]; (f) suppression of action potentials in embryonic chick heart cells by TTX [78].

Since its discovery in 1964 [34], the archetypal biochemical oscillator has been the glycolytic oscillator. Originally observed in cell-free extracts of yeast, oscillations in glycolytic intermediates have since been observed in preparations derived from cardiac and skeletal muscle, cultured fibroblasts and *Ehrlich ascites* tumour cells. (An extensive bibliography is given in [90].) The features of the glycolytic pathway important in analysing oscillations are shown in Fig. 9.2.

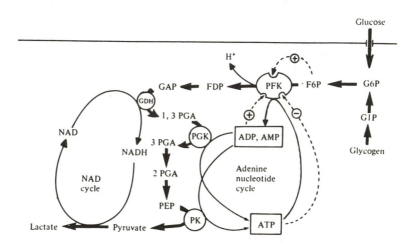

Fig. 9.2 Control structure of the glycolytic pathway. Solid lines denote material transport or chemical reactions. Dashed lines denote allosteric control relationships. PFK = phosphofructokinase, PGK = phosphoglycerate kinase, PK = pyruvate kinase, G6P = glucose 6-phosphate, G1P = glucose 1-phosphate, F6P = fructose 6-phosphate, FDP = fructose diphosphate, GAP = glyceraldehydephosphate, 2-PGA, 3-PGA and 1,3-PGA = the phosphoglyceraldehydes, PEP = phosphoenolpyruvate (from [4]).

Phosphofructokinase is the crucial control enzyme. An early model of the oscillation was constructed using the feedback activation of this enzyme by its product as the crucial control feature [7, 39]. The model takes the form of a two-dimension ordinary differential equation

$$d\alpha/dt = \sigma_1 - \sigma_M \Phi$$

(9.1)
$$d\gamma/dt = \sigma_M \Phi - k_s \gamma$$

$$\Phi = \frac{\alpha e(1+\alpha e)(1+\gamma)^2 + L\theta\alpha ce'(1+\alpha ce')}{L(1+\alpha ce')^2 + (1+\gamma)^2(1+\alpha e)^2}$$

where α is the normalised concentration of glucose, γ is the normalised concentration of ADP, and σ_1 is the rate at which substrate is introduced into the system. σ_1 is strictly positive. Thus, the system is thermodynamically open and an essential condition for sustained oscillations has been met. This differential equation reproduces much of the experimentally observed behaviour. A more systematic mathematical analysis of this model and its extensions to a more detailed representation of the glycolytic pathway has been given by Plesser [85].

In the Goldbeter–Lefever model [39] of the glycolytic oscillator, feedback activation is essential to the destabilisation of the thermodynamic steady state and the resulting oscillation. However, allosteric positive feedback is not sufficient to produce sustained oscillations. It can be shown that simple positive feedback systems fail to display oscillations [91]. Nor is positive feedback necessary for oscillatory behaviour. Negative feedback loops can be periodic. An example of a negative feedback system which may be capable of rhythmic output is the intracellular calcium–cyclic AMP system (Fig. 9.3). In this system calcium inhibits adenylate cyclase, the enzyme that synthesises cyclic AMP, and cyclic AMP produces an increase in intracellular calcium by promoting the desequestration of calcium, possibly via a sequence of intermediates [93, 94].

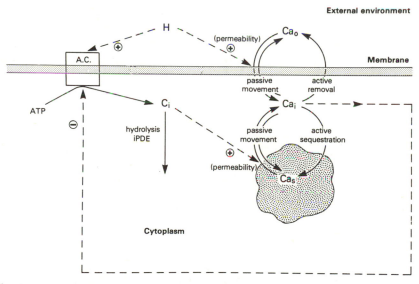

Fig. 9.3 A simplified intracellular second-messenger network. Stimulating hormone, H, has two actions on the network. It activates adenylate cyclase, AC, which results in an increase in intracellular cyclic AMP, C_i. In some preparations the hormone causes a transient increase in membrane permeability to calcium, which results in an increase in intracellular calcium, Ca_i. This increase is augmented by desequestration stimulated by cyclic AMP. The feedback loop is closed by calcium inhibition of adenylate cyclase (modified from [94]).

A detailed mathematical implementation of this system for a specific cell type is given elsewhere [83]. However, some qualitative sense of the behaviour of the system can be obtained by examining a highly idealised phenomenological model [89] of the form

$$x = G(s)f(x)$$

where x is the normalised concentration of intracellular calcium, and $f(x)$ is the calcium dependence of adenylate cyclase in normalised coordinates:

$$f(x) = 1/(1 + x^p)$$

p is a positive integer that determines the degree of sigmoidicity in the inhibition function. $G(s)$, where s is the Laplace transform variable, attempts to depict the sequence of intermediate chemical steps between the synthesis of cyclic AMP and the genesis of the feedback signal:

$$G(s) = \{(b_1 + s)\,(b_2 + s) \ldots (b_n + s)\}^{-1}$$

Parameters b_j, $j = 1, \ldots n$ are positive constants. In the biological literature this system is sometimes termed the Goodwin equations after their use in an early model of the control of protein synthesis [40].

The object of the analysis is to determine the range of parameter values (b's, p and n) which results in periodic solutions. It can be shown that questions about the existence of periodic solutions turn on the stability of the steady state, denoted x^*, which in turn is most expeditiously examined by application of the Nyquist stability theorem [54]. Let $G(i\omega)$ denote the Nyquist locus. It is conventionaly plotted for ω increasing from zero to infinity (Fig. 9.4).

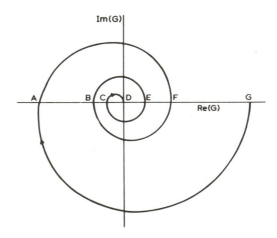

Fig. 9.4 The Nyquist locus for positive ω corresponding to $G(s)$ when $n = 11$, $b_j = 1$. Arrows indicate the direction of increasing ω (from [89]).

For any given value of n there are m intersections of $G(i\omega)$ and the negative real axis for $\omega \geq 0$, where m is the smallest non-negative integer such that $4m \geq n-2$. Let $\omega_1 \ldots \omega_m$ be the frequencies at these intersections. Define the family of m functions $W_j(p, b_1, \ldots b_n)$ by

$$W_j(p, b_1, \ldots b_n) = |f'(x^*)||G(i\omega_j)|$$

Theorem. The following theorem has been demonstrated [79]. Given the system $x = G(s)f(x)$ with $G(s)$ and W_j as defined above, then

(a) the singularity x^* is a local attractor if and only if $W_1 < 1$;
(b) if $W_j > 1 > W_{j+1}$, there are exactly j pairs of eigenvalues with positive real part; if $W_m > 1$, there are m pairs with positive real part;
(c) if $W_j = 1$, there is one pair of eigenvalues with $\lambda = \pm \omega_j$ and $j-1$ pairs with positive real part.

These results on the local stability of the steady state can be related directly to the existence of periodic solutions. Application of the Hopf bifurcation theorem shows that a periodic solution appears at the $W_1 = 1$ bifurcation [79]. Calculation of the curvature of the centre manifold shows that this periodic solution is an attractor. This result carries the limitations of any result obtained by the Hopf bifurcation theorem, namely it is valid only for parameters in a neighbourhood of the bifurcation. A theorem constructed on the contraction mapping theorem shows that there is a periodic solution whenever $W_1 > 1$. The condition that the parameter vector be near bifurcation is no longer necessary [46]. However, this result carries limitations of contraction-mapping theorem constructions. It demonstrates the existence of a periodic solution, but does not show that it is unique and does not offer any information about its stability properties. None the less, these two results and the substantial body of empirical computational evidence strongly argue that the system has a unique globally attracting steady state if $W_1 < 1$, and a unique attracting periodic solution if $W_1 > 1$.

Though this mathematical description is a crude characterisation of the chemical reality of calcium–cyclic AMP control systems, it does offer some qualitative lessons of possibly wider applicability. These results suggest that the tightness of control stability tradeoff long recognised in technological control systems is reflected in biochemical networks. This is seen by examining the function W_1 which, according to the arguments given above, predicts the presence of oscillations. W_1 contains two multiplicative factors, $|f'(x^*)|$, the value of the derivative of the feedback function at the steady state, and $|G(i\omega_1)|$, the magnitude of the Nyquist locus at its first intersection with the negative real axis. Factors that increase their magnitude will promote the appearance of oscillations. The function $f(x)$ gives, in normalised coordinates, the rate of synthesis of cyclic AMP as a function of the concentration of the feedback variable, intracellular calcium. The function for three values of p is shown in Fig. 9.5. As p increases, the transition from high synthesis rate to low synthesis rate becomes sharper and thus the magnitude of the derivative increases. Therefore increasing p, which produces sharp transition functions, the chemical analogue of tight control, can result in the appearance of oscillations. In fact, a value of $p>1$ is essential for periodic behaviour [1]. Chemically, p can correspond to the number of monomers that form the functional allosteric enzyme. Thus, this result is consistent with Goldbeter's in concluding that allostery is an important, possibly crucial determinant of periodic metabolic activity. This point has been made independently by Hess [48].

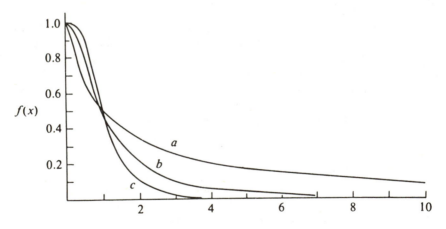

Fig. 9.5 The effect of increasing p on the nonlinearity $f(x) = 1/(1 + x^p)$. (a) $p = 1$; (b) $p = 2$; (c) $p = 3$ (from [91]).

The second factor in W_1 is $|G(i\omega_1)|$. An increase in this factor can also lead to oscillations. Two related mechanisms can do this. The first is to increase n, the number of intermediate steps in the feedback loop. The second dynamically equivalent mechanism is to introduce delays into the loop that correspond to synthesis and transport delays. In these systems $G(s)$ becomes

$$G(s) = e^{-Ds}/\{(b_1 + s) \ldots (b_n + s)\}$$

The Nyquist loci corresponding to increasing values of delay are shown in Fig. 9.6. Since $|G(i\omega_1)|$ is monotone increasing in the total delay, another engineering result is thus obtained for biochemical systems; increasing the delay between a process, in this instance the synthesis of cyclic AMP, and its feedback response can destabilise a network and result in oscillations.

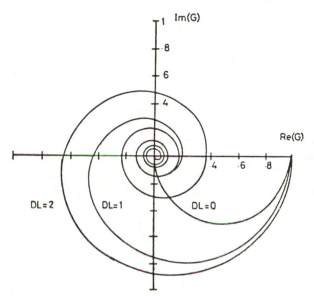

Fig. 9.6 The effect of increasing delay on the Nyquist locus for the case $n = 1$, $b_1 = 1$ (from [91]).

On summarising this cursory examination of high-frequency biological oscillators, it can be concluded that a large experimental literature demonstrates that oscillations of this type are frequently encountered, and that in some cases the techniques of nonlinear control system analysis can be helpful in investigations of these networks. It will be shown in the next section that the study of biochemical oscillators by these methods has established a set of results and a form of analysis that can be extended to the examination of chaotic metabolic systems.

9.3 Chaotic behaviour in biochemical and physiological systems

The most intensively investigated example of chaotic chemical behaviour has not been in a biochemical system but rather in the inorganic Belousov–Zhabotinskii reaction. Early theoretical analyses of chaotic behaviour [112, 114] were anticipated by empirical studies by Hudson and co-workers [99] and Rössler *et al* [97]. In the Rössler experiments, electrochemical potential was measured in a continuously stirred, isothermal system that received a continuous injection of manganese sulphate, sulphuric acid, malic acid and potassium bromate. Thus, the system was thermodynamically open, but autonomous in the sense that it did not receive time-dependent input. In addition to periodic potential signals, seemingly chaotic potential variation was observed. In subsequently performed systematic investigations it has been found that flow rate through the reaction chamber can be used as a parameter to tune reproducibly between periodic and chaotic dynamical regimes. Many of the analysis techniques outlined elsewhere in this volume, such as the graphical construction of the associated attractor, have found application in this chemical system [98]. Periodic doubling bifurcations and chaotic one-dimensional return maps have been observed [102], as has the parameter-dependent appearance of a broad-band component of the Fourier spectrum [116].

Although it is possible to show that even comparatively simple models of metabolic control circuits can display chaos [25, 38], experimental investigations of chaos in biochemical systems have not yet reached the level of detail attained with the Belousov-Zhabotinskii reaction. In part this may reflect a consistency with Goldbeter's theoretical conclusion that chaos is comparatively rare when compared with periodicity [25]. It has been reported that chaos was observed in the oxidation of NADH catalysed by horseradish peroxidase [26]. Using enzyme concentration as the bifurcation parameter it was possible to induce reversible transitions between periodic and chaotic behaviour. An associated mathematical model [81] can reproduce the periodic but not the chaotic regime.

Periodic input in both numerical models of the glycolytic pathway and in an experimental preparation derived from *Saccharomyces cerevisiae* results in a rich variety of behaviours including chaos [75]. The mathematical model of chaos incorporated phosphofructokinase and pyruvate kinase as the principal enzymes of the glycolytic pathway and examined the effect of periodic variation in fructose 6-phosphate. As the result of varying input frequency, integer entrainment ratios of 2^n, 1, 3, 5, 7 and 11 were observed, as were period-doubling bifurcations and chaos. In the experimental system, the NADH fluorescence was observed in response to constant glucose input (the system is autonomously periodic: see Fig. 9.7) and periodic glucose input. By varying input frequency, entrainment ratios of 1, 2, 3, 5 and 7 or chaotic fluorescence could result. A stroboscopic transfer function was constructed by plotting the value of a local maximum in NADH fluorescence against the previous maximum, thus approximating a first return map. The transfer function obtained by this procedure admits a period-3 solution which is consistent with chaos [61]. Using input amplitude as an additional bifurcation parameter, further complex behaviours appear in the mathematical model of glycolysis including transitions between oscillatory modes and intermittency [74]. No examples of chaotic behaviour in autonomous glycolytic systems (theoretical or experimental) have been reported.

In commonly encountered metabolic systems the activity of an allosteric enzyme is a monotonic function (increasing or decreasing) of its allosteric effectors. In the previous sections of this chapter it was argued that oscillations may occur in metabolic feedback systems in which the activity of a crucial control enzyme is uniformly decreasing with the concentration of the feedback effector. It can be shown that in a network with a similar structure in which the feedback activator metabolite is an allosteric activator, switching behaviour can result [91]. This summary motivates the following three questions.

(a) Does enzyme activity inevitably vary monotonically with effector concentration?
(b) If nonmonotonic effector–activity relationships are possible in metabolic feedback systems, what might the dynamical consequences be at the level of cellular metabolism?
(c) What might the dynamical consequences be at the level of systemic physiological regulation?

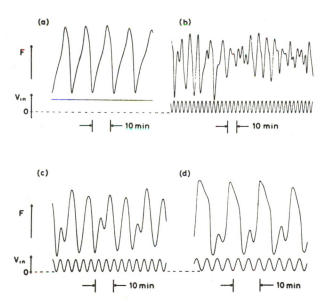

Fig. 9.7 Measured NADH fluorescence (upper trace) for different glucose input functions (lower trace). All ordinate units are arbitrary. (a) Oscillations at constant input flow; (b) chaos; (c) 1 : 5 entrainment; (d) 1 : 3 entrainment (from [75]).

This chapter will conclude with an examination of these questions. It will be shown that the effector–activity relationship is not invariably monotonic; a specific example, the effect of intracellular calcium on the activity of adenylate cyclase, will be considered. The dynamical consequences at the cellular level of these feedback relationships are not yet fully characterised. However, results now available suggest that chaotic variation in intracellular concentrations of calcium and cyclic AMP could result. The ensuing assessment of the consequences of chaotic second-messenger networks is frankly speculative, but when combined with other theoretical and experimental evidence these results suggest that chaotic neural activity may be possible. This in turn may result in clinically observed pathologies of neural regulation.

As indicated previously, intracellular calcium is usually an inhibitor of adenylate cyclase, but this is not invariably the case. In some tissues the action of calcium on adenylate cyclase is mediated by the calcium binding protein calmodulin [19]. Calmodulin is now conventionally described as ubiquitous. It is found in plants [118], in invertebrates (examples include twelve species belonging to eight phyla that include the sea anemone, clam, snail, starfish and earthworm [117]), and in all mammalian tissues examined [19, 118]. Calcium combines stoichiometrically with calmodulin in all tissues, and in some tissues the resulting calcium–calmodulin complex activates adenylate cyclase. Thus, at low concentrations, calcium is an activator of the enzyme. However, at high concentrations of intracellular calcium, all calmodulin binding sites are occupied. Excess calcium, acting directly on the enzyme, is inhibitory. A nonmonotonic, biphasic activation–inhibition curve results. (It should be noted that biphasic effects of calcium on adenylate cyclase had previously been reported in hormone-activated preparations. On reviewing adipocyte data, Bradham and Cheung [8] concluded that calcium was required for the binding of some hormones to receptors, and that calcium in excess of that required for successful binding of agonist to receptor is inhibitory. The consequence of these relationships in *in vitro* preparations is a biphasic adenylate cyclase versus calcium function when hormone is present. When hormone is absent, calcium acts as an inhibitor at all concentrations. Examples are not limited to adipocytes, but also include the ACTH-sensitive adenylate cyclase in the bovine adrenal gland [3] and the oxytocin-sensitive enzyme of frog bladder epithelial cells [6]. In *in vitro* preparations of partially purified enzymes, the functional distinction between intracellular and extracellular calcium is lost. It seems possible that hormone-dependent biphasic systems do not constitute a true example of a biphasic effect of intracellular calcium on adenylate cyclase.) Calmodulin-dependent calcium activation of adenylate cyclase resulting in a biphasic calcium dependence has been observed in bovine brain [20], porcine brain [11], rat cerebral cortex [10], rat corpus striatum [37], rat glial tumour [12], guinea pig brain [86], and a human neuroblastoma [21]. It is quite possible that these cells contain more than one form of adenylate cyclase. These results would indicate only that at least one, but not necessarily all forms, is calmodulin sensitive.

Though all of the many mammalian tissues tested contain calmodulin, the adenylate cyclase in these cell types is not necessarily calcium–

calmodulin sensitive. Indeed this sensitivity and its consequent biphasic activity seem to be exceptional. For example, rat heart, rabbit heart, porcine renal medulla and frog erythrocytes all contain calmodulin, but adenylate cyclase is insensitive to the calcium–calmodulin complex in these cells. Calcium is a monotonic inhibitor of the enzyme in these tissues [21]. The pattern of sensitivity across cell types has encouraged some to speculate that calcium–calmodulin activation of adenylate cyclase is limited to neural and neurally derived tissue [119]. However, this generalisation is evidently unwarranted since activation has been reported in rat pancreatic islet cells [101,115].

The form of the biphasic enzyme activity versus calcium function should be considered since it is important to subsequent dynamical arguments. In the rat cerebral cortex in the absence of calmodulin, calcium has a negligible effect at low concentrations. At concentrations greater than 1.5 $\times$ 10^{-4}M, it was found to be inhibitory. In the presence of calmodulin a biphasic response was observed with a peak at 1.5 $\times$ 10^{-4}M calcium [10,13]. (A general observation should be made concerning cited values of calcium concentration. In some cases, particularly in the earlier literature, high nonphysiological concentrations of calcium were employed, inevitably resulting in the inhibition of the enzyme, even in preparations where calmodulin-dependent activation was subsequently demonstrated. Also, the concentration of ionic calcium, free of chelator and available to act on the enzyme, can be imperfectly calculated. Examples include the failure to calculate the competition of magnesium and calcium for the same EDTA and EGTA ligands. The reaction mixture also contains ATP, which is a calcium chelator [82]. Given the availability of these ligands, the concentration of ionic calcium may be less than the cited value. Computer programs to perform these calculations have been published [31].) A more complex regulatory mechanism was discovered in this preparation when the role of GTP was considered [14]. If calmodulin is absent and GTP is present, a slightly biphasic function with a peak at 1.5 $\times$ 10^{-4}M calcium is observed. If calmodulin is present and GTP absent, a more markedly biphasic response is obtained. When both calmodulin and GTP are present, a strongly biphasic curve is produced with a peak activity at 1.1 $\times$ 10^{-4}M. The function is much more nonlinear than that obtained in the preceding cases [14]. Thus, as the conditions of the experimental preparation more closely approach *in vivo* conditions, a more sharply defined enzyme response is obtained. Very nonlinear responses have been obtained with adenylate cyclase isolated from guinea pig brain ([86], Fig. 9.8).

What might be the effect of this biphasic response on the stability of intracellular calcium–cyclic AMP loops? It has already been argued that to a first approximation these systems can be modelled by a highly idealised control loop of the form $x = G(s)f(x)$, where x is the concentration of intracellular calcium, s is the Laplace transform variable, and G is a

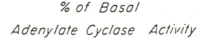

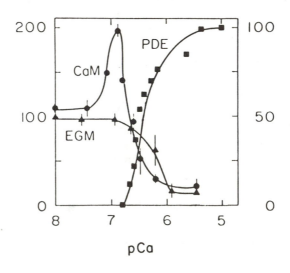

Fig. 9.8 Calcium dependence of guinea pig brain adenylate cyclase and phospho-diesterase. Calcium activation of adenylate cyclase is shown in the curve CaM (marked by circles in the diagram). The calmodulin content of the preparation was 1.57 μg mg^{-1} protein. The role of calmodulin in the calcium-dependent activation is demonstrated in a subsequent experiment in which the preparation was washed with chelators to produce an EGTA-membrane system (labelled EGM, marked by triangles). This process results in dissociating calmodulin bound to the enzyme and abolishes activation of the enzyme. The calcium dependence of soluble calmodulin-dependent phosphodiesterase is shown in the curve marked with squares. Note that the horizontal axis is decreasing in units of pCa. (From [86].)

large-dimensional, low-pass, linear filter. The function $f(x)$ characterises the effect of intracellular calcium on adenylate cyclase. In previous cases where $f(x)$ was monotone decreasing oscillations could result. What can the dynamical consequences be if a nonmonotonic function were intro-duced? Specifically, can chaotic behaviour result in a single-loop feedback system with a nonmonotonic nonlinearity? The answer is now known to be yes. Rössler *et al.* [97] produced a three-dimensional system, $G(s) = (s+1)^{-3}$, where $f(x)$ is a highly nonlinear function. Sparrow [104] con-structed an example that employs a well-behaved biphasic nonlinearity that had been previously studied by May in chaotic difference schemes, $f(x) = rxe^{-x}$ (Fig. 9.9).

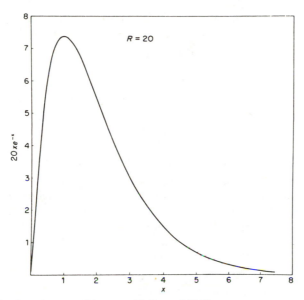

Fig. 9.9 The function rxe^{-x} for $r = 20$ (from [104]).

If $0<r<1$, there is a unique globally attracting non-negative equilibrium at $x=0$. If $r>1$, there are two non-negative equilibria, $x = 0$, which is nonattracting, and $x^* = \ln r$. For $1<r<8.197$, x^* is an attractor. At $r = 8.197$ the system undergoes a Hopf bifurcation and an attracting periodic solution results. Subsequent period-doubling bifurcations appear as r is increased, resulting in increasingly complex periodic solutions. For r slightly greater than 21.0, a periodic orbit was not detected in the numerical results. Empirical computational evidence suggests that the system entered a chaotic regime. As r is increased further, the system appears to undergo a series of reverse bifurcations, compressed in a comparatively narrow range of r values, resulting in a return to a stable periodic orbit. The Sparrow system has very large dimensional, linear filter, $G(s) = (1+s/n)^{-n}$. The numerical values cited here are for the case $n = 50$. It might be supposed that this value of n is far larger than the typical dimension of biochemical feedback systems. Indeed, at this value of n, the system behaves similarly to the finite difference scheme $G(s) = e^{-s}$ obtained in the limit as n approaches infinity. However, this large value of n may not be as objectionable as it at first seems. If a discrete time delay is incorporated in the system, as is frequently done in biological models [64, 72, 73], the value of n resulting in chaos can be reset to lower values if the delay is sufficiently large. Further, $G(s)$ as it appears in Sparrow's system is an exceptionally well behaved element. The introduction of numerator dynamics would further reduce the value of n needed to produce dynamical behaviour. This is a single-loop system. A more realistic representation would incorporate additional loops. It can be shown that single-loop models of biochemical networks that have a globally attracting singular point can lose stability when additional loops, even negative feedback loops, are introduced into the system [79]. Finally, it should be noted that even at large values of r the nonlinearity in this network is a smooth biphasic function. It does not have the abrupt on–off characteristic of the experimentally determined calcium enzyme activity function (Fig. 9.8). A comparison of the Sparrow system and the Rössler system provides a useful example of the trade-off between the nonlinearity $f(x)$ of the system and the dimension needed in $G(s)$ for the appearance of complex dynamics. In the Rössler system an ill-behaved $f(x)$ produces chaos if $n = 3$. In the Sparrow system $f(x)$ is well behaved and large dimensions are required. The limited body of mathematical results describing chaotic control networks makes generalisation difficult. However, the results seem to suggest that the chaotic behaviour observed in this simple model will not necessarily be lost when complications that result in a more faithful depiction of biological systems are introduced.

Combining these mathematical results with the biochemical evidence concerning calmodulin-dependent adenylate cyclase, it is possible to draw some provisional conclusions. Specifically, the elements required for a chaotic feedback system, including a nonmonotonic nonlinearity, are present in intracellular second-messenger networks. It is possible to speculate that this system might enter a region of chaotic behaviour *in vivo*. The heuristic nature of this argument is explicitly recognised, but its dynamical consequences are of interest. Both intracellular calcium and adenylate cyclase are coupled to processes at the cell membrane that are reflected in membrane potential [4, 113]. Thus, if the second-messenger network were to enter a domain of chaotic behaviour, chaotic variation in membrane potential would result. The biphasic feedback function that is essential to this argument has been identified in neural tissue. Therefore, this line of reasoning leads to the speculation that neurones can be chaotic.

This is not the only theoretical argument suggesting that chaotic neural behaviour is possible. Carpenter [16] analysed generalisations of the Hodgkin–Huxley equations and found highly nonlinear cases to be chaotic. Chay [17] constructed a model of a single neurone using Eyring multi-barrier rate theory and subsequently demonstrated [18] that apparently chaotic numerical solutions could be obtained with appropriately selected parameter values. However, in the Chay study the distinction between chaotic solutions and complex periodic solutions with long periods and substantial higher harmonic content was made by qualitative visual assessment. Quantitative measures of chaotic behaviour, such as the dimension of the attractor, Lyapunov exponents or Kolmogorov entropy, were not provided. Chay observed that some model calculations resemble the recordings of *Aplysia* neurone R3 in response to flurazepam [53]. Interestingly, flurazepam, a benzodiazepine-derived hypnotic, includes ataxia among its adverse reactions. Theoretical arguments also indicate that multicellular neural systems can enter domains of chaotic behaviour. Mackey and an der Heiden [70] analysed a system modelling recurrent inhibition in hippocampal CA3 pyramidal cells in response to penicillin. They found evidence for chaotic behaviour. These results are of particular interest since, at excessively high serum levels, the neurotoxic action of penicillin can result in convulsions. Experimentally, a penicillin focus is created in animals by applying penicillin in very high concentrations directly to the cortex. The resulting seizures are commonly regarded as a successful experimental approximation of focal epilepsy. A large-scale brain behaviour, the central dopaminergic system, has been modelled by King *et al* [60]. Changes in the efficacy of dopamine at the post-synaptic membrane can result in seemingly chaotic solutions. They have suggested possible clinical connection with labile behaviour in schizophrenia and disordered motor activity in Parkinsonian patients.

As is frequently the case, the rich diversity and vivid imagination of the theoretical literature is correlated with, and quite possibly a consequence of, a paucity of experimental evidence. However, some suggestive experimental evidence is now available. In both experimental and mathematical systems, chaotic behaviour can result when an autonomous oscillator is subjected to periodic stimulation of appropriate amplitude and frequency. Behaviour that appears to be chaotic has been observed in periodically stimulated neurones. Examples include the molluscs *Lymnaea stagnalis* [51] and *Onchidium verruculatum* [47], and the squid *Dorytenthis bleekeri* [77]. Since this is a common behaviour of nonlinear oscillators, it is the most readily observed form of chaos. Because the stimulating input in these experiments has a comparatively high amplitude, the degree to which this is a physiologically realistic stimulation is open to question. For these reasons the functional significance of these observations is difficult to assess. Chaos in periodically forced nonlinear oscillators, including biological systems, is considered in detail in Part IV of this volume. Experimental literature that explicitly suggests that chaotic behaviour has been observed in neural systems not subject to periodic stimulation is more limited. Holden and his colleagues [52] treated the neurone R.Pe.D.1 of the mollusc *Lymnaea stagnalis* with high concentrations (10 mM) of aminopyridine for prolonged periods of time. Irregular-amplitude modulations, which the authors suggest may be chaotic, were seen only after the preparation was returned to normal saline and the drug was washed away. Care must be exercised in the interpretation of these results since the effect was observed only after prolonged exposure at high concentrations. Also, characterisation of the signal as chaotic was a qualitative, subjective assessment; the dimension of the attractor or the flow's Lyapunov exponents were not calculated. However, the Holden results are very suggestive. 4-Aminopyridine depresses a transient potassium current during depolarisation. This results in prolonging the action potential and thus increases calcium influx into the cell. This increased intracellular calcium is probably responsible for the drug's capacity to increase neurotransmitter release [36, 111]. The drug can counteract the depressant effects of opiates in some systems [22] and, most importantly from the point of view of this analysis, 4-aminopyridine acts as a convulsant in the mammalian central nervous system [109]. Recently, single-unit recordings of the spontaneous activity of neurones in the precentral and postcentral gyri of the squirrel monkey were obtained and the dimensions of the corresponding attractors were calculated. In this preliminary study a small number of units were examined. In so far as it is possible to draw conclusions from such a small sample size, it was observed that two distinct populations of neurones were identified. One group of rapidly firing neurones (the peak of the interspike interval histogram was 1 or 2 ms) obeyed large-dimension dynamics. The procedure used to calculate the dimension, essentially that of Grassberger and Procaccia [41, 42], failed to find a finite dimension up to the last tested embedding dimension, which was either $n=20$ or $n=40$. The other population of neurones was slower; the peak of the interspike interval histogram was greater than 15 ms. The dimension was low, less than 3.5, and noninteger.

The third element in the list of motivating questions concerned the possible consequences of complex dynamics at the level of systemic physiological regulation. It has been proposed that parameter-dependent transitions in dynamical behaviour (fixed-point attractor to periodic attractor, or periodic attractor to strange attractor) may result in clinically observed defects in control. These processes have been termed 'dynamical diseases' [71]. The following discussion will centre on neurophysiological examples. However, this form of analysis is not limited to neurology, and a number of other examples deserve a brief consideration.

Mackey and Glass [35, 71] have proposed that respiratory disorders such as Cheyne–Stokes respiration and Biot breathing occur when the fixed-point attractor governing respiratory feedback control bifurcates to become a periodic attractor. The rate of formation of formed elements of the blood (haematopoiesis) is known to be under feedback control. In a number of clinical disorders the cell count is subject to periodic or highly irregular variation. Examples include cyclic neutropenia (an irregular oscillation in circulating neutrophil count), periodic chronic myelogenous leukaemia (leukocyte and thrombocyte numbers oscillate), cyclic thrombocytopenia (rhythmic variation in platelets and megakaryocyte numbers) and periodic autoimmune haemolytic anaemia (oscillations in erythrocytye count). All of these oscillations are irregular. On the basis of a mathematical analysis of haematopoietic control systems [66–69, 71] it has been proposed that in some cases the feedback system governing haematopoiesis enters a domain of either chaotic or periodic motion. The most systematically studied example of complex behaviour in a biophysical system has been the study of chaotic cardiac dynamics. It has been suggested that some cardiac arrhythmias may occur as a consequence of a chaotic transition [44, 45, 58, 59, 103].

The possible role of dynamical transitions in the aetiology of three neurological disorders will now be considered. They are tremor, dyskinesia and epilepsy. Tremor, involuntary trembling or quivering in an approximately periodic manner, can be roughly classified into two groups [27]. Resting tremors (typically at frequencies of 3–7 Hz) are present at rest but subside or are absent when the affected limb is moved. Intention tremor (typically at frequencies of 7–12 Hz) arises or is intensified when a voluntary movement is attempted. Many investigations of tremor centre on establishing if the control defect is in peripheral neuromuscular feedback systems or in the central nervous system. The paradigm fixed-point attractor to periodic attractor to strange attractor has equal potential applicability in the analysis of either peripheral or central control pathologies. Typically hypothesised peripheral mechanisms propose multicellular systems [108]. A central origin hypothesis is particularly attractive from the point of view of the preceding analysis since it suggests that an investigation of single-neurone transition behaviour could offer useful insights into the origin of the defect.

It might be supposed that tremor was the consequence of a fixed-point attractor to periodic attractor transition of central nervous system neurones. This anticipation is unwarrantably simplistic for two reasons. First, a central origin model can be multicellular and does not necessarily require autonomous oscillatory behaviour in a single neurone. For example, it has been proposed that harmaline-induced tremor may be the consequence of irregular rhythmic discharges of olivo-cerebellar neurones (specifically inferior olivary cells [28]). These authors proposed a multi-cellular origin for the oscillation due to electrotonic coupling and recurrent inhibition in these cells. If this is so, then an understanding of single-neurone transition behaviour would have limited value in elucidating the aetiology of this particular type of tremor. A second argument against the proposition that tremor invariably follows from a fixed-point attractor to periodic attractor transition is recognised when it is noted that olivary neurones show a normal tendency to fire rhythmically. Harmaline only serves to enhance this behaviour [28]. This observation follows a pattern frequently encountered in research in this area. Periodic neural activity is a normal behaviour. The control defects are associated with rhythmic discharges that are somehow abnormal. This qualitative distinction has been expressed by DeLong ([27], page 2172): 'Unfortunately, the data do not allow definitive conclusions about the precise neural mechanisms responsible for the rhythmic firing of thalamic and olivary neurones in these animal models, but one possibility is that this abnormal activity might result from unrestrained intrinsic pacemaker discharge, rather than from oscillation of an internal feedback loop.' DeLong's statement presents two concepts: (a) an understanding of single pacemaker neurones rather than multicellular feedback systems may offer an explanation of the observed pathology; and (b) the defect lies not in rhythmic discharge but in abnormal 'unrestrained' rhythmic output.

The recognition that erratic but near-periodic discharge governed by a strange attractor is a distinct form of dynamical behaviour may provide a quantitative basis for distinguishing between the 'normal' and 'abnormal' rhythms frequently described qualitatively in the neurological literature. The loss of coherent temporal organisation following a periodic attractor–strange attractor transition might have the clinically observed consequences. This is a testable hypothesis since it is sometimes possible to distinguish experimentally between periodic attractors and strange attractors. In the specific case of olivary neurones, harmaline concentration may be a bifurcation parameter that leads to transitions in attractor topology.

Dyskinesias are the class of movement disorders characterised by involuntary movements that are distinguishable from tremor by the absence of a regular period. Specific examples of dyskinesias include chorea (a ceaseless occurrence of rapid, highly complex, involuntary movements), athetosis (mobile spasm), dystonia (involuntary contortions of muscles) and gait disorders. The complementary defects, akinesias (the absence or profound reduction of movement) are not the result of muscular weakness. It has been proposed that akinesias follow as the consequence of the loss of motor programmes that co-ordinate the many individual components of a voluntary movement. Conversely, it has been argued that dyskinesias are the consequence of the inappropriate release of motor programmes [27].

The most completely understood motor programmes are the rhythmic central pattern generators that control locomotion [43]. Gait disorders follow from defects in these control systems. Analysis of normal and abnormal function of central pattern generators probably has a more general significance. Roberts [95] has argued that both learned and innate behavioural sequences are released to function by the disinhibition of central pacemaker neurones. Thus, it has been suggested [27] that defects in pacemaker control may be a causative factor in a larger class of dyskinesias in addition to gait disorders.

Theories of how locomotor control rhythms are generated have been summarised by DeLong [27]. In the first class of model, the rhythm is generated by a network of interconnected neurones. In the second, the rhythm is generated by pacemaker neurones. Neural network models fall into two subcategories: half-centre models and ring models. The half-centre model is the classical model of gait locomotion [15]. Periodic alternating activation of flexors and extensors is achieved by reciprocal inhibition mediated by inhibitory collateral neurones. Ring models consist of a closed circuit of interneurones, some connected to flexors and some to extensors. Periodicity is the consequence of activity propagating through the closed loop. Elaborated ring models have been used to explain movement in both vertebrates and invertebrates [33]. Detailed application of these structures has been made in the analysis of the leech swimming mechanism [32, 107]. Models in which the motor programme is generated by pacemaker neurones include Pearson's [84] identification of pacemaker neurones in the cockroach locomotor system.

The strict division of models into multicellular and single-cell classes is probably more a logical convenience than an accurate reflection of the underlying physiology of these systems. It is probable that individual neurones are able, given the appropriate extracellular environment, to discharge rhythmically. Similarly, networks of interconnected, normally nonoscillatory, neurones can generate rhythmic output. However, the observed behaviour probably results from the collective interaction of individual oscillatory units. If this view is correct, the rigid distinction between single-cell and network models is artificial. This observation also applies to investigations of tremor and, as will be seen presently, seizure disorders. Transitions of dynamical behaviour in either pacemaker neurones or neural networks considered in the discussion of neurological tremor might also be applied to the examination of dyskinesias. Similar experimental protocols would be applicable.

In the examination of tremors and dyskinesias two questions were considered. Does the defect occur at the peripheral or central nervous system level? If the pathology originates in the central nervous system, is it predominantly the consequence of aberrant behaviour that can occur in individual neurones or is it a collective behaviour that can occur only in multineurone networks? In the case of the epilepsies the first question has a definite answer; the seizure is the consequence of disordered behaviour in the central nervous system. The answer to the second question, however, is the object of intense discussion. (Again, we reiterate an earlier observation that a strict division into multicellular and cellular aetiologies may not be realistic. Experimental evidence supporting this view will be cited at the end of this section). The epilepsies encompass a wide range of neurological events. An understanding of the cellular mechanisms involved is limited to small subclasses of these disorders, and it should be recognised that even in these limited cases present understanding is fragmentary. Before considering the possible applications of the dynamical transition paradigm, it is necessary to clearly delineate this restricted domain of application. Only a broad outline is considered here. The present discussion follows that of Martin [76].

Partial or focal epileptic seizures begin in a localised region of the brain. The clinical manifestations will reflect the location of the epileptogenic focus. Typically, the patient remains conscious if the disturbance is confined to one hemisphere. Complex partial seizures (psychomotor epilepsy) result when a focal disturbance spreads to the other hemisphere. Consciousness is lost. Generalised (nonfocal) epilepsy involves large parts of the brain and may be bilateral at initiation. The two principal subcategories are *petit mal* (transient loss of consciousness) and *grand mal* (alternating muscular contraction and relaxation in rapid succession and abrupt loss of consciousness). Seizures analogous to focal epilepsy can be readily induced in animals by direct application of a convulsant (for example: penicillin, bicuculline or aluminium hydroxide) to localised regions of the cortex. The characteristics of the resulting EEG are seemingly identical to those recorded during focal seizures. Also, these drugs can be added to the medium of *in vitro* neural preparations. The intracellular electrical records parallel the seizure behaviour of cortical neurones. Thus, there are both *in vivo* and *in vitro* experimental models of focal epilepsy. The subsequent discussion will be restricted to a consideration of the results from these systems.

The first intracellular electrical response to a convulsant agent is a marked depolarisation of the membrane potential that results in a rapid volley of action potentials. This depolarisation is referred to as the paroxysmal depolarisation shift. Since this is the first event in the genesis of a seizure, research has concentrated on elucidating its mechanism [87].

The most commonly held explanation of the paroxysmal depolarisation shift proposes that it is a network phenomenon [2]. The effects of convulsants on chemical synapses provide a membrane-level mechanism that would account for drug-altered network transmission properties that cause an increase in the gain of cortical feedback circuits, generating the depolarisation. In this model the termination of the depolarisation is caused by network-generated reciprocal inhibition. Much of the evidence supporting network-level models of paroxysmal depolarisation shifts has been obtained from studies with primates and other mammals. However, evidence obtained from the primitive nervous system in a mollusc (*Aplysia californica*), in which specific neurones can be identified, has indicated that, in addition to the previously established effects on chemical synapses, epileptogenic agents can enhance transmission at an identified electrical synapse. This could result in a network instability of the type proposed for focal epilepsy. The simplicity of the *Aplysia* nervous system permits a level of analysis that is impossible in mammals, and subsequent studies with this preparation could lead to further support for network-generated paroxysmal depolarisation shift models.

The alternative view of epileptogenesis attaches greater significance to processes intrinsic to individual neurones. It has been proposed that chronic foci contain 'pacemaker epileptic' neurones, which are responsible for the generation of paroxysmal depolarisation shifts [100, 121]. In support of this hypothesis, it was observed that the proportion of pacemaker epileptic neurones in a given chronic focus is logarithmically correlated with seizure frequency. Results obtained in response to varying the ionic composition of the external medium of *in vitro* preparations suggest that calcium-dependent dendritic action potentials analogous to those observed in reptilian [63, 80] and avian [62] neurones may play an important role in establishing epileptic behaviour at the neurone level [87].

A probable resolution of the network/single neurone dichotomy lies in recognising that convulsants affect both network interactions and intrinsic neurone properties. At a network level, bicuculline [23, 55] and penicillin [24, 65] antagonise GABA-mediated synaptic inhibition, which increases network excitability. At the cellular level both of these drugs prolong calcium-dependent action potentials of mouse dorsal root ganglion and spinal cord neurones in cell culture [50]. The clinical state may well be the consequence of the activity of intrinsically aberrant neurones acting collectively in a network with aberrant transmission properties.

The prospects for experimental tests of the dynamical transition paradigm seem particularly promising in the case of focal epilepsy. *In vitro* slice preparations [29] and cell culture systems are accessible for continuous microelectrode recordings in a controlled environment. Tunable parameters known to effect abnormal behaviour (ionic composition and convulsant concentration in the external medium) can be experimentally controlled. Anticonvulsant drugs may also provide a family of bifurcation parameters. Electroencephalographic *in vivo* tests of the hypothesis are also feasible. The possible role for a transition to chaotic behaviour in epileptogenesis should be specified with care. It is not suggested that the seizure itself is a chaotic event. Indeed, this seems unlikely since a seizure is frequently characterised electroencephalographically by uniform electrical activity over a large part of the brain. However, a seizure may be an automatic corrective response to an earlier transition to chaos. When viewed in this fashion, the convulsion is seen to be the cortical analogue of defibrillation, the cure rather than the disease. It can be hypothesised that a transition from stable behaviour in a local neural network leads to a loss of co-ordinated electrical activity. The massive, rhythmic depolarisations of a seizure could entrain subsystems and re-establish effective communication. Given the emerging mathematical technology that can be used to characterise the complexity of a time series (Lyapunov exponents, Hausdorff dimension and Kolmogorov entropy), this also becomes a testable hypothesis. Activation procedures [5, 110] are methods used to induce a seizure during an EEG recording session. It should be possible to determine if activation results reproducibly in an increase in the disorder of the electroencephalographic signal, particularly in the region of the epileptogenic focus in the case of focal disorders, prior to the appearance of the seizure itself.

To date, the applications of dynamical systems theory to biology have generated more questions than answers, but it is now clear that the dynamical transition paradigm constructed on the concept of parameter-dependent transitions of attractor topology offers a rich language for the phenomenological description of behaviour in complex systems. The associated measures of disorder provide the basis for construction of empirical correlations between dynamical behaviour and physiological function. It is now possible to look forward to the fusion of phenomenological descriptions of processes and empirical correlations that will lead to the development of novel therapeutic protocols of increasing specificity and efficacy.

Acknowledgements

This project is supported in part by NIH Grant NS19716 from the Epilepsy Branch of the National Institute of Neurological and Communicative Disorders and Stroke. The assistance of Richard Latta, Andrew Goldstein and Neil Greenbaun is gratefully acknowledged.

References

[1] Allwright, D. J. A global stability criterion for simple control loops. *J. Math. Biol.* **4**, 363–73 (1977).

[2] Ayala, G. F., Dichter, M., Gumnit, R. J., Matsumoto, H. and Spencer, W. A. Genesis of epileptic interrictal spikes. New knowledge of cortical feedback systems suggests a neurophysiological explanation of brief paroxysms. *Brain Res.* **52**, 1–17 (1973).

[3] Bar, H.-P. and Hechter, O. Adenyl cyclase and hormone action. III. Calcium requirement for ACTH stimulation of adenyl cyclase. *Biochem. Biophys. Res. Commun.* **35**, 681–6 (1969).

[4] Berridge, M. J. and Rapp, P. E. A comparative survey of the function, mechanism and control of cellular oscillators. *J. Exp. Biol.* **81**, 217–80 (1979).

[5] Bickford, R. G. Activation procedures and special electrodes. In *Current Practice of Clinical Electroencephalography*, eds D. W. Klass and D. D. Daly, pp. 269–305. Raven Press, New York (1979).

[6] Blockaert, J., Roy, C. and Jard, S. Oxytocin-sensitive adenylate cyclase in frog bladder epithelial cells. Role of calcium, nucleotides and other factors in hormonal stimulation. *J. Biol. Chem.* **247**, 7073–81 (1972).

[7] Boiteux, A., Goldbeter, A. and Hess, B. Control of oscillating glycolysis of yeast by stochastic, periodic and steady source of substrate: a model and experimental study. *Proc. natl. Acad. Sci. USA* **72**, 3829–33 (1975).

[8] Bradham, L. S. and Cheung, W. Y. Calmodulin-dependent adenylate cyclase. In *Calcium and Cell Function*. Vol. 1, ed. W. Y. Cheung, pp. 109–26. Academic Press, New York (1980).

[9] Brogardh, T. and Johnsson, A. Effects of magnesium, calcium and

lanthanum ions on stomatal oscillations in *Avena sativa. Planta* **124**, 99–103 (1975).

[10] Brostrom, C. O., Huang, Y., Breckenridge, B.McL. and Wolff, D. J. Identification of a calcium-binding protein as a calcium-dependent regulator of brain adenylate cyclase. *Proc. natl. Acad. Sci. USA* **72**, 64–8 (1975).

[11] Brostrom, C. O., Brostrom, M. A. and Wolff, D. J. Calcium-dependent adenylate cyclase from rat cerebral cortex. Reversible activation by sodium fluoride. *J. Biol. Chem.* **252**, 5677–85 (1977).

[12] Brostrom, M. A., Brostrom, C. O., Breckenridge, B.McL. and Wolff, D. J. Regulation of adenylate cyclase from glial tumor cells by calcium and a calcium-binding protein. *J. Biol. Chem.* **251**, 4744–50 (1976).

[13] Brostrom, M. A., Brostrom, C. O., Breckenridge, B.McL. and Wolff, D. J. Calcium dependent regulation of brain adenylate cyclase. *Adv. Cyclic Nucleotide Res.* **9**, 85–99 (1978).

[14] Brostrom, M. A., Brostrom, C. O. and Wolff, D. J. Calcium-dependent adenylate cyclase from rat cerebral cortex: activation by guanine nucleotides. *Arch. Biochem. Biophys.* **191**, 341–50 (1978).

[15] Brown, T. G. On the nature of the fundamental activity of the nervous centres; together with an analysis of the conditioning of rhythmic activity in progression and a theory of the evolution of function in the nervous system. *J. Physiol. Lond.* **48**, 18–46 (1914).

[16] Carpenter, G. A. Bursting phenomena in excitable membranes. *SIAM J. Appl. Maths* **36**, 334–72 (1979).

[17] Chay, T. R. Eyring rate theory in excitable membranes: application to neuronal oscillations. *J. Phys. Chem.* **87**, 2935–40 (1983).

[18] Chay, T. R. Abnormal discharges and chaos in a neuronal model system. *Biol. Cybern.* **50**, 301–11 (1984).

[19] Cheung, W. Y. (ed.) *Calcium and Cell Function. Vol. 1. Calmodulin*, Academic Press, New York (1980).

[20] Cheung, W. Y., Bradham, L. S., Lynch, T. J., Lin, Y. M. and Tallant, E. A. Protein activator of cyclic $3'$, $5'$-nucleotide phosphodiesterase of bovine or rat brain also activates its adenylate cyclase. *Biochem. Biophys. Res. Commun.* **66**, 1055–62 (1975).

[21] Cheung, W. Y., Lynch, T. J. and Wallace, R. W. An endogenous Ca^{2+}-dependent activator protein of brain adenylate cyclase and cyclic nucleotide phosphodiesterase. *Adv. Cyclic Nucleotide Res.* **9**, 233–51 (1978).

[22] Crain, S. M., Crain, B., Peterson, E. R., Hiller, J. M. and Simon, E. J. Exposure to 4-aminopyridine prevents depressant effects of opiates on sensory-evoked dorsal-horn network responses in spinal cord cultures. *Life Sci.* **31**, 235–40 (1982).

[23] Curtis, D. A. and Felix, D. The effect of bicuculline upon synaptic inhibition in the cerebral and cerebellar cortices of the cat. *Brain Res.* **34**, 301–21 (1971).

[24] Davidoff, R. A. Penicillin and presynaptic inhibition in the amphibian spinal cord. *Brain Res.* **36**, 218–22 (1972).

[25] Decroly, O. and Goldbeter, A. Biorhythmicity, chaos and other patterns of temporal self-organization in a multiply regulated biochemical system. *Proc. natl. Acad. Sci. USA* **79**, 6917–21 (1982).

[26] Degn, H., Olsen, L. F. and Perram, J. W. Bistability, oscillation and chaos in an enzyme reaction. *Ann. NY Acad. Sci.* **316**, 623–37 (1979).

[27] DeLong, M. R. Possible involvement of central pacemakers in clinical disorders of movement. *Fed. Proc.* **37**, 2171–5 (1978).

[28] DeMontigny, C. and Lamarre, Y. Effects produced by local applications of

harmaline in the inferior olive. *Can. J. Physiol. Pharmacol.* **53**, 845–9 (1975).

[29] Dingledine, R. (ed.) *Brain Slices*. Plenum, New York (1984).

[30] Drummond, G. I. and Duncan, L. Adenyl cyclase in cardiac tissue. *J. Biol. Chem.* **245**, 976–83 (1970).

[31] Fabiato, A. and Fabiato, F. Calculator programs for computing the composition of the solutions containing multiple metals and ligands used for experiments in skinned muscle cells. *J. Physiol. Paris.* **75**, 463–505 (1979).

[32] Friesen, W. O. and Stent, G. S. Generation of a locomotory rhythm by a neural network with recurrent cyclic inhibition. *Biol. Cybern.* **28**, 27–40 (1977).

[33] Friesen, W. O. and Stent, G. S. Neural circuits for generating rhythmic movements. *Ann. Rev. Biophys. Bioeng.* **7**, 37–61 (1978).

[34] Ghosh, A. and Chance, B. Oscillations of glycolytic intermediates in yeast cells. *Biochem. Biophys. Res. Commun.* **65**, 174–81 (1964).

[35] Glass, L. and Mackey, M. C. Pathological conditions resulting from instabilities in physiological control systems. *Ann. NY Acad. Sci.* **316**, 214–35 (1979).

[36] Glover, W. E. The aminopyridines. *Gen. Pharmacol.* **13**, 259–85 (1982).

[37] Gnegy, M. and Treisman, G. Effect of calmodulin on dopamine-sensitive adenylate cyclase activity in rat striatal membranes. *Molec. Pharmacol.* **19**, 256–63 (1981).

[38] Goldbeter, A. and Decroly, O. Temporal self-organization in biochemical systems: periodic behavior vs chaos. *Am. J. Physiol.* **245**, R478–83 (1983).

[39] Goldbeter, A. and Lefever, R. Dissipative structures for an allosteric model. Application to glycolytic oscillators. *Biophys. J.* **12**, 1302–15 (1972).

[40] Goodwin, B. Oscillatory behaviour in enzymatic control processes. *Adv. Enzyme Regul.* **3**, 425–38 (1965).

[41] Grassberger, P. and Procaccia, I. Estimation of the Kolmogorov entropy from a chaotic signal. *Phys. Rev.* **28A**, 2591–3 (1983).

[42] Grassberger, P. and Procaccia, I. Characterization of strange attractors. *Phys. Rev. Lett.* **50**, 346–9 (1983).

[43] Grillner, S. Locomotion in vertebrates: central mechanisms and reflex interaction. *Physiol. Rev.* **55**, 247–304 (1975).

[44] Guevara, M. R. and Glass, L. Phase locking, period doubling bifurcations and chaos in a mathematical model of a periodically driven oscillator: a theory for the entrainment of biological oscillators and the generation of cardiac dysrhythmias. *J. Math. Biol.* **14**, 1–24 (1982).

[45] Guevara, M. R., Glass, L. and Shrier, A. Phase locking, period-doubling bifurcations and irregular dynamics in periodically stimulated cardiac cells. *Science, Wash.* **214**, 1350–3 (1981).

[46] Hastings, S., Tyson, J. J. and Webster, D. Existence of periodic solutions for negative feedback cellular control systems. *J. Diff. Eqn* **25**, 39–64 (1977).

[47] Hayashi, H., Ishizuka, S., Ohta, M. and Hirakawa, K. Chaotic behavior in the *Onchidium* giant neuron under sinusoidal stimulation. *Phys. Lett.* **88A**, 435–8 (1982).

[48] Hess, B. Oscillations in biochemical and biological systems. *Bull. Inst. Maths Applics* **12**, 6–10 (1976).

[49] Hess, B., Boiteux, A., Busse, H. G. and Gerisch, G. Spatiotemporal organization in chemical and cellular systems. *Adv. Chem. Phys.* **29**, 137–68 (1975).

[50] Heyer, E. J., Nowak, L. M. and MacDonald, R. L. Membrane depolarization and prolongation of calcium-dependent action potentials of mouse

neurons in cell culture by two convulsants: bicuculline and penicillin. *Brain Res.* **232**, 41–57 (1982).

[51] Holden, A. V. and Ramadan, S. M. The response of a molluscan neuron to a cyclic input: entrainment and phase locking. *Biol. Cybern.* **43**, 157–63 (1981).

[52] Holden, A. V., Winlow, W. and Haydon, P. G. The induction of periodic and chaotic activity in a molluscan neurone. *Biol. Cybern.* **43**, 169–73 (1982).

[53] Hoyer, J., Park, M. R. and Klee, M. R. Changes in ionic currents associated with Flurazepam-induced abnormal discharges in *Aplysia* neurons. In *Abnormal Neuronal Discharges*, eds N. Chalazontis and N. Boisson, pp. 301–10. Raven Press, New York (1978).

[54] Hsu, J. C. and Meyer, A. U. *Modern Control: Principles and Applications.* McGraw-Hill, New York (1968).

[55] Johnston, L. A. R., Beart, P. M., Curtis, D. R., Game, C. J. A., McCulloch, R. M. and Maclachlan, R. M. Bicuculline methochloride as a GABA antagonist. *Nature, New Biol.* **240**, 219–20 (1972).

[56] Junge, D. and Stephens, C. L. Cyclic variation of potassium conductance in a burst-generating neurone in *Aplysia*. *J. Physiol. Lond.* **235**, 155–81 (1973).

[57] Kandel, E. R. and Rayport, S. G. Epileptogenic agents enhance transmission at an identified weak electrical synapse in *Aplysia*. *Science, Wash.* **213**, 462–4 (1981).

[58] Keener, J. P. Chaotic cardiac dynamics. In *Mathematical Aspects of Physiology (AMS Lectures in Applied Mathematics. Vol. 19)*, ed. F. C. Hoppensteadt, pp. 299–325. American Mathematical Society, Providence, RI (1981).

[59] Keener, J. P. On cardiac arrhythmias: AV conduction block. *J. Math. Biol.* **12**, 215–25 (1981).

[60] King, R., Bachas, J. D. and Huberman, B. A. Chaotic behavior in dopamine neurodynamics. *Proc. natl. Acad. Sci. USA* **81**, 1244–7 (1984).

[61] Li, T-Y. and Yorke, A. Period three implies chaos. *Amer. Math. Monthly* **82**, 985–92 (1975).

[62] Llinas, R. and Hess, R. Tetrodotoxin-resistant dendritic spikes in avian Purkinje cells. *Proc. natl. Acad. Sci. USA* **73**, 2520–3 (1976).

[63] Llinas, R. and Nicholson, C. Electrophysiological properties of dendrites and somata in alligator Purkinje cells. *J. Neurophysiol.* **34**, 532–51 (1971).

[64] MacDonald, N. *Time Lags in Biological Models. Lecture Notes in Biomathematics. Volume 27*, Springer, Berlin (1978).

[65] MacDonald, R. L. and Barker, J. L. Specific antagonism of GABA-mediated postsynaptic inhibition in cultured spinal neurons: a common mode of convulsant action. *Neurology* **28**, 325–30 (1978).

[66] Mackey, M. C. Periodic autoimmune hemolytic anemia: an induced dynamical disease. *Bull. Math. Biol.* **41**, 829–34 (1979).

[67] Mackey, M. C. Dynamic haematological disorders of stem cell origin. In *Biophysical and Biochemical Information Transfer in Recognition*, eds J. G. Vassileva-Popova and E. V. Jensen, pp. 373–409. Plenum, New York (1979).

[68] Mackey, M. C. Some models in hemopoiesis: predictions and problems. In *Biomathematics and Cell Kinetics*, ed. M. Rotenberg, pp. 23–38. Elsevier/North-Holland Biomedical Press, Amsterdam (1981).

[69] Mackey, M. C. Unravelling the connection between human hematopoietic cell proliferation and maturation. In *Regulation of Reproduction and Aging*, eds E. V. Jensen and J. G. Vassileva-Popova, Plenum, New York (1981).

[70] Mackey, M. C. and an der Heiden, U. The dynamics of recurrent inhibition.

J. Math. Biol. **19**, 211–25 (1983).

[71] Mackey, M. C. and Glass, L. Oscillation and chaos in physiological control systems. *Science, Wash.* **197**, 287–9 (1977).

[72] Mahaffy, J. M. Cellular control models with linked positive and negative feedback and delays. I. The models. *J. Theor. Biol.* **106**, 89–102 (1984).

[73] Mahaffy, J. M. Cellular control models with linked positive and negative feedback and delays. II. Linear analysis and local stability. *J. Theor. Biol.* **106**, 103–18 (1984).

[74] Markus, M. and Hess, B. (1984) Transitions between oscillatory modes in a glycolytic model system. *Proc. natl. Acad. Sci. USA* **81**, 4394–8 (1984).

[75] Markus, M., Kuschmitz, D. and Hess, B. Chaotic dynamics in yeast glycolysis under periodic substrate input flux. *FEBS Lett.* **172**, 235–8 (1984).

[76] Martin, J. H. Properties of cortical neurons, the EEG and mechanisms of epilepsy. In *Principles of Neural Science*, eds E. R. Kandel and J. H. Schwartz, pp. 461–71. Elsevier/North-Holland, New York (1981).

[77] Matsumoto, G., Aihara, K., Ichika, M. and Tasaki, A. Periodic and nonperiodic responses of membrane potential in squid giant axons under sinusoidal current stimulation. *J. Theor. Neurobiol.* **3**, 1–14 (1983).

[78] McDonald, T. F. and Sachs, H. G. Electrical activity in embryonic heart cell aggregates. *Pflugers Arch.* **354**, 165–76 (1975).

[79] Mees, A. I. and Rapp, P. E. Periodic metabolic systems: oscillations in negative feedback biochemical control networks. *J. Math. Biol.* **5**, 99–114 (1978).

[80] Nicholson, C. and Llinas, R. Field potentials in the alligator cerebellum and theory of their relationship to Purkinje cell dendritic spikes. *J. Neurophysiol.* **34**, 509–31 (1971).

[81] Olsen, L. F. and Degn, H. Oscillatory kinetics of the peroxidase–oxidase reaction in an open system. Experimental and theoretical studies. *Biochim. Biophys. Acta.* **523**, 321–34 (1978).

[82] O'Sullivan, W. J. and Perrin, D. D. The stability constants of metal–adenine nucleotide complexes. *Biochemistry* **3**, 18–26 (1964).

[83] Othmer, H. G., Monk, P. B. and Rapp, P. E. A model for signal relay and adaptation in *Dictyostelium discoideum*. II. Analytical and numerical results. *Math. Biosciences* in press (1985).

[84] Pearson, K. G. Central programming and reflex control of walking in the cockroach. *J. Exp. Biol.* **56**, 173–93 (1972).

[85] Plesser, T. Dynamic states of allosteric enzymes. *Proceedings VII Internationale Konferenz uber Nichtlineare Schwingungen. Abhandlungen der Akademie der Wissenschaften der DD6 N6, Vol. 2*, pp. 273–80. Akademie Verlag, Berlin (1977).

[86] Potter, J. D., Piascik, M. T., Wisler, P. L., Robertson, S. D. and Johnson, C. L. Calcium dependent regulation of brain and cardiac muscle adenylate cyclase. *Ann. NY Acad. Sci.* **356**, 220–31 (1980).

[87] Prince, D. A. Neurophysiology of epilepsy. *Ann. Rev. Neurosci.* **1**, 395–415 (1978).

[88] Pye, E. K. Glycolytic oscillations in cells and extracts of yeasts: some unsolved problems. In *Biological and Biochemical Oscillators*, eds B. Chance, E. K. Pye, A. K. Ghosh and B. Hess, pp. 269–300. Academic Press, New York (1973).

[89] Rapp, P. E. Bifurcation theory, control theory and metabolic regulation. In *Biological Systems, Modelling and Control*, ed. D. A. Linkens, pp. 1–93. Peter Peregrinus, Stevenage (1979).

[90] Rapp, P. E. An atlas of cellular oscillators. *J. Exp. Biol.* **81**, 281–306 (1979).

[91] Rapp, P. E. Biological applications of control theory. In *Mathematical Models in Molecular and Cellular Biology*, ed. L. A. Segel, pp. 146–247. Cambridge University Press, Cambridge (1980).

[92] Rapp, P. E. The appearance of complex dynamics in simple metabolic feedback systems. In *Modeling and Simulation. Vol. 12*, eds W. A. Vogt and M. H. Mickle, pp. 1099–1103. Instrument Society of America, Research Triangle Park, NC (1981).

[93] Rapp, P. E. and Berridge, M. J. Oscillations in calcium–cyclic AMP control loops form the basis of pacemaker activity and other high frequency biological rhythms. *J. Theor. Biol.* **66**, 497–525 (1977).

[94] Rapp, P. E., Othmer, H. G. and Monk, P. B. A model for signal-relay and adaptation in *Dictyostelium discoideum*. I. Biological processes and the model network. *Math. Biosciences* in press (1985).

[95] Roberts, E. Disinhibition as an organizing principle in the nervous system. The role of gamma-aminobutyric acid. *Adv. Neurol.* **5**, 127–43 (1974).

[96] Rössler, O. E. and Wegmann, K. Chaos in the Zhabotinskii reaction. *Nature, Lond.* **271**, 89–90 (1978).

[97] Rössler, R., Gotz, F. and Rössler, O. E. Chaos in endocrinology. *Biophys. J.* **25**, 216a (1979).

[98] Roux, J. C., Rossi, A., Bachelart, S. and Vidal, C. Representation of a strange attractor from an experimental study of chemical turbulence. *Phys. Lett.* **77A**, 391–3 (1980).

[99] Schmitz, R. A., Graziani, K. R. and Hudson, J. L. Experimental evidence of chaotic states in the Belousov–Zhabotinskii reaction. *J. Chem. Phys.* **67**, 3040–4 (1977).

[100] Schwartzkroin, P. A. Epilepsy: a result of abnormal pacemaker activity in central nervous system neurons? In *Cellular Pacemakers. Vol. 2. Function in Normal and Diseased States*, ed. D. O. Carpenter, pp. 323–43. Wiley, New York (1982).

[101] Sharp, G. W. G., Wiedenkeller, D. E., Kaelin, D., Siegel, E. G. and Wollheim, C. B. Stimulation of adenylate cyclase and calmodulin in rat islet of Langerhans. *Diabetes* **29**, 74–7 (1980).

[102] Simoyi, R. H., Wolf, A. and Swinney, H. L. One-dimensional dynamics in a multicomponent chemical reaction. *Phys. Rev. Lett.* **49**, 245–8 (1982).

[103] Smith, J. M. and Cohen, R. J. Simple finite-element model accounts for wide range of cardiac dysrhythmias. *Proc. natl. Acad. Sci. USA* **81**, 233–7 (1984).

[104] Sparrow, C. T. Bifurcation and chaotic behavior in simple feedback systems. *J. Theor. Biol.* **83**, 93–105 (1980).

[105] Sparrow, C. T. Chaos in a three-dimensional single loop feedback system with a piecewise linear feedback function. *J. Math. Anal. Applics* **83**, 275–91 (1981).

[106] Steer, M. L. and Levitzki, A. The control of adenylate cyclase by calcium in turkey erythrocyte ghosts. *J. Biol. Chem.* **250**, 2080–4 (1975).

[107] Stent, G. S., Kristan, W. B., Friesen, W. O., Ort, C. A., Poon, M. and Calabrese, R. L. Neuronal generation of the leech swimming movement. *Science, Wash.* **200**, 1348–57 (1978).

[108] Stiles, R. N. and Pozos, R. S. A mechanical-reflex oscillator hypothesis for Parkinsonian hand tremor. *J. Appl. Physiol.* **40**, 990–8 (1976).

[109] Szente, M. and Pongracz, F. Aminopyridine-induced seizure activity. *Monogr. Neural Sci.* **5**, 20–4 (1981).

[110] Takahashi, T. Activation methods. In *Electroencephalography*, eds E. Niedermeyer and F. Lopes da Silva, pp. 179–95. Urban and Schwarzenberg,

Baltimore (1982).

[111] Thompson, S. Aminopyridine block of transient potassium current. *J. Gen. Physiol.* **80**, 1–18 (1982).

[112] Tomita, K. and Tsuda, I. Chaos in the Belousov–Zhabotinsky reaction in a flow system. *Phys. Lett.* **71A**, 489–92 (1979).

[113] Tsien, R. W., Kass, R. S. and Weingart, R. Cellular and subcellular mechanisms of cardiac pacemaker oscillations. *J. Exp. Biol.* **81**, 205–15 (1979).

[114] Tyson, J. J. On the appearance of chaos in a model of the Belousov reaction. *J. Math. Biol.* **5**, 351–62 (1978).

[115] Valverde, I., Vandermeers, A., Anjaneyulu, R. and Malaisse, W. J. Calmodulin activation of adenylate cyclase in pancreatic islets. *Science, Wash.* **206**, 225–6 (1979).

[116] Vidal, C., Bachelart, S. and Rossi, A. Bifurcations en cascade conduisant a la turbulence dans la reaction de Belousov–Zhabotinsky. *J. Physique* **43**, 7–14 (1982).

[117] Waisman, D., Stevens, F. C. and Wang, J. H. The distribution of the Ca-dependent protein activator of cyclic nucleotide phosphodiesterase in invertebrates. *Biochem. Biophys. Res. Commun.* **65**, 975–82 (1975).

[118] Wang, J. H. and Waisman, D. M. Calmodulin and its role in the second messenger system. *Curr. Topics Cell Regul.* **15**, 47–107 (1979).

[119] Weiss, B. and Wallace, T. L. Mechanisms and pharmacological implications of altering calmodulin activity. In *Calcium and Cell Function. Vol. 1. Calmodulin*, ed. W. Y. Cheung, pp. 329–79. Academic Press, New York (1980).

[120] Winfree, A. T. *The Geometry of Biological Time*. Springer, New York (1981).

[121] Wyler, A. R., Fetz, E. E. and Ward, A. A. Injury-induced long-first-interval bursts in cortical neurons. *Exp. Neurol.* **41**, 773–6 (1973).

Part IV

Forced chaos

10

Periodically forced nonlinear oscillators

K. Tomita

Department of Physics, Faculty of Science,
Kyoto University, Kyoto, Japan

10.1 Introduction

It is now widely recognised that chaos is a basic mode of motion underlying almost all natural phenomena, and it is neither exceptional nor peripheral as had been conjectured until quite recently. However, there are a variety of situations under which chaos emerges, and chaos becomes increasingly complex as the dimension of the reference space is increased. To distinguish chaos from a mere kinematical complexity, we have been interested in examining a model of the lowest possible dimensions that leads to definite chaos. A forced nonlinear oscillator clearly suits this purpose in the sense that it provides the simplest nontrivial example exhibiting the full variety of behaviour including chaos. Specific examples may not illustrate all mathematical aspects of chaos in their full generality; however, conceptual clarity and physical insight are to be expected from cases with the lowest dimension.

This means, incidentally, that field theory that involves an infinite number of dimensions is avoided. However, even in this case, tractability may usually be secured after a truncation of modes in the Fourier-transformed space, leaving again only a few degrees of freedom.

Let us now start with a linear dissipative system which is familiar in the neighbourhood of thermodynamic equilibrium. In the simplest case the system has a unique fixed point, which turns out to be a stable node. The possibility of a focus is eliminated through the postulate of microscopic reversibility for thermodynamic equilibrium due to L. Onsager. It follows that a macroscopic clock cannot be expected in thermal equilibrium. Point stability may not be assured in a pathological situation, i.e. in a phase transition when the possibility of a second fixed point appears. In general the stability of one mode is given up for that of another through a phase transition. When there exists an intermediate range, however small, in control parameter space, in which both are stable, the transition is called the *first order*. When such range is vanishing, the transition is called the *second order,* in which simple exchange of stability occurs at the critical point. The state represented by a stable node, i.e. equilibrium phase, stands in general for a *spatial pattern,* which is the only variety expected in thermal equilibrium.

When nonlinearity is introduced into the dynamics, there appear additional new situations. A common feature of the new variety is a large repetitive excursion of the representative point without damping, i.e. an active oscillatory behaviour. In two dimensions it has been proved by Poincaré and by Bendixon [1,16] that the orbit of oscillation asymptotically forms a closed curve, which is often called a *limit cycle*. In this case, point stability is taken over by orbital stability. This self-sustained oscillation has its own natural amplitude, in contrast to the vanishing amplitude of the linear case without external forcing. The transition from a fixed point to a limit cycle, through the change in control parameter, is called a *Hopf bifurcation*. In contrast to the exchange of stability between two fixed points, it is sometimes called *overstability*, with overshooting of the restoring motion.

For dimensions higher than two, it is known that a limit cycle is not a unique alternative for a fixed point. In the new variety of motion, the representative point also comes repeatedly back very close to some of its past locus; however, its orbit is never closed in an accurate manner. In the new variety a sensible definition of the neighbourhood of an orbit is impossible and the orbital stability is lost accordingly. It is, then, associated with a stochastic-like feature such as sensitive dependence on initial conditions, and the prediction of future behaviour becomes impossible in practice. The overall situation, therefore, is often described by the word *chaos,* and the asymptotic orbit is called a *strange attractor*.

It should, however, be remembered that historically the first example of chaos was not the result of a general theory, but came from the practical analysis of a forced nonlinear oscillator, i.e. of the van der Pol type, by Cartwright and Littlewood 40 years ago [2,7]. Levinson took up the phenomenon and succeeded in a clear-cut description of the sporadic behaviour [12]. Some 20 years later, his results gave Smale the incentive to produce a simple intuitive model, which is now called Smale's *horseshoe* [18]. Following Smale's work there finally appeared a general theory [17], which specifies the lowest dimension beyond which the appearance of a strange attractor becomes *generic*.

According to the result of the general theory, the lowest dimension required for chaotic behaviour to be *observed* is three. This implies that a forced oscillation of a simple two-dimensional limit cycle is the simplest possible example exhibiting most of the features expected above. Let us therefore proceed to discuss the behaviour of typical limit cycles under external periodic excitation.

The classical results found for a forced van der Pol oscillator are summarised in section 10.2. In section 10.3 a mathematical interpretation of the classical results, which was elaborated by Mark Levi, is briefly described. This provides us with a certain insight into the general background of chaotic response, typically an embedded Smale's horseshoe structure. In section 10.4, changing the viewpoint slightly, various quantities useful in the physical description of chaos are discussed using an example of the forced Brusselator model. It is pointed out that the global bifurcation structure may be related to the general mechanism. The chapter closes with section 10.5, a discussion that relates in particular to biological applications.

10.2 Forced nonlinear oscillators

10.2.1 *Nonlinear resonance*

Let us start with the noted historic example of the forced van der Pol oscillator [2,7]. It is associated with an electric circuit involving a triode, as shown in Fig. 10.1.

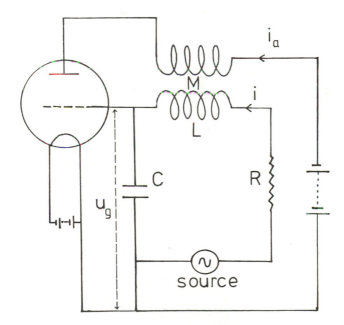

Fig. 10.1 Electrical circuit corresponding to the forced van der Pol oscillator.

According to Kirchhoff's rule, the following differential equation can be written:

(10.1)
$$L\frac{di}{dt} + Ri + u_g - M\frac{di_a}{dt} = P_0 \sin \omega_1 t$$

and

(10.2)
$$C\frac{du_g}{dt} = 1$$

Assuming that the plate current i_a is controlled by the grid voltage u_g through the relation

(10.3)
$$i_a = Su_g(1 - \frac{u_g^2}{3K^2}),$$

and introducing scaled variables by

(10.4)
$$v = u_g/K, \quad \alpha = \frac{MS-RC}{LC}, \quad \gamma = \frac{1}{3}\frac{MS}{LC}$$

$B = P_0/K$ and $\omega_0^2 = 1/LC$, the original equations may be written in the form:

(10.5)
$$\frac{d^2v}{dt^2} + (-\alpha + 3\gamma v^2)\frac{dv}{dt} + \omega_0^2 v = B\omega_0^2 \sin \omega_1 t$$

Setting the solution in the form

(10.6)
$$v(t) = b_1(t) \sin \omega_1 t + b_2(t) \cos \omega_1 t$$

and neglecting higher-order terms in $b_1(t)$ and $b_2(t)$ as small, and introducing new quantities by

(10.7) $a_0^2 = d/(\tfrac{3}{4}\gamma), b^2 = b_1^2 + b_2^2$, and $\Delta = 2(\omega_0 - \omega_1)$

one is led to

(10.8)
$$2\dot{b}_1 + b_2\Delta - \alpha b_1(1 - \frac{b^2}{a_0^2}) = 0$$
$$2\dot{b}_2 - b_1\Delta - \alpha b_2(1 - \frac{b^2}{a_0^2}) = -B\omega_0$$

Rescaling various quantities by $x = b_1/a_0$, $y = b_2/a_0$, $\rho^2 = x^2 + y^2$, $\sigma = \Delta/\alpha$, $F = -B\omega_0/a_0\alpha$ and $\tau = t\alpha/2$, one is finally left with

(10.9)
$$\frac{dx}{dt} = -\sigma y + x(1 - r^2)$$
$$\frac{dy}{dt} = \sigma x + y(1 - r^2) + F$$

Based on this approximate differential equation it is possible to delineate a nonlinear resonance profile and discuss the stability. In Fig. 10.2 the resonance profile is shown as a function of the impressed frequency ω_1.

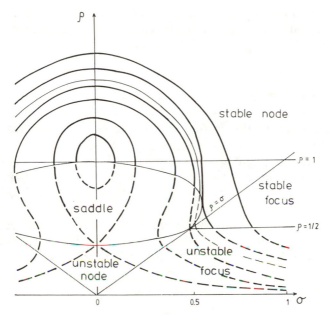

Fig. 10.2 Response profile of the van der Pol oscillator.

Let us discuss the resonance profile of an active nonlinear oscillator in contrast to the more familiar case of a linear oscillator. In the latter case there may appear the effect of induced nonlinearity, e.g. resonance saturation; however, new essentially different features may be pointed out in the former case, which are due to intrinsic nonlinearity. They are associated with new instabilities, described in what follows.

Hard-mode instability (Hopf bifurcation)

In the profile of ordinary linear resonance the wing extends indefinitely off the exact resonance, though the response becomes smaller. This means that a linear oscillator is essentially *passive* in yielding a response. In the present case (cf. Fig. 10.2), however, the entrainment is restricted in a finite frequency range around the exact resonance.

In other words, complete entrainment is observed only above the horizontal line $\rho = 1/2$. The expected wing is bound to become lower than this line, but below the line $\rho = 1/2$ the motion observed is doubly periodic, having frequencies roughly corresponding to the natural frequency ω_0 and the impressed frequency ω_1. On the Poincaré section, in which the entrainment is reduced to a fixed point, a limit cycle now appears instead (Hopf bifurcation). This is an indication of the *active*, or self-sustained, nature of the free oscillation. The tendency that the natural mode presents itself competes with the tendency that the impressed mode entrains the natural mode.

Out in the wing of the response, therefore, the former effect cannot be suppressed, and both ω_1 and ω_0 appear on the stage, leading to quasi-periodicity.

Soft-mode instability (coexistence)

In the case of ordinary resonance it is known that the profile is a unique continuous function of the applied frequency ω_1. In the case of nonlinear resonance, however, the profile may not be unique as a function of the applied frequency ω_1. This corresponds to the fact that in the absence of external excitation there are two distinct modes, i.e. active self-sustained oscillation with finite amplitude and an inactive fixed point, as are indicated in Fig. 10.2 by the points ($\sigma = 0$, $\rho = 1$) and ($\sigma = 0$, $\rho = 0$), respectively.

External excitation tends to mix these two modes, and in fact they are mixed when the amplitude (F) of the impressed excitation is large enough. However, when the amplitude is smaller than a critical value, it fails to mix the two modes, and the critical situation is characterised by the first appearance of a vertical tangent in the response curve.

It is true that the lower branch, which leads to the inactive fixed point of the free system, is mostly unstable owing to the hard-mode instability for $\rho < 1/2$; however, in an intermediate narrow range, i.e. for $1/4 < F^2 < 8/27$, there actually appear two distinct modes depending on the initial condition and the hysteresis phenomenon is observed. The stability limit for each mode is indicated by the ellipse passing ($\sigma = 0$, $\rho = 1.0$) and ($\sigma = 0.5$, $\rho = 0.5$), on which a resonance curve has a vertical tangent, and an exchange of stability (i.e. soft-mode instability) takes place.

10.2.2 *Irregular response*

There is yet a third peculiar aspect in the behaviour of a forced van der Pol oscillator, which was first observed by Cartwright and Littlewood [2, 7] some 40 years ago. This is associated with the situation in which two different modes are coexisting; however, as it involves apparently random behaviour, it cannot be represented in the form of a simple resonance profile.

To facilitate the discussion, let us introduce a simple rescaled version of the original equation [11], i.e.

$$(10.10) \qquad\qquad \varepsilon\ddot{x} + \varphi(x)\dot{x} + \varepsilon x = bp(t)$$

where $p(t)$ is a periodic function with period T, and b and T are considered as control parameters. As is the case in the original equation, the damping $\varphi(x)$ is assumed symmetric around the origin, and satisfies

$$(10.11) \qquad\qquad \varphi(x) \gtrless 0 \quad \text{for} \quad |x| \gtrless 1$$

For small values of $\varepsilon(> 0)$ and $b \neq 0$, it was found that the response has a much longer period than the forcing term, i.e. a larger integer (~ 100) multiple of the forcing period T. In addition, on changing the parameter b (with fixed T), it was seen that two kinds of interval, i.e. type A and type B, alternate. Here type A is associated with a single such mode, and type B is associated with two modes different in period. It was in the interval of type B that an *irregular behaviour* of the response was observed. The two types of interval are separated by small gaps labelled by g_i, in which a complicated series of bifurcation seems to appear. The alternating structure is shown schematically in Fig. 10.3.

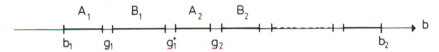

Fig. 10.3 The global structure of alternating two phases, A_k and B_k, shown schematically as a function of control parameter b.

The following are the established behaviours of the solution in different types of interval.

(1) Interval A_k: when $b \in A_k$, eqn (10.10) has a pair of periodic solutions with period $(2n + 1)T$, i.e. a stable node and a saddle, where $n = n(k) \sim \varepsilon^{-1}$ and k specifies the interval A_k.

(2) Interval B_k: in addition to the pair of solutions having period $(2n + 1)T$, there appears a second pair having period $(2n - 1)T$. Moreover, there exists a class of solution which behaves irregularly, as shown in Fig. 10.4.

The solution approaches $x = \pm 1$ in a forced oscillatory fashion, jumps over the interval $-1 < x < 1$ to reach $x = \mp 3$, and then repeats similar behaviours. Let us specify the instant for the consecutive downward crossings with the level $x = 0$ by t_j, then the quantity $\tau_j = t_{j+1} - t_j$, corresponding to the overall period, fluctuates according to the formula

(10.12) $$\tau_j = t_{j+1} - t_j = (2n + \sigma_j)\frac{T}{2}$$

where σ_j may take the value $+ 1$ or $- 1$.

A question arises, then, as to just what kinds of complexity are allowed to appear in the sequence $\sigma = (\ldots\sigma_{-1}\ \sigma_0\ \sigma_1\ldots)$. Unexpectedly it turned out that one cannot restrict or specify the sequence in any way, i.e. σ_is behave as if they are statistically independent of each other. In other words, any irregular sequence corresponds to a possible solution of the original equation. In this sense *randomness*, or *unpredictability*, may be associated with solutions of a deterministic equation, which is certainly remarkable. As a corollary one may expect an infinite number of periodic solutions existing in this case.

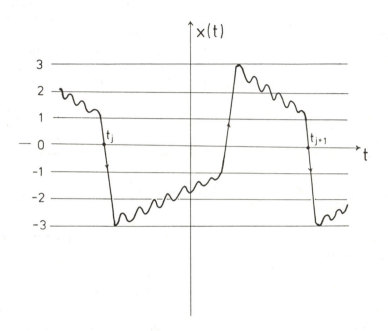

Fig. 10.4 Temporal variation of a forced van der Pol oscillator in region B_k (schematic).

Along with the irregular sporadic behaviour in temporal domain, there appears a strange behaviour in spatial domain. This is the character of the asymptotic invariant manifold in the reference variable space. Owing to the existence of dissipation, the maps on the Poincaré section are generally contracting and the asymptotic invariant manifolds are described as *attractors*. In the more familiar case of a fixed point or a limit cycle, an attractor usually has its own basin of finite size, inside which there is no dependence on initial conditions. In contrast, in the present case of irregular sequences there is nothing like a basin of attraction associated with an orbit, although the maps are contracting on the Poincaré section. An aggregate of such isolated orbits packed into a small space is called a *strange attractor*, because no orbital stability is expected and the future behaviour has a sensitive dependence on initial conditions.

One should, however, be aware that the appearance of a strange attractor is not due to a pathology, for it cannot be destroyed by a small perturbation of the dynamics (structural stability).

Having described empirical facts, it is natural that the theoretical background is sought for the irregular response. Recent attempts along this line are described in the next section.

10.3 Mathematical interpretation

In this section a mathematical analysis and interpretation are attempted of the empirical results on the forced oscillator sketched in the previous section. It should be admitted at the outset, however, that mathematical understanding is not yet complete, and one must be content with a plausible insight. As an insight, however, the *horseshoe* model proposed by Smale, for instance, is very enlightening and convincing. The only difficulty lies in the connection between the actual system and the premises required for the model.

10.3.1 *Forced Van de Pol oscillator*

For the particular case of the forced van der Pol oscillator, Levi [11] and Guckenheimer [5] attempted to bridge the gap in a plausible way.

It is known that a nonlinear map in a one-dimensional interval may easily become noninvertible, which leads to mixing or stochastic-like behaviour [14,15]. On the other hand, the forced van der Pol oscillator, of which the original dimension is three, may be reduced to a map in two dimensions by taking a stroboscopic representation. Is it possible to associate the resulting two-dimensional map with any one-dimensional map, through any deformation or contraction? It may not always be possible; however, in the present example one can do pretty well in making a bridge between the two, and it will be a good help in understanding *deterministic chaos*.

To start with, let us rewrite the original equation (10.10) in the following form [11]:

$$\dot{x} = \frac{1}{\varepsilon}(y - \Phi(x))$$

(10.13)

$$\dot{y} = -\varepsilon x + bp(t)$$

where $\Phi(x) = \int_0^x \varphi(\xi) d\xi$.

In order to simplify the argument, henceforth ε is assumed to be fairly small and figures are drawn almost as if [11]

$$\varphi \to \varphi_0 \, \text{sgn}\,(x^2 - 1)$$

(10.14)

$$p \to p_0 \, \text{sgn}\,(\sin\frac{2\pi t}{T})$$

In Fig. 10.5 $y = \Phi(x)$ is shown by a thick line. Horizontal arrows attracting the flow towards the outside indicate the fast manifold in the phase plane, but the motion along $y = \Phi(x)$ is much slower; hence it is called the slow manifold. Without the external excitation the solution is a simple relaxation oscillation around the parallelepiped $A \rightarrow \bar{B} \rightarrow \bar{A} \rightarrow B \rightarrow A \rightarrow \ldots$, and the stroboscopic phase portrait at natural period is simply a fixed point.

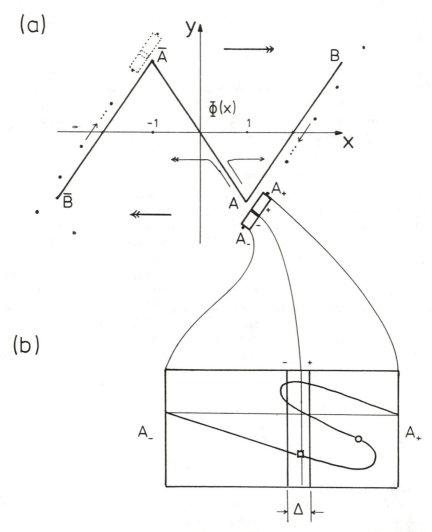

Fig. 10.5 Stroboscopic phase portrait and the conjugate annulus map (schematic). The vertical strip, having width Δ, is mapped into the quasi-horizontal folded string.

On applying a periodic external excitation the period of the strobo-scopic portrait is chosen to equal the forcing period T, which defines map F, and the phase portrait starts to drift along the parallelepiped $A\bar{B}\bar{A}B$

in a circulating manner (dots in Fig. 10.5a). For example, in the range A_k, described in the previous section, it takes $(2n + 1)$ steps to come back to the starting point, which is a fixed point with respect to $(2n + 1)$-fold mapping $F^{(2n+1)}$. On the other hand, in the range B_k, the map does not exactly come back to the starting point after one circulation around $A\overline{B}\overline{A}B$, thus leading to an irregular behaviour which one now wishes to analyse. In order to proceed further a contracted map is introduced with reference to the slow and fast manifolds. Namely, one wishes to contract the integral multiple of period T steps along slow manifolds and concentrate on the way in which the positive domain of x is mapped on to the negative domain, and vice versa, along the fast manifolds.

To do this, let us skip all the stroboscopic image points along slow manifold $BA(\overline{BA})$ except the last two, i.e. A_- and A_+; then they define a narrow two-dimensional strip A_-, A_+ $(\overline{A}_-, \overline{A}_+)$, for the result of global circulation in fact does not coincide each time. On identifying the A_+ $(\overline{A}_+)$ end of the strip with the A_- $(\overline{A}_-)$ end, one finally obtains an annulus $A(\overline{A})$, located around the minimum (maximum) point of $\Phi(x)$. Let us suppose that A and $\overline{A}$ are mapped on to each other through a contracted map N. Finally a map of A on to itself is defined by $M = N:N$.

The gist of this annulus mapping M lies in that the central narrow strip $\Delta(-, +)$ of annulus A is elongated along the annulus direction a great deal, because Δ bridges two different sides of the watershed of $\Phi(x)$, which flows fast in directions opposite to each other. As the circumference of the annulus is constant, however, a certain angular range is multiply covered by the elongated (and squeezed transversally at the same time) image of original Δ. The image of the strip Δ is schematically shown in Fig. 10.5b.

Considering the transverse contraction, one is finally led to a close parallelism between the two-dimensional annulus map M and a simple one-dimensional map f with a modulus shift in the middle, as is indicated in Fig. 10.6.

10.3.2 *Levi's theorem*

After the foregoing preparatory considerations on the parallelism between annulus map M and a one-dimensional map f with shift, one may now state the mechanism for the appearance of irregularity in the form of a general theorem according to Mark Levi [11]. A short illustration of the theorem in physical terms will be given here.

The premises for the theorem are stated in two ways, one in terms of annulus map $M(\rightarrow\text{'A'})$ and the other in terms of a one-dimensional projection $f(\rightarrow\text{'B'})$ of the annulus map.

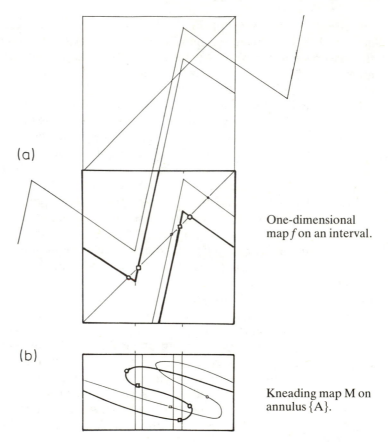

(a)

One-dimensional
map f on an interval.

(b)

Kneading map M on
annulus $\{A\}$.

Fig. 10.6 An annulus map related to an interval map f. Thin line corresponds to region A_k, and thick line corresponds to region B_{k-1}.

'A': *Kneading character of annulus map M*

Suppose a vertical narrow strip V_i in the annulus domain covers the watershed region, it will be mapped into a long horizontal strip H_i with folding by M, i.e.

$$(10.15) \qquad\qquad M(V_i) = H_i$$

as shown in Fig. 10.5.

In addition, it is assumed that through forward mappings the horizontal strip H_i is contracted vertically and nested in itself on each step, and through backward mappings the vertical strip V_i is contracted horizontally and nested in itself on each step. Namely,

$$(10.16) \qquad\qquad \begin{aligned} dM\,(H_i) &\subset H_i \\ dM^{-1}\,(V_i) &\subset V_i \end{aligned}$$

'B': *Stretching and folding in circular map f*

As shown in Fig. 10.6 the circular map f is thought to be obtained from the annulus map M through projection. Corresponding to the overriding on watershed, f should have a steep slope in a narrow range Δ. Δ is stretched to a length satisfying $c < f(\Delta) < 2c$, where c is the circumference of the circular domain. In the remaining range f should be moderately contracting. The overall length of the image is definitely longer than the circumference, and inevitably there appears folding. On changing the control parameter (here the amplitude of the forcing oscillation), the image is simply rotated clockwise. In other words, the map f is shifted upwards with little change in shape, which is structurally stable.

Under the assumptions 'A' and 'B', the main results of the theorem are stated as follows. Namely, for small enough Δ and deviation between M and f, there exists a certain interval $[b_1,b_2]$ which consists of alternating two types of intervals A_k and B_k, separated by short gaps g_k or $g_k{}^+$

$$(10.17) \qquad [b_1,b_2] = A_1 \cup g_1 \cup B_1 \cup g_1^+ \cup \cdots \cup A_n \cup g_n \cup B_n$$

and the behaviour in each interval is expected as follows.

(1) *The case $b \in A_k$.* The map M has exactly one pair of fixed points, a sink and a saddle, and M is an ordinary diffeomorphism (Morse–Smale). This situation corresponds to Fig. 10.5 and to the thin curve in Fig. 10.6

(2) *The case $b \in B_k$.* In this case the map M possesses two pairs of fixed points and an invariant Cantor set S covering the neighbourhood of two saddles, as shown by thick curves in Figure 10.6. The behaviour of map M on S is homeomorphic to that of *left shift* θ on a sequence Σ of numbers specifying different strips corresponding to different saddles, and the former can be described in terms of the latter. As the choice of the sequence is arbitrary, except for the selection rule imposed by the model, the behaviour on the Cantor set S is stochastic-like, as it is considered an embedded Smale's *horseshoe* map. Finally, the measure of S is expected to be zero.

(3) *The case $b \in g_k$ or g_k^+.* In every gap g_k there exists a point $b_k^* \in g_k$ for which the map M has a *homoclinic tangency*. If this tangency is nondegenerate, there are uncountably many values of $b \in g_k$, for each of which M has infinitely many sinks corresponding to periodic motions. The boundary between A_k and g_k corresponds to a homoclinic tangency, but that between g_k and B_k is not understood very well. The total length of the gaps

$$\sum_{k=1}^{n-1} (g_k + g_k^+)$$

is of the order ε and the details in the gaps are hard to see when ε is small.

For the detailed proof and justification of the premises A and B in the case of the forced van der Pol oscillator, the reader should refer to the original paper by Levi [11].

The behaviour of the forced van der Pol oscillator described in the previous section has acquired a mathematical interpretation in a general form. It is now clear that the stochastic-like delocalisation in region B_k is due to Smale's horseshoe structure. Owing to the monotonic contraction of strips, both towards the past and the future (assumption A), S forms a Cantor set with vanishing measure, i.e. at most a countably infinite variety of complexities are involved. However, for physical preparations of the initial condition, this is dense enough and the stochastic-like motions are inevitably observed.

10.4 Physical description of chaos

Up to the previous section our discussion was centred around a single historic example, i.e. the forced van der Pol oscillator in which the existence of irregular motion (chaos) was first discovered. This does not mean, however, that this system is the most convenient system to study. In fact, to the writer's knowledge, there seems to exist no exhaustive phase diagram on this system, including chaotic phases.

In this section, therefore, a number of methods or quantities are introduced, which are of help in physical description and application of chaos, even if the mathematical background is yet to be clarified.

10.4.1 *Phase diagram*

Although the behaviours of individual bifurcations should receive careful study, no less important in physical understanding and applications is the global bird's-eye view of the whole bifurcation structure in the control parameter space, i.e. the *phase diagram*.

In Fig. 10.8 an example of a phase diagram is shown with the forced Brussels model which was investigated by us [19,22]. The model is based on the flow chart given in Fig. 10.7, and is described by the differential equation

(10.18)
$$\frac{dx}{dy} = x^2y - Bx + A - x$$
$$\frac{dy}{dt} = Bx - x^2y$$

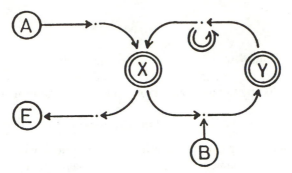

Fig. 10.7 Formal scheme for the Brussels model.

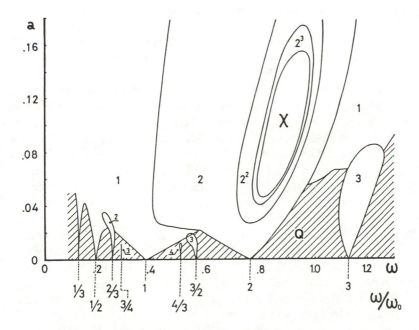

Fig. 10.8 Phase diagram (the forced Brussels model). The number indicates the harmonic periods appearing in the respective regions in the unit of forced periods. A limit cycle of nonintegral period appears in the shaded region Q, and a chaotic response is found in the region χ.

By changing

(10.19) $A \rightarrow A + a \cos \omega t$

the forced oscillation was investigated by simulation, and the result is shown as a world map in the a–ω plane in Fig. 10.8, where the forced frequency is scaled by the natural period under the condition $A = 0.4$ and $B = 1.2$

All the modes of motion described in 10.1 appear in Fig. 10.8.

(a) *Entrainment.* In the stroboscopic representation with the forcing period T, the fundamental entrainment corresponds to a fixed point which is indicated by number 1 in Fig. 10.8. Similarly, in the region marked with the number k, $k = 1,2, \ldots$, there appears k point periodicity. In fact, however, in the region touching the abscissa at $\omega = k\omega_0/l$, there appears entrainment by frequency $l\omega/k$, $l = 1,2, \ldots$.

(b) *Quasi-periodic oscillation.* This corresponds to the shaded region with label Q. Quasi-periodic oscillation appears as a result of a Hopf bifurcation in the stroboscopic representation.

(c) *Period-doubling cascade.* This is the layer-like region indicated by $2^n (n = 2,3, \ldots)$ which does not touch the abscissa and is surrounded by a crust-like layer of subharmonic resonance of the order 1/2. Also it does not include the central core indicated by χ. Although it is not possible to indicate separately, there appear successive regions 2^n, where integer n increases without bound, each being nested in the region one step earlier. The corresponding region of stability converges to a certain limit in a geometrical way, thus leaving the space for χ. As a result of period doubling (without bounds), an oscillation with an arbitrarily long period is expected in the limit, being in touch with the region χ. An analogous period-doubling sequence appears in the case of a one- dimensional map, which was treated by Feigenbaum [3,4] by the renormalisation technique. A similar theory seems to apply to the present case, and the observed shrinking ratio of the consecutive parameter range is consistent with Feigenbaum's ratio 4.669. . . .

(d) *Chaos—aperiodic recurrence.* This appears in the core region labelled by χ. An example of the phase portrait is shown in Fig. 10.9a, and its fine detail is magnified in Fig. 10.9b. In the latter figure, the strobo-scopic portrait is also shown. It bears a basic four-point periodicity; however, the result of iteration emerges as four islands, i.e. ①, ②, ③ and ④, instead of four points, indicating the chaotic divergence in recurrence. In fact in four steps indicated by ① → ② → ③ → ④, the island is stretched and twisted roughly by π (① → ②), folded (② → ③), twisted by π again (③ → ④), and finally squeezed into the original island (④ → ①). This indicates clearly the process of kneading, supposed in foregoing sections.

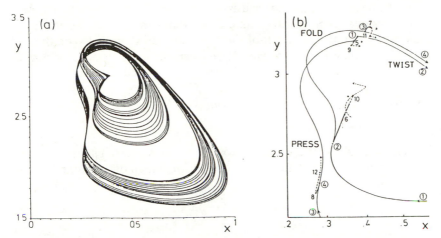

Fig. 10.9 Phase portrait of a chaos ($a = 0.05$, $\omega = 0.80$). (a) Orbital profile of a chaotic response. (b) Extraction of the stroboscopic representation. (Four islands indicated by thick curves correspond to the asymptotic invariant manifolds in the stroboscopic representation. Thin curves visiting these islands in the order of attached numbers indicate the behaviour of the system between the stroboscopic illumination.)

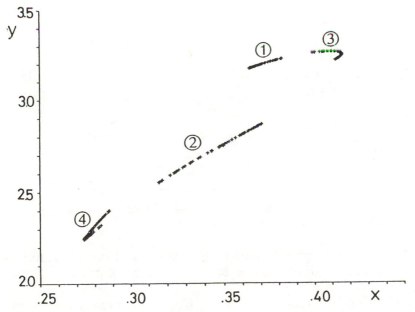

Fig. 10.10 Stroboscopic phase portrait of a chaos ($a = 0.05$, $\omega = 0.80$), which corresponds to the thick curve in Fig. 10.9b.

The stroboscopic phase portrait is separately shown in Fig. 10.10. As a two-dimensional section it is remarkably thin, and looks like an inter-rupted single curve, as a result of the big contrast in rate between the fast and the slow manifolds. This result suggests strongly the use of a one-dimensional interval map in place of a two-dimensional mapping.

10.4.2 *Use of one-dimensional terminology*

(a) *Transfer function [19,22]*

Choosing the island ②, which almost resembles a straight line, as the basic interval, one may define a one-dimensional map of this interval on to

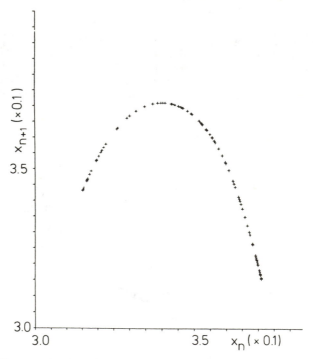

Fig. 10.11 The transfer function $f(x)$ which governs the mapping of island ②. This particular plot is obtain for a chaos ($\omega = 0.80$).

itself. (Actually the projection on to the x-axis was used.) In Fig. 10.11 an example is shown of the transfer function $f(x)$, thus obtained, which has the following features.

First, the shape of $f(x)$ is surprisingly simple, and almost quadratic in x. Secondly, the supporting interval is well defined ($x_m < x < x_M$). This means that the interval $[x_m, x_M]$ is repeatedly mapped on to itself, being stretched and folded each time. Thirdly, the shape of $f(x)$ is almost structurally invariant, including purely periodic phases alternating with chaotic ones. Based on one-dimensional transfer function $f(x)$, several useful concepts or quantities may be introduced in what follows.

(b) *Rate of expansion λ (x) [19,22]*

An obvious point of concern is the rate of local deformation which is defined by

$$(10.20) \qquad \lambda(x) = \left| \frac{df}{dx} \right|$$

$\lambda(x) > 1$ corresponds to expansion, and $\lambda(x) < 1$ to contraction. If one chooses x at the position of a fixed point, it is clear that $\lambda(x) > 1$ corresponds to instability and $\lambda(x) < 1$ to stability of the fixed point.

For a k-fold map $f^{(k)}$ one may define $\lambda^{(k)}$, which is related to λ ($= \lambda^{(1)}$) by the *chain rule:*

$$(10.21) \qquad \lambda^{(k)}(x) = \left| \frac{d}{dx} f(f^{(k-1)}(x)) \right| = \prod_{i=1}^{k-1} \left| \left(\frac{df}{dy} \right)_{y = f^{(i)}(x)} \right|$$

The stability of a k-periodic point $x_i^{(k)}$ is assured only in the range

$$(10.22) \qquad +1 \geq \lambda^{(k)}(x_i^{(k)}) \geq -1$$

The left-hand equality corresponds to the condition for a *tangent* bifurcation, where a pair of periodic points come into existence, one stable and the other unstable. The right-hand equality corresponds to a *pitchfork* bifurcation or a *period doubling,* where the periodic point loses its stability without losing its existence. When the chain rule is applied to this particular situation, one finds

$$(10.23) \qquad \lambda^{(2k)}(x_i^{(k)}) = \{\lambda^{(k)}(x_i^{(k)})\}^2 = 1$$

therefore a mode having a double period emerges through a tangent bifurcation at this point, and the number of periodic points is doubled, hence the term *pitchfork* bifurcation. It is then not very difficult to imagine that the period doubling occurs repeatedly when the control parameter is changed in a definite direction. Feigenbaum [3,4] found that the range for the stability is shortened in a geometrical way with a *universal* ratio, i.e. 1/4.669.... The stability range for a basic frequency plus a period-doubling cascade, derived from it, was called a *window* by May [14,15] who observed many windows and also the convergence limits of doubling.

(c)   *Lyapunov number and invariant measure [10,19,22]*

The rate $\lambda\,(x)$ of expansion may lead to another convenient concept, the *Lyapunov number,* which is defined by a time average of $\lambda\,(x)$, i.e.

(10.24)
$$\Lambda\,(x_0) = \lim_{n\to\infty} \frac{1}{n} \log \lambda^{(n)}(x_0)$$
$$= \lim_{n\to\infty} \frac{1}{n} \sum_{k=0}^{n-1} \log \lambda\,(x{=}x_k)$$

where $x_k{=}f^{(k)}\,(x_0)$.

When the motion in the long run is *ergodic,* then it is expected that the average $\Lambda\,(x_0)$ does not depend on the initial point x_0, and a constant Λ characterises the invariant manifold as a whole. For the case in which $a = 0.05$ and $\omega = 0.81$ (chaos), the Lyapunov number is computed to be $\Lambda = 0.536$.

Under the same situation a measure supported by the interval may be defined by

(10.25)
$$\mu_{x_0}(x) = \lim_{N\to\infty} \frac{1}{N} \sum_{i=0}^{N-1} \delta[x{-}f^{(i)}(x_0)].$$

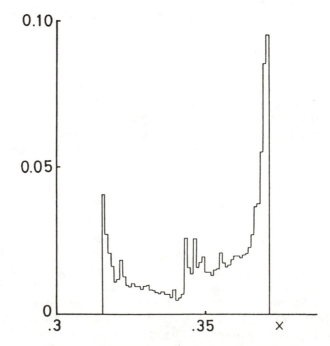

Fig. 10.12 Invariant measure based on histogram construction $(a = 0.05,$ $\omega = 0.81)$.

When this quantity is independent of x_0, it is called an invariant measure, which is associated with every locality of the invariant manifold. An example of invariant measure is shown in Fig. 10.12 which corresponds to the above Lyapunov number.

(d) Entropy (Kolmogorov-Sinai) and variation principle [10]

The invariant measure $\mu\,(x)$, found empirically in the previous section, may alternatively be obtained by a variation principle. Let us, for this purpose, introduce the entropy associated with a partition. Suppose a map f with a maximum is given (cf. Fig. 10.11): one may introduce a partition which divides the interval with reference to the position of the maximum. Iterating the map f many times, then, the partition $\alpha = \{A_i\}$ ($i = 1, 2, \ldots n(\alpha)$) becomes increasingly fine grained. Associated with the partition α one may introduce an associated entropy H by

$$(10.26) \qquad H_\mu(\alpha) = - \sum_{i=1}^{n(\alpha)} \mu\,(A_i) \log \mu\,(A_i)$$

where $\mu(A_i)$ is the invariant measure associated with interval A_i. One may further introduce an nth refinement $\alpha^{(n)}$ of the partition α, i.e.

$$(10.27) \qquad \alpha^{(n)} = \sum_{i=0}^{n-1} f^{(-i)}(\alpha)$$

and define an entropy per unit step of refinement by

$$(10.28) \qquad h_\mu(f,\alpha) = \lim_{n \to \infty} \frac{1}{n} H_\mu(\alpha^{(n)})$$

when $A_i^{(n)}$ includes at most one fixed point. If the size of every lap tends to zero with increasing n, it is shown that

$$(10.29) \qquad h_\mu(f,\alpha) = h_\mu(f) \qquad \text{(Kolmogorov–Sinai)}$$

In a number of typical cases it may be shown that the invariant measure $\mu(x)$ may be found as a result of a variation principle. Namely, choose the variational function as

$$(10.30) \qquad \Psi_v = h_v(f) - \int v(\mathrm{d}x) \log \lambda\,(x)$$

and maximise Ψ_v with respect to $v(x)$.

As a result, then, one finds $v(x) \to \mu(x)$, and the corresponding maximum value of Ψ_v is zero. The last statement indicates that the Kolmogorov–Sinai entropy of the invariant measure is closely related to the Lyapunov number of the dynamics.

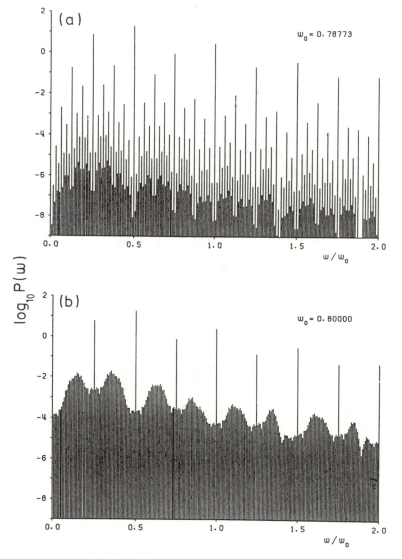

Fig. 10.13   Power spectral density $\phi^{(2)}(l\omega)$ (a = 0.05) (a) ω = 0.787 73
(period = 2^7); (b) ω = 0.800 00 (chaos).

(e) *Correlation spectrum* [9]

As an independent quantity useful in identifying chaos the time correlation
and its Fourier transform, i.e. spectrum, are often invoked. The spectrum
is generally defined by

(10.31) $$\varphi^{(k)}(l\omega) = \left| Y(l\omega) \right|^k = \left| \frac{\omega}{2\pi} \int_0^{2\pi/\omega} y(t) e^{-jl\omega t} dt \right|^k$$

Examples in the case of the forced Brusselator model are shown in Fig. 10.13. In contrast to the collection of line spectra (a) which are associated with periodic motions, chaos is characterised by the existence of a continuous spectrum (b), even if several separate peaks are recognised at the same time.

10.4.3 *Global window structure*

In the previous section, i.e. in subsection 10.4.2(b), a single window based on a certain periodicity was described in terms of $\lambda(x_i^{(k)})$. In fact a window is embedded in a *wall*, which corresponds to the region for chaos.

Coming now back to the phase diagram (Fig. 10.8) of the forced Brusselator model, the region χ assigned to chaos is not a single continuum and a systematic window structure was found in it as shown in Fig. 10.14 [10].

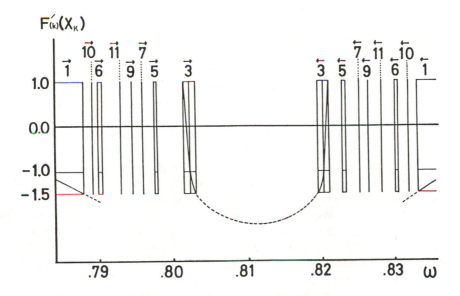

Fig. 10.14 The global window structure. Characteristic multiplier i.e. the rate of expansion, $(F'_{(k)}(X_{(k)})$ is plotted against the input frequency (a control parameter). The window specified by k corresponds to a region of stability for a series of harmonic bifurcations based on the k-point periodicity with respect to a mapping of island ② on to itself. The arrow points to its conjugate window having the same basic periodicity. (k-point periodicity appears in a parenthetic pair in the present case).

Examining the structure of Fig. 10.14, one becomes aware that the appearance of alternating periodic and chaotic regions resembles closely the structure of Fig. 10.3, which was found for the forced van der Pol oscillator in section 10.2.2. The only difference lies in the fact that in Fig. 10.3 the forced amplitude is swept, keeping the frequency constant, whereas in Fig. 10.12 the forced frequency is swept, keeping the amplitude constant. This resemblance in structure strongly suggests that the underlying mechanisms are rather similar, namely, that the mathematical interpretation in section 1.3 may also apply to the present example.

Alternating periodic and chaotic structure in the phase diagram is known to appear also in the case of autonomous systems, e.g. in the Lorenz system, the Zhabotinskii reaction, etc. Recalling the fact that a nonautonomous system can mathematically be transformed into an autonomous system with increased dimension, it is natural to expect that the theoretical considerations, given in the foregoing sections, may also apply to the case of autonomous systems.

10.5 Discussion

Although theorists have long avoided a thorough study, nonlinear oscillations have many examples at various levels of nature. Stellar rotations and orbital revolutions are the oldest examples, in the investigation of which Poincaré invented various concepts and tools in dynamical system theory. At the other extreme, the characteristic stability of living organisms is often associated with an underlying biorhythm, which is another example of a self-sustained oscillation. When a small oscillator (e.g. a living organism) is placed under the influence of a large oscillator (e.g. the Earth's rotation), the coupling between the two oscillators is duly approximated by a forced oscillation, provided the change induced in the motion of the large oscillator is neglected. This implies that there are numerous cases in our surroundings to which the foregoing theoretical considerations apply. A few examples of recent topics might be mentioned. In the field of biochemistry, chaotic behaviour in yeast glycolysis[13] has been observed under periodic substrate input flux. In the field of medicine the generation of cardiac dysrhythmias[6] was related to a chaos in a forced oscillator model. The phase resetting in insect population dynamics[8,23] seems also to belong to the same category, and irregular behaviours may well be related to chaos.

In view of the contents of other chapters of this section of the book, in this chapter I have tried to concentrate on the theoretical understanding and the way of describing *irregular response*, or chaos, in a rather abstract way, though I am much interested in the implications of chaos in practical contexts. Having practical applications in mind, a few comments are due here.

First, a conceptual proposal[20]: it seems a good idea to discriminate 'chaos' from 'randomness' in the following sense. Kolmogorov gave an unambiguous definition to 'randomness', which belongs to a completely tractable range with respect to a certain processor. In contrast, it is proposed here that the concept 'chaos' should be reserved as belonging strictly to the unprocessible range of a given processor. The reader will find that this discrimination will be a great help in avoiding the confusion hitherto existing. It should be noted that in this way 'randomness' and 'chaos' are both defined with respect to a given processor, but they belong to different ranges which are never overlapping.

Secondly, a question [21]: is *chaos* an exceptional or pathological phenomenon in nature? Traditionally chaos has been considered a rather peripheral phenomenon from the point of view of exact science. In fact, however, it is neither a special nor pathological phenomenon, and forms a matrix for all the regularities found in nature hitherto. In practice, irregularity is often treated as noise, or disease, which should be avoided or suppressed as much as possible. The existence of such cases are admitted; however, along the previously stated line of thought, chaos is not just a pathological phenomenon but can be equally or more important in physiology. The characteristic flexibility in response of living organisms to external excitation will never be fully understood without invoking chaos as a fundamental mode of motion.

Acknowledgement

It is our great pleasure to thank Professor Chihiro Hayashi for the stimulating discussions we have enjoyed with him while preparing this chapter. He also kindly showed us recent results of his study on the route to chaos found in the forced Duffing–van der Pol system.

References

Description of forced nonlinear oscillations may be found in *standard textbooks,* such as the following:

Hayashi, C. *Nonlinear Oscillations in Physical Systems.* McGraw-Hill, New York (1964).
Minorsky, N. *Nonlinear Oscillations.* Van Nostrand, London (1962).
Stoker, J. *Nonlinear Vibrations.* Wiley Interscience, New York (1950).

Although irregular responses had been noticed, they were not given serious attention in standard texts.

[1] Bendixson, I. Sur les courbes défines par les équations différentielles. *Acta Math.* **24,** 1–88 (1901).
[2] Cartwright, M.L. Forced oscillations in nearly sinusoidal systems. *J. Inst. Elect. Engng.* **95,** 223 (1948).
[3] Feigenbaum, M.J. The onset spectrum of turbulence. *Phys. Lett.* **74A,** 375–8 (1979).
[4] Feigenbaum, M.J. The transition to aperiodic behaviour in turbulent systems. *Comm. Math. Phys.* **77,** 65–86 (1980).

[5] Guckenheimer, J. Symbolic dynamics and relaxation oscillations. *Physica 1–D,* 227–35 (1980).
[6] Guevara, M.R. and Glass, L. Phase-locking, period doubling bifurcations and chaos in a mathematical model of a periodically driven oscillator: a theory for the entrainment of biological oscillators and the generation of cardiac dysrhythmias. *J. Math. Biol.* **14,** 1–23 (1982).
[7] Holmes, P.J. and Rand, D. Bifurcations of the forced van der Pol oscillator. *Quart. Applied Math.* **35,** 495–509 (1978).
[8] Hoppenstaedt, F.C. and Keener, J.P. Phaselocking of biological clocks. *J. Math. Biol.* **15,** 339–49 (1982).
[9] Kai, T. Universality of power spectrum of a dynamical system with an infinite sequence of period doubling bifurcations. *Phys. Lett.* **86A,** 263–6 (1981).
[10] Kai, T. and Tomita, K. Statistical mechanics of deterministic chaos. *Progr. Theor. Phys.* **64,** 1532–50 (1980).
[11] Levi, M. Qualitative analysis of periodically forced relaxation oscillations. *Mem. Amer. Math. Soc.* **32** (244), 147 pp. (1981).
[12] Levinson, N. A second order differential equation with singular solutions. *Ann. Math.* **50,** 127–53 (1949).
[13] Marcus, M., Kuschmitz, D. and Hess, B. Chaotic dynamics in yeast glycolysis under periodic substrate input flux. *FEBS Lett.* **172,** 235–8 (1984).
[14] May, R. Biological populations obeying difference equations: stable points, stable cycles and chaos. *J. Theor. Biol.* **51,** 511–24 (1975).
[15] May, R. Simple mathematical models with very complicated dynamics. *Nature, Lond.* **261,** 459–67 (1976).
[16] Poincaré, H. Memoire sur les courbes définies par les équations différentielles. I–IV, Oeuvre I, Gauthier-Villas, Paris (1880–90).
[17] Ruelle, D. and Takens, F. On the nature of turbulence. *Comm. Math. Phys.* **20,** 167–92 (1971), added note *Comm. Math. Phys.* **23,** 343–4 (1971).
[18] Smale, S. Differential dynamical systems. *Bull. Am. Math. Soc.* **73,** 747–817 (1967).
[19] Tomita, K. Chaotic response of a nonlinear oscillator. *Phys. Rep.* **86,** 114–67 (1982).
[20] Tomita, K. Coarse graining revisited—the case for macroscopic chaos. In *Chaos and Statistical Methods,* ed. Y. Kuramoto, pp. 2–13. Springer, Berlin (1984).
[21] Tomita, K. The significance of the concept 'Chaos'. *Progr. Theor. Phys. Suppl. No. 79,* 1–25 (1985).
[22] Tomita, K. and Kai, T. Chaotic response of a nonlinear oscillator. *J. Stat. Phys.* **21,** 65 (1979).
[23] Winfree, A. T. *The Geometry of Biological Time.* Springer, Berlin (1980).

11
Chaotic cardiac rhythms

Leon Glass[1], Alvin Shrier[1] and Jacques Bélair[2]

[1]*Department of Physiology, McGill University,*
3655 Drummond Street, Montreal, Quebec, Canada H3G 1Y6, and
[2]*Département de Mathématiques et de Statistique, Université de Montréal,*
Montréal, Quebec, Canada H3C 3J7

> The term chaotic heart action should be reserved to denote irregular heart action caused by frequent premature systoles of multiple points of origin, occurring singly or in short paroxysms, and especially when the ectopic foci are not confined to the ventricles and auricles alone, but come from both chambers or, in addition from the A-V node. The significance of this condition lies in its always being taken as a prelude to sudden death.
>
> *L.N. Katz* [36]

11.1 Introduction

'Chaos' is a term that is used to denote dynamics in deterministic mathematical equations in which the temporal evolution is aperiodic in time and sensitive to the initial conditions. But, as the above quotation illustrates, in electrocardiography 'chaotic' dynamics have been recognised for a much longer period of time than the recent surge of interest among mathematicians and physicists. Of course, the use of the term 'chaotic' by cardiologists is purely descriptive, and does not reflect a detailed theoretical analysis of the underlying mechanisms. In this chapter we describe experimental studies on the effects of periodic electrical stimulation of spontaneously beating cells from embryonic chick heart. The irregular dynamics observed in this experimental system are interpreted using theoretical results from the theory of bifurcations of finite difference equations. The close correspondence between many of the rhythms observed in the experimental system, and abnormal rhythms observed clinically in electrocardiography, leads us to speculate that clinically observed 'chaotic' rhythms may indeed be associated with deterministic 'chaos' in mathematical models of the intact human heart.

In the normal heart the cardiac rhythm consists of a regular periodic contraction of the atria (auricles) followed a short time (0.08–0.12 s) later by contraction of the ventricles. The normal cardiac rhythm (frequency at rest of 60–100 min^{-1}) is set by a pacemaking site in the right atrium in a small region of specialised tissue called the sinoatrial (SA) node. From the

SA node the cardiac activity spreads sequentially through the atrial musculature, the atrioventricular (AV) node separating the atria and ventricles, and then through specialised conduction tissue to the ventricles. The electrochemical events associated with the heartbeat can be monitored noninvasively on the electrocardiogram (ECG), which is a record of potential differences between different points on the surface of the body. On the ECG well-characterised deflections or waves are associated with the excitation of the atria and the ventricles. The morphology and timing of these waves reflect the location of the sites of initiation of the atrial and ventricular excitation and the conduction pathways of the electrical activity of the heart. In a great many pathological conditions, abnormal cardiac rhythms (called either arrhythmias or dysrhythmias) are observed and classified from an analysis of alterations in the timing and morphology of waves on the ECG [8].

The arrhythmia described above as 'chaotic heart action' is believed to arise as a consequence of multiple pacemaking foci (called ectopic foci) located in different regions of the heart. The interactions of spreading electrical activity originating from such foci lead to frequent premature systoles (ventricular contractions). The appearance of the ECG in such circumstances is extremely complex and irregular. Contemporary workers still sometimes use the term 'chaotic heart rhythm' to designate multifocal atrial and ventricular rhythms [8,49]. The grave prognosis for patients who display these complex rhythms is amply supported by clinical data. Recordings of the ECG in which ventricular arrhythmias display multiform extrasystoles (which may arise from multiple foci) are frequently observed prior to sudden death [46]. A theoretical model of the postulated mechanism for 'chaotic heart action' would necessarily assume several independent pacemakers (the ectopic foci) in an excitable medium (the cardiac muscle) of complex geometry. The temporal evolution of such a system could then be analysed. At the current time we do not attempt to analyse this difficult problem.

We study a greatly oversimplified model system consisting of an electronic stimulator coupled unidirectionally to an independent cardiac oscillator. In section 11.2 we sketch out the main ideas of the theory. In section 11.3 we describe the experimental system and present experimental evidence for chaotic dynamics. The results are discussed in section 11.4.

11.2 Theory

11.2.1 *Periodic forcing and the bifurcations of circle maps*
In an early paper, van der Pol and van der Mark [55] proposed that the cardiac rhythm could be modelled by coupled nonlinear oscillators. By changing relative frequencies of the oscillators it was possible to reproduce many different cardiac arrhythmias. It is, however, extremely difficult to analyse mathematically the original equations proposed by van der Pol and

van der Mark, and subsequent workers have carried out computer simulations to determine the dynamics [35,52]. Indeed, theoretical analysis of the sinusoidally forced van der Pol equation is a difficult problem which lies at the heart of much current research in nonlinear mathematics [40]. The difficulties inherent in the analysis of sinusoidal forcing have led us to consider the effects of brief, pulsatile stimuli delivered periodically to the spontaneous cardiac oscillator. Under certain well-defined approximations, which we outline below, the dynamics in these situations can be analysed by consideration of circle maps.

Assume that the dynamics of a cardiac pacemaker can be represented by an ordinary differential equation $dy/dt = f(y)$ where $y \in \mathbb{R}^n$, and f represents nonlinear functions describing evolution in time (e.g. the McAllister *et al* equations [41]). The cardiac oscillation is given by a stable limit cycle oscillation with period T_0. Let $y(0)$ be a point on the cycle (typically taken as the depolarisation or upstroke of the action potential) and define the phase, ϕ, ($0 \leq \phi < 1$) of any point $y(t)$ on the cycle to be t/T_0 (mod 1). The locus of all points which asymptotically approach the limit cycle in the limit $t \rightarrow \infty$ is called the basin of attraction of the limit cycle. Let the trajectories of two points in the basin of attraction of the limit cycle be given by $y(t)$, $y'(t)$. Then if $\lim d[y(t), y'(t)] = 0$, where d is the Euclidean distance, $y(0)$ and $y'(0)$ have the same eventual phase. A locus of points with the same eventual phase is called an isochron [18,19,20,37,56].

The effect of a stimulus delivered at some phase ϕ is to shift the oscillator to a new point in phase space with eventual phase ϕ';

$$(11.1) \qquad \phi' = g(\phi)$$

where g is called the phase transition curve (PTC). The PTC can be experimentally measured. A stimulus is delivered to the system at phase ϕ, resulting in a perturbed cycle of length T. Then, provided the return to the limit cycle is very rapid,

$$(11.2) \qquad g(\phi) = 1 + \phi - T/T_0$$

In situations in which the return to the limit cycle is not rapid or the oscillator is switched out of its basin of attraction, eqn (11.2) does not apply [18,37].

Now consider the effect of a periodic train of stimuli with an inter-stimulus interval t_s. Assume that following each stimulus there is a rapid relaxation back to the limit cycle, and that the stimuli do not change the intrinsic properties of the system. Then if ϕ_i is the phase of the oscillator before the ith stimulus

$$(11.3) \qquad \phi_{i+1} = f(\phi_i) = g(\phi_i) + \tau$$

where $\tau = t_s/T_0$. Equation (11.3) determines a circle map often called the first return or Poincaré map.

Equations (11.2) and (11.3) are the basic equations for theoretical studies. These equations were first derived by Perkel *et al.* [48] and have been used subsequently by many workers [2,17,19,24,25,29,30,44,53, 59,60], but notation differs between different groups. The PTC can be experimentally determined from eqn (11.2), and then the Poincaré map of eqn (11.3) can be numerically iterated to compute expected dynamics. The stable fixed points of eqn (11.3) correspond to stable phase-locked dynamics in which there is a stable repeating pattern consisting of N stimuli and M cycles of the cardiac oscillator ($N:M$ phase locking). Since the PTC is a function which maps the unit circle into itself, analysis of the bifurcations of eqn (11.3) requires a knowledge of the bifurcations of two-parameter circle maps. One parameter, τ, corresponds physically to the period of the periodic stimulus, and the second parameter reflects the strength of the perturbing current passed through the microelectrode (implicit in the PTC). There has been recent interest in the bifurcations of two-parameter circle maps [3,5–7,13,15,17,33,34,47]. We summarise some of the main findings of this work with particular emphasis on points related to the experimental observation of 'chaotic' dynamics.

Consider the map given in eqn (11.3). Defining

$$(11.4) \qquad\qquad \Delta\phi_i = g(\phi_i) + \tau - \phi_i$$

then the rotation number ρ is given by

$$(11.5) \qquad\qquad \rho = \lim_{n \to \infty} \sup \sum_{i=1}^{n} \Delta\phi_i$$

Clearly, ρ is rational for cycles of the finite difference equation, and hence is rational for stable phase locking. The topological degree of the PTC counts the number of times g winds around the unit circle as ϕ winds around the unit circle once. Winfree [56] has given examples to show that the PTC is generally of degree 1 for low-amplitude stimuli and is of degree 0 for higher stimulus amplitude. For monotonic degree 1 circle maps, quasi-periodic dynamics arise for irrational ρ. If a critical point of the Poincaré map is a point on a cycle, the cycle will be stable and is called a superstable cycle.

To make matters concrete, we display theoretically computed phase-locking zones using a PTC which was fit to experimental data [20]. The PTC is given by

$$(11.6) \qquad g(\phi) = \phi - C \exp\left[-\frac{(\phi - \phi_{max})^2}{\sigma^2} \right] - \frac{S(\phi-1)\phi^n}{\theta^n + \phi^n}$$

where

$$(11.7) \qquad \begin{array}{l} C = 0.125 + 0.025\overline{A}, \quad \phi_{max} = 0.34 + 0.12 \times 2^{-\overline{A}} \\ \sigma^2 = 0.04 \times 2^{-\overline{A}}, \quad \theta = 0.34 + 0.48 \times 2^{-\overline{A}} \\ n = 1.875 \times 2^{\overline{A}}, \quad S = 0.92 \end{array}$$

The parameter $\overline{A}$ is given by $\overline{A} = 50A$, where A (in arbitrary units) is related to the amplitude of the current pulse stimulus. The PTC is of topological degree 1, monotonic for $A < 0.039$ and nonmonotonic with a single maximum and minimum outside of this range. Figure 11.1 displays

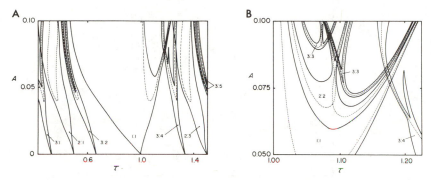

Fig. 11.1 (A) Principal phase-locking regions (enclosed by solid lines) and associated superstable cycles (dashed lines) numercially computed from eq. (11.6). For values of A between 0 and 0.02, ϕ_{max}, σ^2, θ and n were held at their value for $A = 0.02$, and C and S were linearly interpolated between 0 and the values of these parameters at $A = 0.02$. (B) Enlargement of a portion of (A) to show period-doubling and higher-order bifurcations with rotation number $\rho = 1$. Note the 1:1, 2:2, and 3:3 phase locking zones associated with period -1, -2 and -3 superstable cycles. From ref. [20] with permission.

boundaries of theoretically computed phase-locking zones with the dashed lines giving the superstable cycles. Figure 11.2 gives the superstable cycles up to period 4 over a still narrower region of parameter space. Many of the features in Figs 11.1 and 11.2 are well understood and can be expected to arise in the bifurcations of any circle map of degree 1 as there is a change from monotonicity to nonmonotonicity.

In the monotonic region, the Poincaré map is a one-to-one invertible map of the circle. For this case for A and τ fixed, ρ is independent of the initial condition. For fixed A as τ varies, the rotation number is a Cantor function piecewise constant on the rationals. The zones of stable phase locking are nonoverlapping cusp-shaped regions (called Arnold tongues) extending to $A = 0$. Arnold tongues exist for all rational rotation numbers. Such zones form an open dense set in parameter space (i.e. they are structurally stable), but the irrational rotation numbers exist on a set of positive Lebesgue measure. All of the above properties will be found in general for two-parameter invertible maps of the circle [3].

For the region in which the PTC is nonmonotonic, the structure of the phase-locking zones is much more complex and not as well understood. In the nonmonotonic region it is known that, for A fixed, as τ varies one must find at least two distinct values of τ at which there exist two

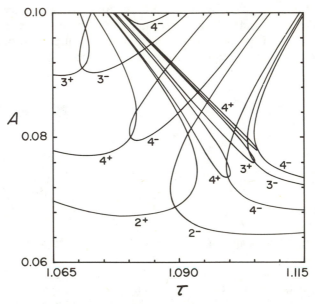

Fig. 11.2 Superstable cycles with rotation number $\rho = 1$. From ref. [20] with permission.

superstable cycles for each rational rotation number (see the dashed lines in Fig. 11.1) [17]. Consequently, the Arnold tongues must extend into the region in which the PTC is nonmonotonic. For A and τ fixed in the nonmonotonic region, the rotation number may depend on the initial condition. Such a situation arises, for example, at the point of intersection of the superstable cycles associated with 1:1 and 3:4 phase locking in Fig. 11.1. At this point the maximum of the PTC is on a cycle of period 1 ($\rho = 1$) and the minimum is on a cycle of period 3 ($\rho = 4/3$). In general, if the maximum is attracted to a cycle with rotation number ρ, and the minimum is attracted to a cycle with rotation number ρ', then some initial condition can be found for which the rotation number is any value in the closed interval $[\rho', \rho]$ [22,45]. There can also be two different stable cycles both having the same rotation number. This situation arises for example at the upper intersection point of the 2:2 superstable cycles (at $\tau = 1.09$, $A = 0.08$) or at the intersection points of the period 2 with the period-4 and period-3 superstable cycles in Fig. 11.2. Another prominent feature is the complex sequence of bifurcations found in the two-parameter space in some of the Arnold tongues (Fig. 11.2). A topologically equivalent bifurcation structure has been found in other maps of the circle and interval in which there are two parameters and two extrema [5,6,14, 17]. Thus, we have conjectured that the bifurcations represent a 'universal' structure, that is, a generalisation of the U-sequence found in one-dimensional maps with one parameter and one extremum [42]. We believe that the reason this complex structure does not arise in all the Arnold tongues in this map reflects the slow growth of the extrema of the PTC in the parameter range considered. In other situations in which extrema grow linearly with parameter increase, this structure is apparently found in all the Arnold tongues [6,7,17].

11.2.2 *Routes to chaos in circle maps*

One of the main interests of current theoretical research in nonlinear dynamics is in characterising the nature of the transition from regular periodic dynamics to irregular, turbulent or chaotic dynamics [12]. Several characteristic types of transition have been observed mathematically and in physical and biological systems as parameters are changed. Many of these main 'routes to chaos' can be observed in different regions of the frequency –amplitude parameter space of Fig. 11.1. We briefly describe three routes to chaos, intermittency, period-doubling bifurcations, and quasi-periodicity (overlapping of resonances) and show that all three routes can be found in the two-parameter space of circle maps.

(a) *Intermittency* arises as a result of tangent bifurcations (when $\partial f^n/\partial \phi|_{\phi^*} = 1$, where ϕ^* is a point of period n). In the region in which the PTC is monotonic, the only bifurcations that are found are tangent bifurcations. The boundaries of all of the Arnold tongues are associated with tangent bifurcation in this monotonic region [34]. As well, tangent bifurcations are present throughout the region in which the circle map is nonmonotonic.

(b) *Period doubling* arises as a consequence of the nonmonotonicity of the PTC (when $\partial f^n/\partial \phi|_{\phi^*} = -1$ where ϕ^* is a point of period n). In Fig. 11.2 as A increases for $\tau = 1.065$ or $\tau = 1.115$, one observes superstable orbits of period 2, 4, 3. These are the periods arising in the U-sequence for one-dimensional, one-parameter interval maps with one extremum [42]. Finer examination of the sequence of periodic orbits reveals the other orbits of the U-sequence, including the period-doubling cascades. In addition to these sequences in the region en-closed in the square in Fig. 11.2, there is a complex pattern of overlapping of stable orbits of different periodicities displaying period-doubling and tangent bifurcations and showing a self-similar structure [5,6]. The extensions of the outer boundaries of the Arnold tongues are still associated with tangent bifurcations whereas the inner bound-aries are associated with period-doubling bifurcations. Since the Arnold tongues overlap, as a single parameter changes it is possible to observe a cascade of period-doubling bifurcations associated with one ex-tremum of the PTC and a tangent bifurcation associated with the other extremum.

(c) *Quasi-periodicity (overlapping of resonances)*. Quasi-periodic dynamics are associated with irrational rotation numbers in the region in which the PTC is monotonic. In this situation, the locus of points associated with any given irrational rotation number is a curve extending from $A = 0$ to $A = 0.39$. For the region in which the PTC is nonmonotonic, there is a wedge-shaped region in which some initial condition can be found associated with any given irrational rotation number. This wedge-shaped region arises as a direct consequence of the overlapping of Arnold tongues (leading to bistability) and of the observation that the rotation number covers an interval in nonmonotonic circle maps [7, 19,33,47]. The existence of aperiodic orbits in such circumstances has been carefully described by Levi [40] and was anticipated by the seminal paper of Levinson in 1949 (which described aperiodic orbits in a situation with bistability but did not explicitly deal with circle maps). Full descriptions of topological properties of aperiodic orbits in the nonmonotonic region have not yet been given. It is also of some interest to determine whether the set of initial conditions associated with irrational rotation numbers has non-zero measure for the case in which there are two stable periodic orbits with different rotation number; to our knowledge this is not now known.

In summary, in the region in which the Poincaré map is monotonic, the phase-locking zones have a comparatively simple structure which is well understood. On the other hand, in the region in which the Poincaré map is nonmonotonic, the phase-locking zones overlap, and in the internal region of each phase-locking zone there are complex bifurcation sequences. Consequently, although one can identify some regions of parameter space in which simple routes to chaos can be isolated in the nonmonotonic region, in general, as one parameter is changed, extremely complex sequences of bifurcations will be observed. The resulting dynamics will be difficult to interpret unless the underlying two-dimensional bifurcation structure in Figs. 11.1 and 11.2 is understood. A similar observation has been made by Holmes [28] in a related context.

11.3 Experimental observations

Experiments were performed on a preparation of spontaneously beating aggregates of embryonic chick heart cells (100–200 μm in diameter) which were electrically stimulated by passing a brief current pulse through an intracellular microelectrode, a preparation initially developed by De Haan [10,11]. The results of the experiments are reported in detail in ref. [23] with partial results in refs. [19, 20 and 24a]. Here we present a summary of the results with emphasis on the experimental observation of chaos. In a given aggregate, single pulses of current were delivered at various phases of the spontaneous cycle to determine the PTC using eqn (11.2). Then, at the same current amplitude, periodic stimuli were delivered over a range of stimulation frequencies with interstimulus period, t_s, mainly in the range $T_0/3 < t_s < 2T_0$ where T_0 is the intrinsic time interval between beats in the absence of stimulation (generally T_0 is between 0.5 and 1.0 s). The interstimulus interval was first changed in comparatively coarse steps of about 50 ms, and then, in regions in which the dynamics were more sensitive to stimulation frequency, t_s was varied in 10 ms steps (the finest resolution possible in most of our experiments). The dynamics at any given stimulation frequency were recorded for a minimum of about 20 s, when the dynamics were regular, and for as long as several minutes when the dynamics were irregular. Between each stimulation trial, the stimulator was turned off for at least 30 s.

There are several difficulties in performing and interpreting these experiments, which reflect the complex biological nature of the preparation. One of the main difficulties is to maintain the intracellular impalement of the microelectrode over extended periods of time. In practice, the longest it has been possible to maintain an impalement in a single aggregate is about 5 h. Since the diameters and intrinsic periods of beating of the aggregates differ, comparison of results from two different aggregates requires a normalisation procedure [20]. Also, there are fluctuations in the intrinsic frequency of any one aggregate (coefficient of variation about 2%). We take these fluctuations as a reflection of intrinsic biological 'noise' [9]. Therefore, in comparison with physical systems and numerical simulations of deterministic mathematical models, the experimental system is 'noisy' and there is uncontrollable variability between experiments done in different preparations.

For a fixed current amplitude, as the stimulation frequency is decreased, periodic rhythms are observed at some frequencies and aperiodic dynamics at other frequencies. The periodic rhythms covering the largest areas in parameter space, the 2:1, 1:1 and 2:3 rhythms (Fig. 11.3), are readily observed in all preparations and can be maintained indefinitely. Phase-locked dynamics in a number of other patterns can be observed, but the number of different patterns that can be obtained depends on the particular preparation and the stimulation strength. Guevara [23] has observed phase-locked patterns, stably maintained for at least several cycles in the ratios 5:1, 4:1, 7:2, 3:1, 8:3, 5:2, 7:3, 2:1, 7:4, 5:3, 8:5, 3:2, 7:5, 4:3, 5:4, 1:1, 4:5, 3:4, 5:7, 2:3, 5:8, 3:5, 4:7, 1:2, where the order from left to right corresponds to decreasing stimulation frequency. Not all of these different rhythms have been observed in a single preparation, and these rhythms are not all observed at each stimulation strength. The sequence of observed phase-locking patterns follows the Farey sequence as expected, and shows a monotonic increase of rotation numbers [16,26]. In general, the higher the periodicity of the orbit the more difficult the orbit is to observe experimentally.

If one observes two different stable rhythms at different stimulation

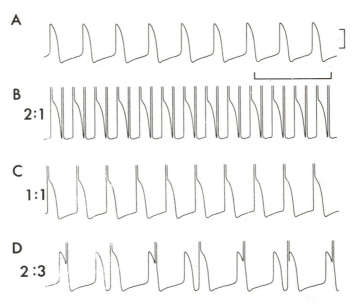

Fig. 11.3 Recordings of transmembrane potential showing regular dynamics. The average interbeat interval t_{av} is computed from the five spontaneous interbeat intervals immediately preceding the start of each stimulation run. Aggregate number 1, diameter $= 114\,\mu m$. (A) Control unperturbed activity; (B) $A = 0.10$, $t_{av} = 469\,ms$, $t_s = 150\,ms$, $\tau = 0.32$, 2:1 phase locking; (C) $A = 0.10$, $t_{av} = 437\,ms$, $t_s = 400\,ms$, $\tau = 0.91$, 1:1 phase locking; (D) $A = 0.10$, $t_{av} = 471\,ms$, $t_s = 600\,ms$, $\tau = 1.27$, 2:3 phase locking. Vertical scale represents $50\,mV$; horizontal scale represents $1\,s$. Adapted from ref. [20] with permission.

frequencies keeping stimulation strength fixed, then, by probing with intermediate stimulation frequencies, it is often possible to obtain dynamics which are aperiodic. The range of parameters over which aperiodic dynamics are observed varies with stimulation frequency and amplitude. However, over large ranges of parameter space, for example with stimulation parameters in the range $0.06 < A < 0.1$, $1.05 < \tau < 1.4$, it is difficult to maintain stable entrainment for extended periods of time.

We now offer an interpretation of the irregular rhythms observed experimentally, based on the preceding discussion of chaotic dynamics in circle maps. In particular, we have obtained evidence that the three routes to chaos described above can be experimentally observed.

(a) *Intermittency.* The high-frequency boundary of the 1:1 zone in Fig. 11.1 corresponds to a tangent bifurcation. By increasing the stimulation frequency starting at the 1:1 zone, a situation eventually arises in which there are occasional dropped beats. These dropped beats occur at approximately equal time intervals, but the pattern is irregular (Fig.

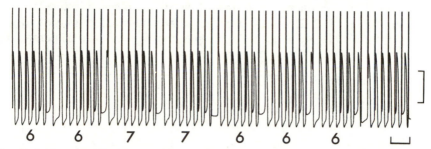

Fig. 11.4 Nonperiodic Wenckebach cycles, $n + 1:n$, with $n = 6$ or 7. Numbers below the recordings represent the number of action potentials during each series. Aggregate number 1, $A = 0.06$, $t_{av} = 470\,\text{ms}$, $t_s = 350\,\text{ms}$, $\tau = 0.75$. Vertical scale represents 50 mV; horizontal scale represents 1 s.

11.4). This intermittent dropping of beats is expected theoretically near the tangent bifurcation in a noisy system.

(b) *Period doubling.* Provided the stimulation strength is sufficiently large (but not too large), period-doubling bifurcations are expected as the

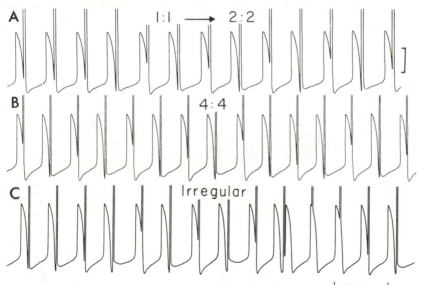

Fig. 11.5 Period-doubling bifurcations and irregular dynamics. (A) 1:1 phase locking spontaneously changing to 2:2 phase locking. Note that during 2:2 phase locking there are two distinct phases of the cycle at which the stimuli fall. Aggregate number 2, diameter = 181 μm, $A = 0.10$, $t_{av} = 519\,\text{ms}$, $t_s = 550\,\text{ms}$, $\tau = 1.06$. (B) 4:4 phase locking. There are four distinct phases of the cycle at which the stimuli fall. Aggregate number 1, $A = 0.10$, $t_{av} = 470\,\text{ms}$, $t_s = 490\,\text{ms}$, $\tau = 1.04$. (C) Irregular dynamics. Aggregate number 1, $A = 0.10$, $t_{av} = 471\,\text{ms}$, $t_s = 500\,\text{ms}$, $\tau = 1.06$. Vertical scale represents 50 mV; horizontal scale represents 1 s. Adopted from [24a] with permission (part (C) has been changed).

stimulation frequency is decreased, starting from within the 1:1 zone. In Fig. 11.5, 2:2, 4:4, and irregular dynamics are displayed. In Fig. 11.6, we display ϕ_{i+1} vs ϕ_i as measured using experimental data

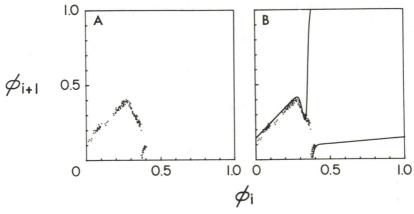

Fig. 11.6 (A) ϕ_{i+1} vs ϕ_i from the experimental data shown in Fig. 11.5C. (B) The Poincaré map, calculated from eqn (11.6) superimposed on the experimental points.

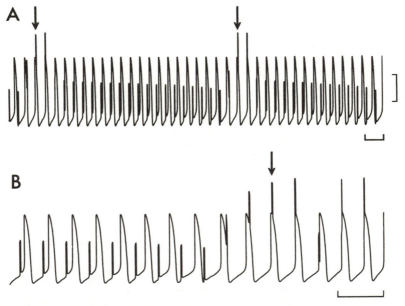

Fig. 11.7 (A) 'Quasi-periodic' dynamics demonstrating the Wenckebach phenomena. The arrows indicate blocked stimuli. Aggregate number 3, diameter $= 95\,\mu\text{m}$, $t_{av} = 540\,\text{ms}$, $t_s = 500\,\text{ms}$, $\tau = 0.92$. A is very weak (<0.02). (B) Expanded time scale for a portion of (A). Vertical calibration represents $50\,\text{mV}$; horizontal calibration represents $1\,\text{s}$.

from which the sequence shown in Fig. 11.5c was extracted. Super-imposed on these data is the plot of ϕ_{i+1} vs ϕ_i based on the PTC curve from eqn 11.2. The nonmonotonic curve with the unstable period-1 fixed point in the nonmonotonic region is typical of finite difference equations in the period-doubling route to chaos. The close correspondence between the experimental data derived from periodic stimulation (points), and the curves based on single pulse perturbation experiments (solid curve) give strong support for the theory described in section 11.2.

(c) *Quasi-periodicity*. At low stimulation intensities, the zones of entrain-ment are narrow and stable phase locking is difficult to obtain. A typical tracing of the resulting quasi-periodic dynamics is shown in Fig. 11.7. The plot of the experimentally measured ϕ_{i+1} vs ϕ_i (Fig. 11.8) is a monotonic circle map, almost parallel to the 45° line. The periodic electrical stimulation in this case barely perturbs the ongoing spon-taneous rhythm.

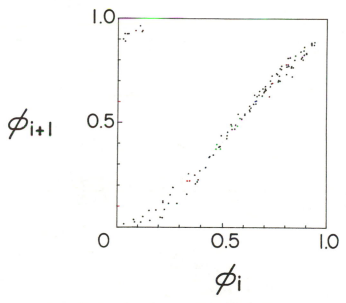

Fig. 11.8 ϕ_{i+1} vs. ϕ_i from the experimental data shown in Fig. 11.7.

At intermediate stimulation strengths, the dynamics should be described by a nonmonotonic circle map of degree 1. In Fig. 11.9 we show irregular dynamics and in Fig. 11.10 we show ϕ_{i+1} vs ϕ_i experimentally measured for this situation along with the superimposed curve based on the PTC, eqn 11.2. The dynamics for this situation are not appreciably (at least from a coarse qualitative perspective) different from the dynamics that are experimentally observed at a lower stimulation strength but the same

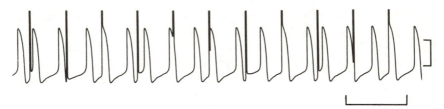

Fig. 11.9 Irregular dynamics with apparently randomly inserted extra or escape beats. Aggregate number 1, $A = 0.06$, $t_{av} = 484\,ms$, $t_s = 700\,ms$, $\tau = 1.44$. Vertical calibration represents $50\,mV$; horizontal calibration represents $1\,s$. Adopted from ref. [20] with permission.

stimulation frequency. Indeed, there are no striking qualitative differences between the experimentally observed dynamics just beneath or just above the critical intensity at which the circle map becomes nonmonotonic. We hypothesise that the inability to experimentally observe such differences is

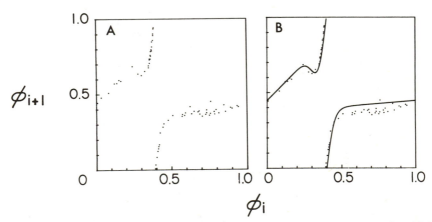

Fig. 11.10 (A) ϕ_{i+1} vs. ϕ_i from the experimental data shown in Fig. 11.9. (B) The Poincaré map, calculated from eqn (11.6), superimposed on the experimental points. Adopted from ref. [20] with permission.

due to small amounts of noise which destroy the higher-order resonances that overlap near the transition from monotonicity to nonmonotonicity.

At higher stimulation intensities, as lower-order resonances overlap, the effects on the dynamics are expected to be more striking. In Fig. 11.11 we show three traces taken at a single stimulation frequency in a single preparation. Figure 11.11A contains a sequence of 1:1 locking; Fig. 11.11B contains a sequence of 3:4 locking; and Fig. 11.11C contains an irregular sequence combining elements of both of the other patterns. This dynamics arises near the neighbourhood in which there is an overlapping

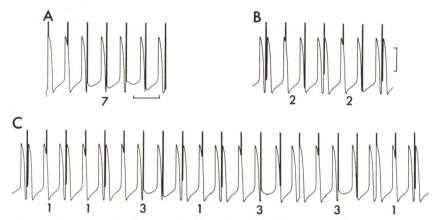

Fig. 11.11 Complex dynamics and bistability. The three panels in this figure represent dynamics observed during a given train of pulses during which stimulus frequency and intensity were being held constant. (A) 1:1 phase-locking dynamics; (B) 3:4 phase locking; and (C) irregular sequences combining elements of (A) and (B). Integers below the recordings represent the number of 1:1 events during each series of action potentials between couplets. Each couplet consists of an escape beat followed by a stimulated beat. Aggregate number 1, $A = 0.10$, $t_{av} = 470\,\text{ms}$, $t_s = 560\,\text{ms}$, $\tau = 1.19$. Vertical calibration represents $50\,\text{mV}$; horizontal calibration represents $1\,\text{s}$.

of 1:1 and 3:4 zones. We believe that the complex dynamics in Fig. 11.11 is a manifestation of the bistability arising as a consequence of overlapping of the resonances in a noisy system. The plot of ϕ_{i+1} vs ϕ_i, and the fitted PTC (Fig. 11.12) provides evidence in support of this hypothesis. In

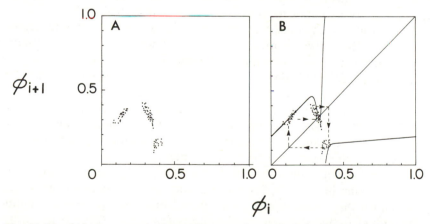

Fig. 11.12 (A) ϕ_{i+1} vs. ϕ_i from the experimental data shown in Fig. 11.11. (B) The Poincaré map, calculated from eqn (11.6), superimposed on the experimental points. The staircase method for graphically iterating the map shows a stable cycle corresponding to a 3:4 phase locking.

Fig. 11.12B we show the period-3 orbit corresponding to 3:4 entrainment. The local minimum at about $\phi_i = \phi_{i+1} = 0.37$ is associated with the intermittent 1:1 sequences. Thus both 1:1 and 3:4 sequences can be found along with more irregular sequences due to the effects of 'noise' in mixing the basins of attraction. A definitive demonstration of the bistability such as finding one or another stable rhythm by initiating stimulation at different phases in the cardiac cycle has not been performed (and may be impossible in this noisy system).

11.4 Discussion

The periodic stimulation of periodically beating cells from chick heart gives rise to a wide range of different regular and irregular dynamics. By theoretically analysing this experiment, the results can be understood based on the analysis of one-dimensional circle maps. Such a model system can be used to interpret a number of regular and irregular cardiac rhythms which may be due to the periodic forcing, by an impulse originating at the SA node or a secondary focus, either in the AV node [24,35,52,55] or ventricular tissue [2,29,32,44,59]. Indeed, practically all of the great variety of experimentally observed rhythms have counterparts in clinically observed arrhythmias [23].

In the intact human heart, more complex situations can also arise. For example, there can be interactions between two beating foci, so that activity from each focus acts to reset the other. Theoretical analysis of the dynamics in such situations shows the possibility for a wide range of regular and irregular dynamics [29,30,31,58]. The existence of 'chaotic' dynamics arising from the periodic stimulation of an excitable but not spontaneously oscillating medium has also been discussed [38].

In contrast to rhythms that are believed to be due to interactions between multiple discrete foci is fibrillation. During fibrillation there are low-voltage comparatively rapid fluctuations on the ECG. These are believed to be due to disorganised re-entrant wave propagation throughout the cardiac tissue, without independent pacemaker sites [1, 8]. Although fibrillation is sometimes described as 'chaotic', frequency analysis of fibrillation has shown strongly peaked amplitudes over comparatively narrow frequency ranges [4, 21, 27], and the use of the adjective 'chaotic' to describe such rhythms has been questioned [21]. It has been shown that 'fibrillatory' activity can arise in spatially distributed systems with variable refractory times [39, 43, 51, 54]. Recent studies using computer simulation have shown that period-multupling can be found during periodic stimulation at high frequencies prior to the onset of fibrillatory activity in a spatially distributed system with variable refractory times [51, 54]. However, this dynamics may at least partially reflect the details of the spatial discretisation of the network of excitable cells and the finite difference approximation used in the simulation [51]. The observation of alternans during electrical stimulation of cardiac tissue at high-stimulation frequency [25, 50] may be due to a period-doubling bifurcation which is not part of a cascade of period-doubling bifurcations [25]. Thus it is not clear if fibrillation in continuous systems is associated with cascades of period-multupling bifurcations displayed in the discrete models of fibrillation. A different theoretical model for the initiation of fibrillatory activity in a network of spatially distributed self-oscillatory elements based on analysis of phased resetting due to a single shock has been proposed by Winfree [57]. Finally, we reiterate our earlier observation that 'chaotic' heart rhythms, perhaps due to the interaction of multiple pacemaking foci, are often observed prior to fibrillation and sudden death.

To summarise, a theoretical analysis of cardiac arrhythmias can be developed based on the following observations:

(1) Cardiac arrhythmias are normally classified on the basis of qualitative properties of the dynamics.
(2) Simple biological and theoretical models for the intact heart display extremely rich dynamics as parameters in the model representing stimulation frequency and amplitude are varied.
(3) There is a correspondence between many of the rhythms observed in the model systems and rhythms observed clinically. Thus, a theoretical understanding of the mechanisms underlying cardiac arrhythmias, and particularly 'chaotic' cardiac rhythms, may be possible.

Acknowledgements

This work has been supported by grants from the Canadian Heart Foundation and the Natural Sciences Engineering and Research Council. We thank M. R. Guevara for invaluable contributions to the experimental and theoretical work presented in this paper, and R. Brochu for excellent technical assistance. LG was a resident in the Institute for Nonlinear Science at the University of California, San Diego, during the 1984–5 academic year.

References

[1] Allesie, M.A., Lammers, W.J.E.P., Bonke, F.I.M. and Hollen, J. In *Cardiac Electrophysiology and Arrhythmias*, eds D. P. Zipes and J. Jalife, pp. 265–75. Grune & Stratton, Orlando (1985).
[2] Antzelevich, C., Jalife, J. and Moe, G.K. Electrotonic modulation of pacemaker activity. *Circulation* **66**, 1225–32 (1982).
[3] Arnold, V.I. *Geometrical Methods in the Theory of Ordinary Differential Equations*, Section 11. Springer, New York (1983).

[4] Battersby, E.J. Pacemaker periodicity in atrial fibrillation. *Circ. Res.* **17**, 296–302 (1965).

[5] Bélair, J. and Glass, L. Self similarity in periodically forced oscillations. *Phys. Lett.* **96A**, 113–16 (1983).

[6] Bélair, J. and Glass, L. Universality and self-similarity in the bifurcation of circle-maps. *Physica 16D*, 143–54 (1985).

[7] Boyland, P. Bifurcations of circle maps; Arnold tongues bistability and rotation intervals. 1984 preprint.

[8] Chung, E.K. *Principles of Cardiac Arrhythmias*, 2nd edn. Williams and Wilkins, Baltimore (1977).

[9] Clay, J.R. and DeHaan, R.L. Fluctuations in interbeat interval in rhythmic heart cell clusters—role of membrane voltage noise. *Biophys. J.* **28**, 377–89 (1976).

[10] DeHaan, R.L. The potassium sensitivity of isolated embryonic heart cells increases with development. *Dev. Biol.* **23**, 226–40 (1970).

[11] DeHaan, R.L. and De Felice, L.J. Oscillatory properties and excitability of the heart cell membrane. *Theor. Chem.* **4**, 181–233 (1978).

[12] Eckmann, J.P. Roads to turbulence in dissipative dynamical systems. *Rev. Mod. Phys.* **53**, 643–54 (1981).

[13] Feigenbaum, J.P., Kadanoff, L.P. and Shenker, S.J. Quasiperiodicity in dissipative systems: a renormalization group analysis. *Physica 5D* 370–86 (1982).

[14] Fraser, S. and Kapral, R. Analysis of flow hysteresis by a one-dimensional map. *Phys. Rev.* **25A**, 3223–33 (1982).

[15] Fraser, S. and Kapral, R. Universal vector scaling in one dimensional maps. *Phys. Rev.* **30A**, 1017–25 (1984).

[16] Glass, L., Graves, C., Petrillo, G.A. and Mackey, M.C. Unstable dynamics of a periodically driven oscillator in the presence of noise. *J. Theor. Biol.* **86**, 455–75 (1980).

[17] Glass, L. and Perez, R. Fine structure of phase locking. *Phys. Rev. Lett.* **48**, 1772–5 (1982).

[18] Glass, L. and Winfree, A.T. Discontinuities in phase resetting experiments. *Am. J. Physiol. (Regulatory Integrative Comp. Physiol.* **15**) **246**, R251–8 (1984).

[19] Glass, L., Guevara, M.R., Shrier, A. and Perez, R. Bifurcation and chaos in a periodically stimulated cardiac oscillator. *Physica 7D* 89–101 (1983).

[20] Glass, L., Guevara, M.R., Bélair, J. and Shrier, A. Global bifurcations of a periodically forced biological oscillator. *Phys. Rev.* **29A**, 1348–57 (1984).

[21] Goldberger, A.L., Bhargava, V., West, B.J. and Mandell, A.J. Ventricular fibrillation is not chaos. *Clin. Res.* **32**, 169A (1984).

[22] Guckenheimer, J. Bifurcations of dynamical systems, in *Progress in Mathematics, vol. 8, Dynamical Systems*, Birkhauser, Boston (1980).

[23] Guevara, M.R. Chaotic cardiac dynamics. PhD thesis, McGill University, Montreal (1984).

[24] Guevara, M.R. and Glass, L. Phase-locking, period doubling bifurcations and chaos in a mathematical model of periodically driven oscillators. *J. Math. Biol.* **14**, 1–23 (1982).

[24a] Guevara, M.R. Glass, L. and Shrier, A. Phase-locking, period-doubling bifurcations and irregular dynamics in periodically stimulated cardiac cells. *Science* **214**, 1350–3.

[25] Guevara, M.R., Ward, G., Shrier, A. and Glass, L. Electrical alternans and period-doubling bifurcations. *Comp. Cardiol.* 167–70 (1984).

[26] Hardy, G.H. and Wright, E.M. *An Introduction to the Theory of Numbers*, 4th edn, Clarendon Press, Oxford (1960).

[27] Herbschleb, J.N., Van der Tweel, I. and Meijler, F.L. The apparent repetition frequency of ventricular fibrillation. *Comp. Cardiol.* 249–52 (1982).

[28] Holmes, P. Bifurcation sequences in horseshoe maps: infinitely many routes to chaos. *Phys. Lett.* **104A**, 299–302 (1984).

[29] Honerkamp, J. The heart as a system of coupled nonlinear oscillators. *J. Math. Biol.* **18**, 69–88 (1983).

[30] Honerkamp, J. and Strittmatter, W. Fibrillation of a cardiac region and the tachycardia mode of a two-oscillator system. 1984 preprint.

[31] Ikeda, N. Model of bidirectional interaction between myocardial pacemakers based on the phase response curve. *Biol. Cybernetics* **43**, 157–67 (1982).

[32] Ikeda, N., Tsurutu, H. and Sato, T. Difference equation model of the entrainment of myocardial pacemaker cells based on the phase response curve. *Biol. Cybernetics* **42**, 117–28 (1981).

[33] Jensen, M.H., Bak, P. and Bohr, T. Transition to chaos by interaction of resonances in dissipative systems. I. Circle maps. *Phys. Rev.* **30A**, 1960–9 (1984).

[34] Kaneko, K. On the period adding phenomena at the frequency locking in a one-dimensional mapping. *Progr. Theor. Phys.* **68**, 669–72 (1982).

[35] Katholi, C.R., Urthaler, F., Macy, J. and James, T.N. A mathematical model of automaticity in the sinus node and AV junction based on weakly coupled relaxation oscillators. *Comp. Biomed. Res.* **10**, 529–43 (1977).

[36] Katz, L.N. *Electrocardiology*, 2nd edn. Lea and Febiger, Philadelphia (1946).

[37] Kawato, M. Transient and steady state phase response curves of limit cycle oscillations. *J. Math. Biol.* **12**, 13–30 (1981).

[38] Keener, J.P. Chaotic cardiac dynamics, in *Mathematical Aspects of Physiology*, ed. F.C. Hoppenstaedt, pp. 299–325. American Mathematical Society, Providence, RI (1981).

[39] Krinskii, V. Fibrillation in excitable media. *Problemi Kyburnetikii*, **20**, 59–80 (1968).

[40] Levi, M. Qualitative analysis of the periodically forced relaxation oscillations. *Mem. Am. Math. Soc.* **32** (244), 147 pp. (1981).

[41] McAllister, R.E., Noble, D. and Tsien, R.W. Reconstruction of the electrical activity of cardiac Purkinje fibres. *J. Physiol. (Lond.)* **251**, 1–59 (1975).

[42] Metropolis, N., Stein, M.L. and Stein, P.R. On finite limit sets for transformations on the unit interval. *J. Comb. Theor.* **15**, 25–44 (1973).

[43] Moe, G.K., Rheinboldt, W.C. and Abildskov, J.A. A computer model of atrial fibrillation. *Am. Heart J.* **67**, 200–20 (1964).

[44] Moe, G.K., Jalife, J., Mueller, W.J. and Moe, B. A mathematical model of parasystole and its application to clinical arrhythmias. *Circulation* **56**, 968–79 (1977).

[45] Newhouse, S., Palis, J. and Takens, F. Bifurcations and stability of families of diffeomorphisms. *Publ. Inst. Hautes Etud. Sci.* **57**, 5–71. (1983).

[46] Nikolic, G., Bishop, R.L. and Singh, J.B. Sudden death recorded during Holter monitoring. *Circulation* **66**, 218–25 (1982).

[47] Ostlund, S., Rand, D., Sethna, J. and Siggia, E. Universal properties of the transition from quasiperiodicity to chaos in dissipative systems. *Physica 8D*, 303–42 (1983).

[48] Perkel, D.H., Schulman, J.H., Bullock, T.H., Moore, G.P. and Segundo, J.P. Pacemaker neurons: effects of regularly spaced synaptic input. *Science,* **145**, 61–3 (1964).

[49] Phillips, J., Spano, J. and Burch, G. Chaotic atrial mechanism. *Am. Heart J.* **78**, 171–9 (1969).

[50] Ritzenberg, A.L., Adam, D.R. and Cohen, R.J. Period multupling: evidence for nonlinear behaviour in canine heart. *Nature, Lond.* **307**, 159–61 (1984).

[51] Ritzenberg, A.L., Smith, J.M., Grumbach, M.P. and Cohen, R.J. Precursor to fibrillation in cardiac computer model. *Comp. Cardiol.* 171–4 (1985).

[52] Roberge, F.A., Nadeau, R.A. and James, T.N. The nature of the P–R interval. *Cardiovasc. Res.* **2**, 19–30 (1968).

[53] Scott, S.W. Stimulating simulations of young yet cultured beating chick hearts. PhD thesis, State University of New York, Buffalo (1979).

[54] Smith, J.M. and Cohen, R.J. Simple finite element model accounts for wide range of cardiac dysrhythmias. *Proc. Natl Acad. Sci. USA* **81**, 233–7 (1984).

[55] van der Pol, B. and van der Mark, J. The heart beat considered as a relaxation oscillator and an electrical model of the heart. *Phil. Mag.* **6**, 763–75 (1928).

[56] Winfree, A.T. *The Geometry of Biological Time.* Springer, New York (1980).

[57] Winfree, A.T. Fibrillation as a consequence of pacemaker activity, in *Heart Rate and Rhythm,* eds L.N. Bouman and H.J. Jongsma, 447–70. Martin Nijhoff, The Hague (1982).

[58] Ypey, D.L., Van Meerwijk, W.P.M., Ince, C. and Groos, G. Mutual entrainment of two pacemaker cells. A study with an electronic parallel conductance model. *J. Theor. Biol.* **86**, 731–55 (1980).

[59] Ypey, D.L., Van Meerwijk, W.P.M. and de Bruin, G. Suppression of pacemaker activity by rapid repetitive phase decay. *Biol. Cybernetics* **45**, 187–94 (1982).

[60] Ypey, D.L., Van Meerwijk, W.P.M. and De Haan, R.L. Synchronization of cardiac pacemaker cells by electrical coupling: a study with embryonic heart cell aggregates, in *Heart Rate and Rhythm,* eds L.N. Bouman and H.L. Jongsma, pp. 363–95. Martin Nijhoff, The Hague (1982).

12
Chaotic oscillations and bifurcations in squid giant axons

K. Aihara[1] and G. Matsumoto[2]

[1]Department of Electronic Engineering, Faculty of Engineering,
Tokyo Denki University, 2-2, Nishiki-cho, Kanda, Chiyoda-ku,
Tokyo 101, Japan, and [2]Electrotechnical Laboratory,
Section of Analogue Information Science, Division of Information Science,
Tsukuba Science City, Niihari-gun, Ibaraki 305, Japan

12.1 Nonlinear dynamics in nerve membranes

The fundamental functions of neurones, such as the generation and propagation of action potentials, are supported by nonlinear dynamics peculiar to the nerve membranes. The nonlinear neural dynamics produces different attractors and bifurcations in far from equilibrium conditions. For example, stable limit cycles with Hopf bifurcations and multiple equilibrium points with saddle-node bifurcations have been analysed both experimentally and theoretically [1,2,11,14,15,17,19,20,24,33].

Self-sustained oscillations, or the spontaneous repetitive firing of action potentials in squid giant axons, can be understood in terms of a dissipative structure that has spatio-temporal order and behaves as a nonlinear neural oscillator [1,24,26]. In this chapter, we study experimentally and numerically the various responses of the neural oscillator of squid giant axons to periodic forcing by a sinusoidal current [3–6,25,27].

12.2 Experimental approaches to squid giant axons

Intact giant axons of squid *(Doryteuthis bleekeri)* were used in the electrophysiological experiments. A mixture of natural sea water and 550 mM NaCl was used as the external medium to induce the self-sustained oscillations in the squid giant axonal membrane [1,24]. The membrane potential was recorded through a pair of glass pipette Ag–AgCl electrodes filled with 550 mM KCl.

A sinusoidal current $A \sin 2\pi f_S t$ that provided the periodic forcing was applied to the axons that were in the state of self-sustained oscillation through an internal current electrode with the conducting length of 5 mm, where A and f_S are the amplitude and the frequency of the stimulating sinusoidal current. The membrane potential of the axon was spatially clamped over the conducting portion of the current electrode. This space-clamping condition makes it possible to describe the forced oscillations by 'ordinary' differential equations as described in section 12.3.

We analysed the oscillating membrane potentials by stroboscopic plots [29,34,36]. Namely, the membrane potential $V(t)$ and its time-differential $dV(t)/dt$ were observed at a fixed phase of the stimulating sinusoidal current and plotted successively on the V–dV/dt plane [27].

12.3 The Hodgkin–Huxley oscillator

It is well known that the Hodgkin–Huxley equations [13] can describe phenomenologically the various phenomena of nerve excitation in squid giant axons. We used the Hodgkin–Huxley ordinary differential equations that have four variables (the membrane potential, V, the sodium activation, m, the sodium inactivation, h, and the potassium activation, n) for numerical analyses of the periodically forced oscillations in squid giant axons surrounded by a mixture of natural sea water and 550 mM NaCl. (See [1,5,26] for the detailed procedures taking the variation of the external ionic concentrations into the Hodgkin–Huxley equations.)

The self-sustained oscillations start through a backward (or subcritical) Hopf bifurcation as the mixing ratio of 550 mM NaCl to natural sea water is increased [1]. We analysed the responses of a Hodgkin–Huxley oscillator in the soft-oscillation mode to the forcing sinusoidal current $A \sin 2\pi f_S t$. The natural frequency f_N, the peak-to-peak amplitude of the action potentials, and the membrane potential of the unstable resting state in the Hodgkin–Huxley oscillator are 174.6 Hz, 126.9 mV and −60.0 mV, respectively, and approximate to the corresponding values of squid giant axons [5]. The forced oscillations were analysed by stroboscopic plots on the two-dimensional planes V–m, V–h, V–n, V–m^3h (the normalised sodium conductance). The Hodgkin–Huxley equations were numerically solved on a HITAC-M280H computer at the University of Tokyo Computer Center.

12.4 Forced oscillations in squid giant axons

The periodically forced oscillations in both squid giant axons and the Hodgkin–Huxley equations are classified into (1) synchronised oscillations, (2) quasi-periodic oscillations, and (3) chaotic oscillations.

12.4.1 *Synchronised oscillations*

When f_S (the stimulating frequency) is close to n/m (a simple rational number) times f_N (the natural frequency), f_F (the fundamental frequency

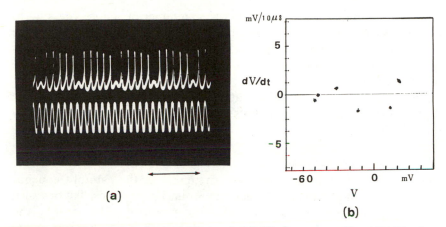

Fig. 12.1 A 5/6-synchronised oscillation in squid giant axons ($f_N = 181\,$Hz, $f_S = 265\,$Hz and $A = 3.27\,\mu$A). (a) The wave forms of the membrane potential (above) and the stimulating current (below). The length of the bar corresponds to 120 mV (for the membrane potential), 12 μA (for the stimulating current) and 30 ms (for the time). (b) The stroboscopic plot on the V-dV/dt plane.

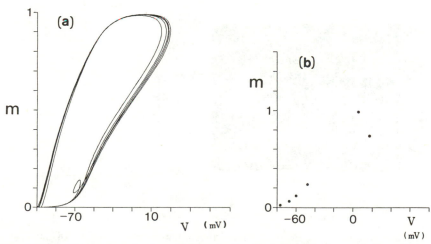

Fig. 12.2 A 5/6 synchronised oscillation in the Hodgkin–Huxley equations ($f_N = 174.6\,$Hz, $f_S = 216.5\,$Hz and $A = 40.0\,\mu$A cm^{-2}). (a) The projection of the trajectory on to the V–m plane. (b) The stroboscopic plot on the V–m plane.

of the forced oscillation) is entrained to f_S/n and the stroboscopic plot is composed of n points. Examples of the synchronised oscillations are shown in Fig. 12.1 (squid giant axons) and Fig. 12.2 (the Hodgkin–Huxley equations).

The rotation number, or the ratio of the number of action potentials to the number of the cycles of the stimulating sinusoidal current [10] in the oscillation synchronised to f_S of nf_N/m is m/n in the Hodgkin–Huxley equations and m/n or $(m-1)/n$ in squid giant axons. The term 'the m/n-synchronised oscillation' in Figs. 12.1 and 2 means that the rotation number, or the average firing rate, is m/n and that f_F is f_S/n. The distribution of the rotation numbers corresponds to a portion of the Farey series [5, 35, 38].

12.4.2 *Quasi-periodic oscillations*

Coexistence of two incommensurable rhythms, i.e. the natural frequency f_N and the forcing frequency f_S, generates quasi-periodic oscillations such as pulse–amplitude–modulation and pulse–phase–modulation [5]. The quasi-periodic oscillations can easily be identified by an invariant closed curve that emerges asymptotically in the stroboscopic plot. Examples of the quasi-periodic oscillations are shown in Fig. 12.3 (squid giant axons) and Fig. 12.4 (the Hodgkin–Huxley equations).

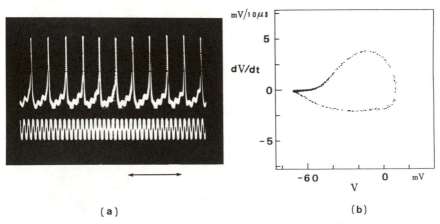

(a)   (b)

Fig. 12.3 A quasi-periodic oscillation in squid giant axons ($f_N = 187\,\text{Hz}$, $f_S = 800\,\text{Hz}$ and $A = 2\,\mu\text{A}$). (a) The wave forms of the membrane potential (above) and the stimulating current (below). The length of the bar corresponds to $60\,\text{mV}$, $12\,\mu\text{A}$ and $15\,\text{ms}$. (b) The stroboscopic plot on the V–dV/dt plane.

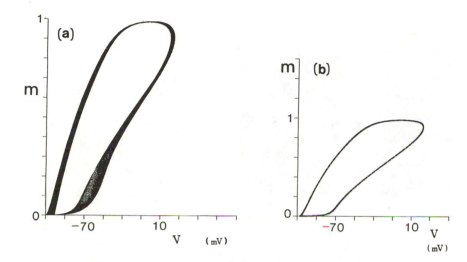

Fig. 12.4 A quasi-periodic oscillation in the Hodgkin–Huxley equations ($f_N =$ 174.6 Hz, $f_S = 800.0$ Hz and $A = 50.0\,\mu\text{A cm}^{-2}$). (a) The projection of the trajectory on to the V–m plane. (b) The stroboscopic plot on the V–m plane.

12.4.3 *Chaotic oscillations*

Chaotic oscillations have been found in the forced Hodgkin–Huxley oscillators for some values of A and f_S [3, 4, 5, 16, 21]. The stroboscopic plots of the chaotic oscillations depict the so-called strange attactors. Examples of the chaotic oscillations are shown in Fig. 12.5 (squid giant axons) and in Fig. 12.6 (the Hodgkin–Huxley equations). The strange attractors in the chaotic oscillations are evidently different from the points of the synchronised oscillation and from the closed curve of the quasi-periodic oscillation.

Figure 12.7 shows another example of the strange attractors, which is obtained before full development of the strange attractor in the Hodgkin–Huxley equations. The strange attractor in Fig. 12.7 is composed of four islands, A, B, C and D, and has four-periodicity. Namely, the four islands are transferred one after another by the stroboscopic mapping such that A→B→C→D→A. Figure 12.8 is a magnification of the islands in Fig. 12.7. Figure 12.9 shows a one-dimensional transfer function which maps successively the membrane potential of a point on the island A to that of the next visiting point on the same island. The transforming dynamics of Fig. 12.8 such as folding (A→B) and pressing (D→A) and the approximately smooth curve of Fig. 12.9 are quite similar to the dynamics of the forced Brusselator reviewed by Tomita [34].

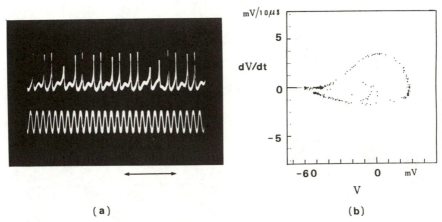

(a)   (b)

Fig. 12.5 A chaotic oscillation in squid giant axons (f_N = 179 Hz, f_S = 270 Hz, and
A = 2.37 μA). (a) The wave forms of the membrane potential (above) and the
stimulating current (below). The length of the bar corresponds to 120 mV, 12 μA
and 30 ms. (b) The stroboscopic plot on the V–dV/dt plane.

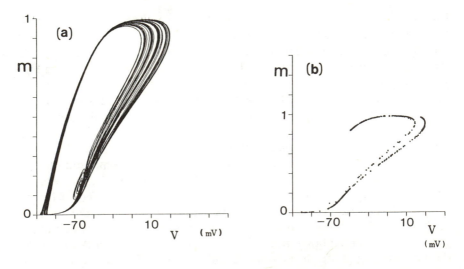

Fig. 12.6 A chaotic oscillation in the Hodgkin–Huxley equations (f_N =
174.6 Hz, f_S = 100.0 Hz and A = 40.5 μA cm^{-2}). (a) The projection of the trajectory on to the V–m plane. (b) The stroboscopic plot on the V–m plane.

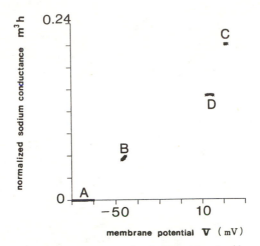

Fig. 12.7 A strange attractor composed of four islands ($f_N = 174.6\,\text{Hz}$, $f_S = 100.0\,\text{Hz}$ and $A = 41.394\,\mu\text{A}\,\text{cm}^{-2}$).

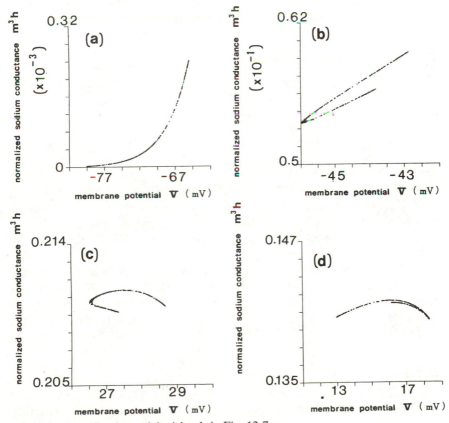

Fig. 12.8 Magnifications of the islands in Fig. 12.7.

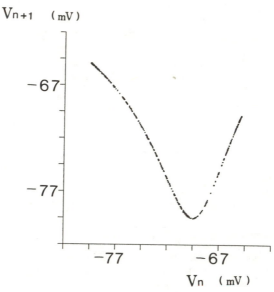

Fig. 12.9 A transfer function on the island A in Fig. 12.7.

12.5 Routes to the chaotic oscillations

The routes or the bifurcations to the strange attractors of the chaotic oscillations so far observed in squid giant axons and the Hodgkin–Huxley equations are (1) period-doubling bifurcations, (2) intermittency, and (3) collapse of quasi-periodicity [4–6].

12.5.1 *Period-doubling bifurcations*

A route to the chaotic oscillation through the period-doubling bifurcations (e.g. [7, 28]) was observed in squid giant axons. Figure 12.10 shows an example of the route, where Figs. 12.10a, b and c correspond to the 5/7-synchronised oscillation, the 10/14-synchronised oscillation, and the chaotic oscillation, respectively.

There also exist successive period-doubling bifurcations into the chaotic oscillation in the Hodgkin–Huxley oscillator [5]. For example, when A is changed from $70.0\ \mu\text{A cm}^{-2}$ to $40.0\ \mu\text{A cm}^{-2}$ with the fixed f_S of 100.0 Hz, the following sequence of the period of the forced oscillation is obtained: $T(= 10.0\ \text{ms}) \to 2T \to 4T \to 8T \to 16T \to \text{chaos} \to 12T \to 24T \to \text{chaos}$.

12.5.2 *Intermittency*

Figure 12.11 shows intermittent chaos (e.g. see [23]) in squid giant axons. The neural oscillator fails to generate the action potentials intermittently in this case. Another example of intermittent chaos was also observed in

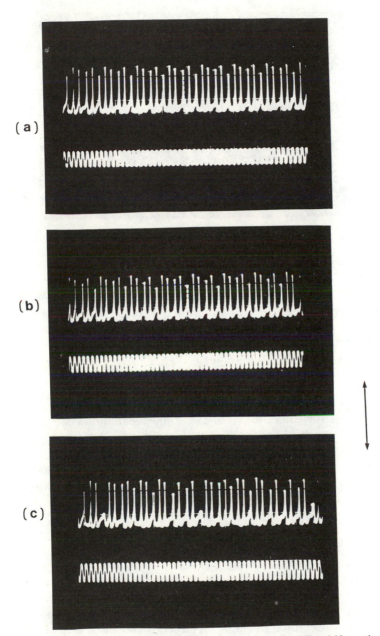

Fig. 12.10 A route to a chaotic oscillation via a period-doubling bifurcation (f_N = 177 Hz and f_S = 250 Hz). The upper and lower traces in (a – c) represent the membrane potential and the stimulating current, respectively. The length of the bar corresponds to 120 mV, 12 μA and 60 ms. (a) A 5/7-synchronised oscillation (A = 1.19 μA). (b) A 10/14-synchronised oscillation (A = 1.33 μA). (c) A chaotic oscillation (A = 1.43 μA).

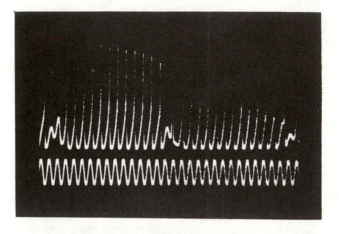

Fig. 12.11 An intermittent chaos in squid giant axons (f_N =228 Hz, f_S = 303 Hz and A = 2 μA). The length of the bar corresponds to 60 mV, 12 μA and 30 ms.

squid giant axons where the chaotic oscillation occurred intermittently between almost synchronised oscillations. Though this may result from unsteadiness of the real membranes, the existence of the same type of intermittent chaos in the Hodgkin–Huxley equations [5] implies that the intermittency is produced deterministically.

12.5.3 *Collapse of quasi-periodicity*

In recent years the route from a quasi-periodic state to a chaotic state has been studied extensively [8, 30, 31]. Usually this route is observed as collapse of a two-dimensional torus via frequency lockings [22, 32, 37]. The collapse of a torus in the Hodgkin–Huxley oscillator is demonstrated in Fig. 12.12. It is a further problem to trace this route in squid giant axons, adjusting two bifurcation parameters of A and f_S.

12.6 **Discussion**

The periodically forced oscillations in both squid giant axons and the Hodgkin–Huxley equations have been analysed by the approximate ω-limiting set of the stroboscopic mapping, or the stroboscopic plot. A synchronised oscillation, a quasi-periodic oscillation and a chaotic oscillation have been identified by points, a closed curve and a strange attractor in the stroboscopic plot, respectively. The chaotic oscillations in the Hodgkin–Huxley equations are also confirmed by the existence of a positive one-dimensional Lyapunov number.

Two rhythms, i.e. the natural frequency and the forcing frequency, create abundant temporal patterns of action potentials in the chaotic

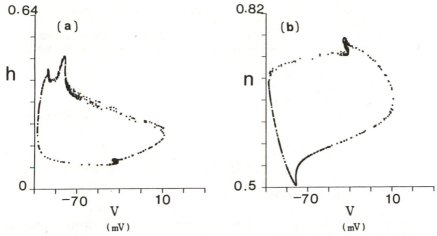

Fig. 12.12 Collapse of quasi-periodicity in the Hodgkin–Huxley equations ($f_N =$ 174.6 Hz, $f_S = 500.0$ Hz and $A = 99\,\mu\mathrm{A\,cm^{-2}}$). (a) The stroboscopic plot on the V–h plane. (b) The stroboscopic plot on the V–n plane.

oscillations. Similar chaotic oscillations have been found in many biological membranes [9, 10, 12, 18]. It is a future problem to clarify whether the chaotic oscillations relate to physiological functions or not.

Acknowledgements

The authors wish to thank M. Ichikawa for his help in the experiments. They would also like to express their cordial thanks to S. Amari and A. V. Holden for their valuable comments.

References

[1] Aihara, K. and Matsumoto, G. Temporally coherent organization and instabilities in squid giant axons. *J. Theor. Biol.* **95**, 697–720 (1982).
[2] Aihara, K. and Matsumoto, G. Two stable steady states in the Hodgkin–Huxley axons. *Biophys. J.* **41**, 87–9 (1983).
[3] Aihara, K., Utsunomiya, T., Matsumoto, G. and Hoshino, H. A chaotic behavior of the membrane potential in squid giant axons. *Proc. World Congress on Medical Physics and Biomedical Engineering* **2**, 34 (1982).
[4] Aihara, K., Kotani, M. and Matsumoto, G. Strange attractors, bifurcations and noise-effects in nerve membranes. *The Institute of Electronics and Communication Engineers of Japan, Technical Report*, MBE84–40 (1984).
[5] Aihara, K., Matsumoto, G. and Ikegaya, Y. Periodic and nonperiodic responses of a periodically forced Hodgkin–Huxley oscillator. *J. Theor. Biol.* **109**, 249–69 (1984).
[6] Aihara, K., Utsunomiya, T., Matsumoto, G. and Kotani, M. Nonlinear oscillations and period-doubling bifurcations in squid giant axons, *Proc. Biological Engineering Society 6th Nordic Meeting*, PMC8,4 (1984).

[7] Feigenbaum, M. J. Quantitative universality for a class of nonlinear transformation. *J. Stat. Phys.* **19**, 25–52 (1978).

[8] Feigenbaum, M. J., Kadanoff, L. P. and Shenker, S. J. Quasiperiodicity in dissipative systems: A renormalization group analysis. *Physica 5-D* 370–86 (1982).

[9] Guevara, M. R., Glass, L. and Shrier, A. Phase locking, period-doubling bifurcations, and irregular dynamics in periodically stimulated cardiac cells. *Science* **218**, 1350–3 (1981).

[10] Guttman, R., Feldman, L. and Jakobsson, E. Frequency entrainment of squid axon membrane. *J. Memb. Biol.* **56**, 9–18 (1980).

[11] Hassard, B. Bifurcation of periodic solutions of the Hodgkin–Huxley model for the squid giant axons. *J. Theor. Biol.* **71**, 401–20 (1978).

[12] Hayashi, H., Nakao, M. and Hirakawa, K. Chaos in the self-sustained oscillation of an excitable biological membrane under sinusoidal stimulation. *Phys. Lett.* **88A**, 265–6 (1982).

[13] Hodgkin, A. L. and Huxley, A. F. A quantitative description of membrane current and its application to conduction and excitation in nerve. *J. Physiol. (London)* **117**, 500–44 (1952).

[14] Holden, A. V. Autorhythmicity and entrainment in excitable membranes. *Biol. Cybern.* **38**, 1–8 (1980).

[15] Holden, A. V. Hopf bifurcation and the repetitive activity of excitable cells, In *Lecture Notes in Biomathematics*, **57**, 335–40. Springer, New York (1985).

[16] Holden, A. V. and Muhamad, M. A. Chaotic activity in neural systems, In *Cybernetics and Systems Research 2*, ed. R. Trappl, pp. 245–50. Elsevier, North-Holland, Amsterdam (1984).

[17] Holden, A. V. and Winlow, W. Neuronal activity as the behavior of a differential system. *IEEE Trans. SMC* **13**, 711–19 (1983).

[18] Holden, A. V., Winlow, W. and Haydon, P. G. The induction of periodic and chaotic activity in a molluscan neurone. *Biol. Cybern.* **43**, 169–73 (1982).

[19] Holden, A. V., Haydon, P. G. and Winlow, W. Multiple equilibria and exotic behaviour in excitable membranes. *Biol. Cybern.* **46**, 167–72 (1983).

[20] Huxley, A. F. Ion movements during nerve activity. *Ann. NY Acad. Sci.* **81**, 221–46 (1959).

[21] Jensen, J. H., Christiansen, P. L., Scott, A. C. and Skovgaard, O. Chaos in nerve. *Proc. Iasted Symposium, ACI* **2**, 15/6–15/9 (1983).

[22] Kaneko, K. Collapse of tori in dissipative mappings, In *Chaos and Statistical Methods*, ed. Y. Kuramoto, pp. 83–8. Springer, Berlin (1984).

[23] Pomeau, Y. and Manneville, P. Intermittent transition to turbulence in dissipative dynamical system. *Comm. Math. Phys.* **74**, 189–97 (1980).

[24] Matsumoto, G. Long-range spatial interactions and a dissipative structure in squid giant axons and a proposal of physical model of nerve excitation. In *Nerve Membrane, Biochemistry and Function of Channel Proteins*, eds G. Matsumoto and M. Kotani, pp. 203–20, University of Tokyo Press, Tokyo.

[25] Matsumoto, G., Kim, K., Uehara, T. and Shimada, J. Electrical and computer simulations upon the nervous activities of squid giant axons at and around the state of spontaneous repetitive firing of action potentials. *J. Phys. Soc. Japan* **49**, 906–14 (1980).

[26] Matsumoto, G., Aihara, K. and Utsunomiya, T. A spatially ordered pacemaker observed in squid giant axons. *J. Phys. Soc. Japan* **51**, 942–50 (1982).

[27] Matsumoto, G., Aihara, K., Ichikawa, M. and Tasaki, A. Periodic and nonperiodic responses of membrane potentials in squid giant axons during sinusoidal current stimulation. *J. Theor. Neurobiol.* **3**, 1–14 (1984).

[28] May, R. M. Simple mathematical models with very complicated dynamics. *Nature, Lond.* **261**, 459–67 (1976).
[29] Minorsky, N. *Nonlinear Oscillations.* Van Nostrand, New York (1962).
[30] Ostlund, S., Rand, D., Sethna, J. and Siggia, E. Universal properties of the transition from quasi-periodicity to chaos in dissipative systems, *Physica 8-D*, 303–42 (1983).
[31] Ruelle, D. and Takens, F. On the nature of turbulence. *Comm. Math. Phys.* **20**, 167–92 (1971).
[32] Sano, M. and Sawada, Y. Transition from quasi-periodicity to chaos in a system of coupled nonlinear oscillators, *Phys. Lett.* **97A**, 73–76 (1983).
[33] Tasaki, I. Demonstration of two stable states of the nerve membrane in potassium-rich media. *J. Physiol. (London)* **148**, 306–31 (1959).
[34] Tomita, K. Chaotic response of nonlinear oscillators, *Phys. Rep.* **86**, 113–67 (1982).
[35] Tomita, K. and Tsuda, I. Towards the interpretation of Hudson's experiment on the Belousov–Zhabotinsky reaction, *Prog. Theor. Phys.* **64**, 1138–60 (1980).
[36] Ueda, Y. and Akamatsu, N. Chaotically transitional phenomena in the forced negative-resistance oscillator. *IEEE Trans. CAS-28*, 217–24 (1981).
[37] Yahata, H. Onset of chaos in some hydrodynamic model systems of equations, In *Chaos and Statistical Methods*, ed. Y. Kuramoto, pp. 232–41, Springer, Berlin (1984).
[38] Yoshizawa, S., Osada, H. and Nagumo, J. Pulse sequences generated by a degenerate analog neuron model. *Biol. Cybern.* **45**, 23–33 (1982).

Part V

Measuring chaos

13
Quantifying chaos with Lyapunov exponents

A. Wolf

*The Cooper Union, School of Engineering,
Cooper Square, New York NY 10003, USA*

13.1 Chaos, orbital divergence and the loss of predictability

Chaos has been discovered both in the laboratory and in the mathematical models that describe a wide variety of systems [1, 3]. In common usage chaos is taken to mean a state in which chance prevails. To the nonlinear dynamicist the word chaos has a more precise and rather different meaning. A chaotic system is one in which long-term prediction of the system's state is impossible because the omnipresent uncertainty in determining its *initial* state grows exponentially fast in time. The rapid loss of predictive power is due to the property that orbits (trajectories) that arise from nearby initial conditions diverge exponentially fast on the average. Nearby orbits correspond to almost identically prepared systems, so that systems whose differences we may not be able to resolve initially soon behave quite differently. In non-chaotic systems, nearby orbits either converge exponentially fast, or at worst exhibit a slower than exponential divergence: long-term prediction is at least theoretically possible.

Rates of orbital divergence or convergence, called Lyapunov exponents [2, 9, 13, 16], are clearly of fundamental importance in studying chaos. Positive Lyapunov exponents indicate orbital divergence and chaos, and set the time scale on which state prediction is possible. Negative Lyapunov exponents set the time scale on which transients or perturbations of the system's state will decay. In this chapter we define the spectrum of Lyapunov exponents, describe the well-known technique for computing a system's spectrum from its defining equations of motion, and outline a new technique for estimating non-negative exponents from experimental data.

13.2 Quantifying chaos in a one-dimensional map

The simplest chaos machine, the one-dimensional (1-D) map, is useful for illustrating the properties of Lyapunov exponents. One such map is the logistic equation, $x(n+1) = r*x(n)*(1-x(n))$, discussed in Chapter 3 [11]. $x(0)$ is an initial condition chosen in the interval $(0,1)$, and r is a tunable parameter in $[0, 4]$. At $r = 4$ the trajectory (the sequence of map iterates $x(i)$, $i = 0, \ldots, \infty$) is known to be chaotic. In Fig. 13.1 the trajectories from two nearby initial conditions are seen to be diverging after only three iterations. If the two initial conditions are viewed as defining the error bar of some single *experimentally determined* initial condition, one's ability to pinpoint the system's state is clearly impaired after a few iterations. The reason for the growth of uncertainty is easily determined from the figure; the average slope of the map, as sampled by the trajectory, must be larger than one. An equivalent viewpoint is to consider the propagation of error through the map with a linear stability analysis. The error in specifying $x(n)$ is defined to be $dx(n)$, whereby

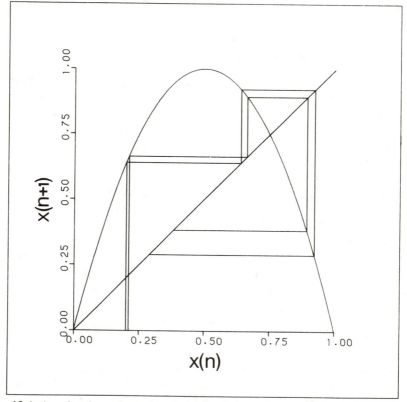

Fig. 13.1 A pair of nearby points is iterated through the logistic map, $x(n + 1) = 4*x(n)*(1-x(n))$. The resulting orbits diverge exponentially fast on the average. Map iteration consists of passing between the map and the line $x(n + 1) = x(n)$.

(13.1) $$x(n + 1) + dx(n + 1) = f[x(n) + dx(n)]$$
$$\sim f[x(n)] + dx(n)*f'[x(n)]$$
$$dx(n + 1) = dx(n)*f'[x(n)] = dx(n)*r[1 - 2x(n)]$$

or, in terms of the initial uncertainty,

(13.2) $$|dx(n)| = |dx(0)| * \prod_{i=0}^{n-1} |f'(x(i))| = |dx(0)| * \prod_{i=0}^{n-1} |r(1 - 2x(i))|$$

For any value of r uncertainty tends to grow in time where the long-term product of the local stretching factors, $|r(1-2*x(i))|$, is greater than one.

If the uncertainty is to grow exponentially fast, eqn (13.2) must be consistent with

(13.3) $$dx(n) = dx(0)*2^{\lambda n}$$

where λ is defined as the Lyapunov exponent. This requires that [13]

(13.4) $$\lambda = \lim_{n \to \infty} \frac{1}{n} \sum_{i=0}^{n-1} \log_2|f'(x(i))|.$$

The limit of large n is necessary if we are to obtain a quantity that both describes long-term behaviour and is independent of initial condition. The limit effectively averages over all initial conditions, except perhaps for a negligible set of points (e.g. all of the points that eventually arrive at the unstable fixed point at the origin in the logistic equation). The probability density of the map is simply the normalised collection of delta functions that mark the locations $x(i)$ visited by a trajectory

(13.5) $$p(x) = \lim_{n \to \infty} \frac{1}{n} \sum_{i=0}^{n-1} \delta(x - x(i)).$$

When combined with eqn (13.4) we obtain

(13.6) $$\lambda = \int p(x)*\log_2|f'(x)| \, dx$$

where the integral is taken over the domain of the map. The Lyapunov exponent is most easily understood in this form: local stretching, determined by the logarithm of the magnitude of the slope, is weighted by the probability of encountering that amount of stretching.

In Fig. 13.2 the Lyapunov exponent for the logistic equation is shown for r in (3.4, 4.0). As r is increased from 0 to approximately 3.57, the system exhibits a period-doubling sequence. At each r, the iterates $x(i)$ converge to a repeating sequence of period 2^n; n increasing by one at bifurcation points such as $r \approx 3.45$. Stable periodic orbits are characterised by negative Lyapunov exponents; bifurcation points correspond to orbits of marginal stability and therefore have zero exponents. For any specified

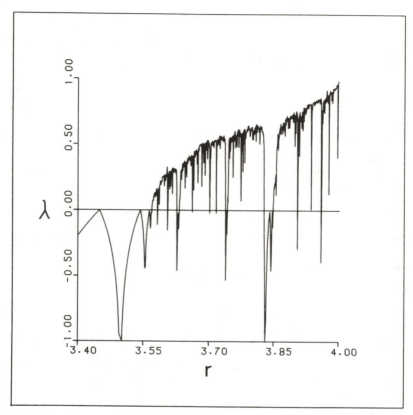

Fig. 13.2 The Lyapunov exponent for the logistic map is shown as a function of the parameter r. The curve for r < 3.4, which exhibits only the first period-doubling bifurcation, is not shown. The units of λ are bits of information lost per map iteration.

period there is a value of r where convergence is faster than exponential, and λ takes the value negative infinity. Only a few such superstable points, such as r ≈ 3.50, can be resolved in Fig. 13.2. As r grows from 3.57 to 4.0, a trend towards increasingly chaotic behaviour and therefore a growing positive exponent is interrupted by an infinite number of 'windows' of periodic behaviour. Figure 13.2 concisely summarises the dynamics of the logistic equation as a function of its tunable parameter, though certain properties, such as the period of the periodic states, cannot be determined from the graph alone.

The units of λ are bits of information per map iteration. (The reader is cautioned that some authors define λ with $\log_e$ rather than $\log_2$ and then incorrectly use the units of (binary) bits per iteration.) For any r, and any specified precision of the initial condition, the value of λ read from the graph quantifies the average rate of loss of predictive power. For example, at $r = 4$, $\lambda = 1.0$ bits per iteration. If an initial condition can be specified to 16 bits of precision, only 8 bits of state information remain after 8 iterations, 4 bits after 12 iterations, and predictive power is completely lost after 16 iterations. (The qualification that λ defines the *average* rate of loss of predictive power is an important one because the uncertainty interval may occasionally shrink for a few iterates if the slope of the map is not everywhere larger than one.) Knowledge of the system's state may be thought of as residing in a 16-bit shift register. At $r=4$, each map iterate has the effect of shifting one bit to the left, past the decimal point and into the void. Bits that come in from the right end to take their place are 'garbage' bits that depend only on the manner in which the iterate is determined. The Lyapunov exponent for a 1-D map is thus the rate at which bits are shifted through the 'state knowledge register' [15]. The analogy of a shift register is not an idle one; with a suitable coordinate transformation, the logistic equation for $r = 4$ becomes the 'bit shift' map $x(n + 1) = 2*x(n)$ (modulus 1).

For the 1-D map, exponential separation is incompatible with motion confined to the unit interval unless a 'folding' process merges widely separated points. In the logistic equation, folding occurs when a pair of simultaneously iterated points fall on opposite sides of $x = 1/2$. These points may be thrown very close together at which time orbital divergence loses (and then regains!) its exponential character. In the repeated stretching and folding that produces chaos, it is a local property of the flow, the stretching, that determines λ, and a larger scale property, the folding, that should never directly appear in exponent calculation. Avoiding the fold proves to be an important consideration in estimating Lyapunov exponents from finite quantities of experimental data. The existence of a fold suggests a modification of our earlier statement on orbital divergence; in chaotic systems, nearby orbits diverge exponentially fast on the average, *so long as their separation remains infinitesimal.* When analysing experimental data, the word 'infinitestimal' must be replaced with 'small compared to the range of dynamical motion'.

Given the functional form of a 1-D map or a long sequence of experimentally obtained map iterates, eqns (13.4) and (13.6) provide a means of estimating λ. This calculation is performed with experimental data for a chaotic chemical reaction governed by a 1-D map in Wolf and Swift [18], where the problems with this approach to estimating λ are discussed.

13.3 Defining the Lyapunov spectrum

For systems whose dimensionality is larger than one, there is a set or spectrum of Lyapunov exponents, each one characterising orbital divergence in a particular direction. The spectrum of exponents is first described for the equations that arise from Lorenz's simple model of fluid convection [8].

$$\frac{dx}{dt} = 16.0\,(y-x)$$

(13.7)
$$\frac{dy}{dt} = 45.92x - xz - y$$

$$\frac{dz}{dt} = xy - 4.0z$$

In the three-dimensional 'phase space' defined by coordinates $(x(t), y(t), z(t))$, post-transient behaviour of almost all trajectories takes place on an 'attractor' whose appearance is locally nearly planar. In Figs. 13.3a–c the solution to these equations is shown for about 15 orbits as a dotted

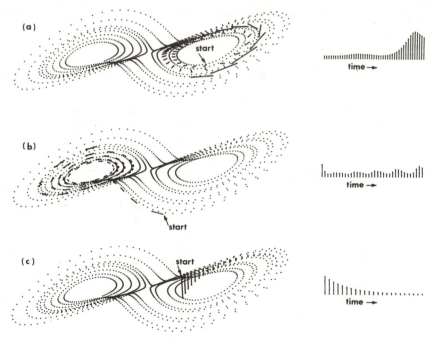

Fig. 13.3 The short-term evolution of the separation vector between three pairs of nearby points is shown for the Lorenz attractor. The true magnitude of the evolving vector appears to the right of each figure. (a) An expanding direction $(\lambda_1 > 0)$. (b) A slower than exponential direction $(\lambda_2 = 0)$. (c) A contracting direction $(\lambda_3 < 0)$.

line. The Lorenz attractor is a 'strange' attractor as it is a chaotic system, possessing a positive Lyapunov exponent. In these figures three kinds of phase-space behaviour are displayed. In Fig. 13.3a an exponential divergence of two nearby points *on different orbits* in the attractor is shown. (It is possible to consider two points in the single long orbit defining the attractor as being on different orbits, provided that their temporal separation is greater than a mean orbital period.) This chaotic motion is characterised by a positive Lyapunov exponent, λ_1. In Fig. 13.3b the behaviour of nearby points *on the same orbit* in the attractor is shown. The separation of these points neither grows nor shrinks exponentially fast so the associated Lyapunov exponent, λ_2, is zero. Finally, in Fig. 13.3c, the decay of a transient or perturbation to the attractor is illustrated. A point displaced off the attractor approaches a (carefully chosen!) point on the attractor, exponentially fast on the average. The associated exponent, λ_3, is negative. Lyapunov exponents involve long time averages, so the short segments of Fig. 13.3 will not accurately characterise the exponents; nevertheless, the qualitative behaviour is already visible. The complete spectrum of exponents for the Lorenz attractor is approximately (2.16, 0.0, −32.4) bits per second.

The three-dimensional phase space of the Lorenz attractor has three Lyapunov exponents, each describing the behaviour of one class of pairs of nearby orbits. In the general case, there are as many exponents as phase-space dimensions, though a particular Lyapunov exponent is not associated with a *unique* direction in phase space, such as a coordinate axis. For example, λ_3 describes orbital decay to the Lorenz attractor, motion orthogonal to the locally nearly planar attractor. However, the attractor is neither globally flat, nor exactly two-dimensional, so this direction varies in a complicated way over the attractor. λ_3 involves a long time average of the contracting nature of phase space over these directions.

The information theory interpretation of the Lyapunov exponents in the Lorenz attractor is a simple extension of the discussion for the 1-D map. If an initial condition is specified to 16 bits in each coordinate, the positive exponent of 2.16 bits per second means that all knowledge of the system's state (except that it still lies within the attractor) is lost after about 8 s, or about 16 mean orbital periods. If a perturbation appears in the next to least significant bit, the negative exponent of −32.4 bits per second means that the orbit will return to the attractor (to our resolution of 16 bits) in about one-sixteenth of an second, or about one-eighth of a mean orbital period.

All strange attractors in a three-dimensional phase space have the same spectral type, $(+,0,-)$: a positive exponent indicating chaos within the attractor, a zero exponent for the slower than exponential motion along an orbit (ref. [5] contains a general proof of the existence of a zero exponent in continuous systems with strange attractors), and a negative exponent so that the phase space contains an attractor.

Even if the magnitudes of the Lyapunov exponents are not known, the spectral type provides a qualitative picture of a system's dynamics, as we have already seen for the logistic equation and the Lorenz attractor. A 1-D map has a single exponent which is positive, negative or zero for chaotic, periodic, and marginally stable behaviour, respectively. In a three-dimensional continuous dissipative system, the only possible spectral types, and the attractors they describe, are: (+,0,−), a strange attractor; (0,0,−), a two-torus; (0,−,−), a limit cycle; and (−,−,−), a fixed point. In four dimensions there are three distinct types of strange attractor with spectral types: (+,+,0,−), (+,0,−,−), and (+,0,0,−).

The spectrum of Lyapunov exponents is now defined in a manner that is particularly useful for the algorithms later presented. Given a continuous dissipative dynamical system in an n-dimensional phase space, the long-term evolution of an infinitesimal n-sphere of initial conditions is monitored. Because of the deforming nature of the flow, the sphere will evolve into an n-ellipsoid. It is assumed that the centre of the sphere is *on* the attractor at $t = 0$, and that the principal axes of the ellipsoid have been ordered from most rapidly to least rapidly growing. The ith Lyapunov exponent is then defined in terms of the growth rate of the ith principal axis, $p_i(t)$

$$(13.8) \qquad\qquad \lambda_i = \lim_{t \to \infty} \frac{1}{t} \log_2 \left[\frac{p_i(t)}{p_i(0)} \right]$$

Notice that the linear extent of the ellipsoid grows as $2^{\lambda_1 t}$, the area defined by the first two principal axes grows as $2^{(\lambda_1 + \lambda_2)t}$, the volume defined by the first three axes grows as $2^{(\lambda_1 + \lambda_2 + \lambda_3)t}$, and so on. This property provides an alternate definition of the spectrum of exponents: the sum of the first j exponents is given by the long-term exponential growth rate of the j-volume defined by the first j principal axes. This alternative definition will prove to be quite useful in spectral calculations. Whichever definition is employed, it is only necessary to follow the motion of as many points on the sphere as there are principal axes, although it may be easier to visualise phase-space behaviour if we consider the evolution of all points on the sphere's surface.

The existence of the limit in eqn (13.8) is not guaranteed for most of the model dynamical systems one is likely to encounter [2,9,13,16], and the situation for experimental data is even murkier. Calculations of orbital divergence rates necessarily characterise the properties of the given data set, and not necessarily the underlying dynamical system. We hope for some correspondence between these two sets of 'Lyapunov exponents', but in general it is not possible to independently confirm exponents determined from experimental data (however, see ref. [3]).

The behaviour of the evolving sphere of states is now examined. The centre of the sphere moves along the trajectory of some particular initial condition, while points on the surface of the sphere move along neighbouring trajectories. The sphere can therefore be expected to rotate and deform as it moves through phase space. Individual axes may grow or shrink exponentially fast, or may show slower than exponential behaviour. The volume of the sphere decreases exponentially fast (as the sum of all of the Lyapunov exponents), but its linear extent grows exponentially fast (as λ_1). These two processes imply that the sphere becomes skewed exponentially fast. Should the sphere initially be finite in extent, each pass through the mandatory folding structure in the attractor would result in its being folded over on itself, ultimately evolving into an infinitely sheeted structure – an object with fractional dimension. Kaplan and Yorke [7] have established a method for estimating the fractional dimension of an attractor from a subset of the spectrum of Lyapunov exponents. When applied to the Lorenz attractor, a fractional dimension of 2.07 is obtained, confirming its not quite planar nature. If the initial sphere of states is infinitesimal in extent, the probability of encountering the fold is zero and the collection of states remains ellipsoidal for all time.

We now outline a proof that the sum of a system's Lyapunov exponents is the time-averaged divergence of its phase space, a quantity that is negative for the dissipative systems considered in this chapter. If a small volume element in a d-dimensional phase space, $\Delta V(t)$, grows exponentially fast with an exponent equal to the sum of all of the Lyapunov exponents, then

$$(13.9) \qquad \frac{\mathrm{d}}{\mathrm{d}t}\,\Delta V(t) = \frac{\mathrm{d}}{\mathrm{d}t}\Delta V(0)2^{[\Sigma\lambda_i]t} = \Delta V(t)[\Sigma\lambda_i]$$

where a factor of $\log_e 2$ is ignored. The change in volume of the element is also given by

$$(13.10) \quad \frac{\mathrm{d}}{\mathrm{d}t}\,(\Delta x(t)\Delta y(t)\ldots) = \Delta x(t)\Delta y(t)\ldots(\frac{1}{\Delta x(t)}\frac{\mathrm{d}\Delta x(t)}{\mathrm{d}t} + \frac{1}{\Delta y(t)}\frac{\mathrm{d}\Delta y(t)}{\mathrm{d}t}+\ldots)$$

$$= \Delta x(t)\Delta y(t)\ldots(\Delta\dot{x}/\Delta x + \dot{y}/\Delta y +\ldots)$$

$$= \Delta V(t)\,(\nabla\cdot\mathbf{v})$$

Comparing eqns [13.9] and [13.10] and noting that the former already involves a long time average, we obtain

$$(13.11) \qquad \sum_{i=1}^{d} \lambda_i = \overline{(\nabla\cdot\mathbf{v})}$$

where the long time average may be taken over a single trajectory. We note that conservative systems may be chaotic, but cannot possess phase-space attractors.

13.4 Computing the Lyapunov spectrum from equations of motion

The calculation of the Lyapunov spectrum from the evolution of an infinitesimal state sphere may not be directly implemented on a computer, as computers cannot represent infinitesimal quantities. In a chaotic system, if the state sphere is initially finite, the exponentially rapid growth of its linear extent means that the fold will be encountered long before the spectrum has converged: we fail to probe only the local structure of the attractor. An additional problem is the exponentially fast growth of the skewness of the sphere: principal axis vectors all collapse to the direction associated with the largest Lyapunov exponent, their directions becoming indistinguishable whatever the precision of one's computer. In Fig. 13.4 the evolution of a pair of initially orthogonal principal axis vectors (their size relative to the attractor greatly exaggerated) is shown in the Hénon strange attractor[6], which is generated by the two-dimensional mapping

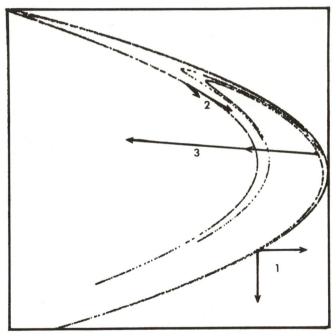

Fig. 13.4 The action of the Jacobian on an initially orthonormal pair of vectors is illustrated for the Hénon map: (1) initial vectors, (2) first iterate, and (3) second iterate. By the second iteration the divergences in extent and orientation are quite apparent. The angular orientations of the two vectors may be *numerically* resolved for a few more iterates.

$$(13.12) \qquad \begin{aligned} x(n+1) &= 1 + y(n) - 1.4{*}x(n)^2 \\ y(n+1) &= 0.3{*}x(n) \end{aligned}$$

After only two iterates the linear extent and the skewness of the collection of states defined by the principal axis vectors are seen to be diverging. A solution to these problems was discovered independently by Bennetin *et al.*[2] and Shimada and Nagashima [16] in 1979. The result is a straightforward technique for computing a complete Lyapunov spectrum to any desired precision for systems whose equations of motion are available. This technique is now described in detail.

The solution to the divergent axis behaviour lies in the use of a technique from linear algebra, the Gram–Schmidt reorthonormalisation procedure, henceforth referred to as GSR [12]. Given a set of linearly independent vectors, GSR provides a new set of vectors that are orthonormal and preserve the orientation of various subspaces of the original set: the first vector in each set is identical in direction, the first two vectors in each set define the same plane, the first three vectors in each set define the same three-volume, and so on. These properties allow us to periodically replace the evolving ellipsoid with a new ellipsoid that is smaller (no problem of diverging extent) and whose principal axes have their orientation 'preserved' (we keep track of each phase-space direction) though the new axis vectors are all orthogonal (no orientation collapse).

We now review the Gram–Schmidt procedure. The first replacement vector is simply the first old vector, normalised. The second replacement vector is the second old vector with its component along the first new vector removed, then normalised. The third new vector is the third old vector with its components along the first two new vectors removed, then normalised, and so on. The reader can easily confirm the orientation preserving property of the replacement vector set from this description.

The importance of the orientation-preserving property of GSR is seen from the alternate definition of the spectrum of Lyapunov exponents, where the rate of length growth determines λ_1, the rate of area growth determines $\lambda_1 + \lambda_2$, and so on. Successive principal axis vectors define volume elements of all dimensions from one to that of the phase space, whose evolution may be simultaneously monitored. GSR allows the state sphere to be evolved for the long times required for exponent convergence because, though changing the directions of all but the first principal axis vector each time it is invoked, it never changes the orientation of any of the volume elements spanned by successive principal axis vectors. When GSR is used the initial and final volumes of the elements of each dimension $(L(t_j),\ L'(t_{j+1});\ A(t_j),\ A'(t_{j+1});\ \dots\ )$ are recorded and used to update running exponential growth rates. If m replacement elements spanning a time t have been used, the exponential growth rate of the first principal axis is

$$(13.13) \qquad (\lambda_1)_m = \frac{1}{t} \sum_{j=1}^{m} \log_2 \left[\frac{L'(t_{j+1})}{L(t_j)} \right]$$

which is identical to the growth rate

$$(13.14) \qquad \lambda_1 = \frac{1}{t} \log_2 \left[\frac{L'(t_1)}{L(t_0)} \right]$$

that would have obtained from the evolution of a single length element had we been able to follow it for time t. Similarly,

$$(13.15) \qquad (\lambda_1 + \lambda_2)_m = \frac{1}{t} \sum_{j=1}^{m} \log_2 \left[\frac{A'(t_{j+1})}{A(t_j)} \right]$$

which is identical to the exponential growth rate of the area element defined by the first two principal axis vectors had we been able to follow it for a long time t. The expressions for the remaining exponents follow similarly.

To summarise the use of the Gram–Schmidt procedure, Lyapunov exponents will be computed from the long-term growth rates of volume elements of various dimensions. Elements may be replaced with smaller elements whose defining vectors are orthogonal, so long as the new elements have the same phase-space orientation as the ones they replace.

The technique described thus far requires that the nonlinear equations of motion be solved once for the centre of the sphere, and once for the end point of each principal axis vector, GSR being invoked when necessary. In a numerical implementation, GSR corresponds to integrating the differential equations for a new set of initial conditions that define the end points of the replacement vectors. The remaining practical difficulty is ensuring that GSR is performed frequently enough that only the local structure of the attractor is being probed.

The problem is avoided if we solve the nonlinear equations for the centre of the sphere, and simultaneously solve the *linearised* equations of motion about that point for each axis vector [2,16]. The linear system can only sample infinitesimal perturbations from the 'fiducial' trajectory of the centre point. This ensures that nearby orbits remain (relatively) near by for long times.

The use of the linear system appears to eliminate the need for GSR, at least for the divergence in the extent of the state ellipsoid. However, the linear equations diverge whenever there exists a positive Lyapunov exponent, just as the nonlinear equations do. The advantage of the linear system is that its solutions continue to represent infinitesimal deviations from the fiducial trajectory, even as they grow numerically large. The divergence in the linear system is simply a problem of exceeding the word size of one's computer: the solution remains small relative to the attractor. The role of GSR in the linear system is to prevent the orientation divergence, and simply to keep numbers manageable in size. More details about this calculation, as well as FORTRAN code for its implementation, may be found in ref. [19].

Although this calculation has been described for a set of ordinary differential equations, it is essentially unchanged if the system is defined by a discrete mapping such as the Hénon map. A linear stability analysis provides us with the linearised equations of motion: the Jacobi matrix for the map, whose evaluation requires iterating the nonlinear equations for the fiducial trajectory. The evolution of nearby orbits is determined by the action of the Jacobian on principal axis vectors. Figure 13.4 shows the action of the Jacobian on a pair of principal axis vectors — not the evolution of three nearby initial conditions by the nonlinear mapping as was previously stated. The evolution of either principal axis vector in a single iteration is given by

$$(13.16) \qquad \begin{pmatrix} dx\,(n+1) \\ dy\,(n+1) \end{pmatrix} = \mathbf{J}_n \begin{pmatrix} dx\,(n) \\ dy\,(n) \end{pmatrix} = \begin{pmatrix} -2.8*x(n) & 1.0 \\ 0.3 & 0 \end{pmatrix} \begin{pmatrix} dx\,(n) \\ dy\,(n) \end{pmatrix}$$

so that

$$(13.17) \qquad \begin{pmatrix} dx\,(n) \\ dy\,(n) \end{pmatrix} = \mathbf{J}_{n-1} \left[\mathbf{J}_{n-2} \cdots \mathbf{J}_1 \begin{pmatrix} dx\,(1) \\ dy\,(1) \end{pmatrix} \right]$$

or, by regrouping the terms,

$$(13.18) \qquad \begin{pmatrix} dx\,(n) \\ dy\,(n) \end{pmatrix} = \left[\mathbf{J}_{n-1}\ \mathbf{J}_{n-2} \cdots \mathbf{J}_1 \right] \begin{pmatrix} dx\,(1) \\ dy\,(1) \end{pmatrix}$$

In eqn (13.17) the latest Jacobi matrix acts on the *current* axis vector, that vector reflecting the action of all previous Jacobi matrices on the *initial* axis vector. When divergences arise in the current vector pair, the pair is replaced. This is the same interpretation of the procedure of 'evolution and replacement' as was presented for the continuous system. In eqn (13.18) the action is contained in the product Jacobian which acts on the *initial* pair of axis vectors. Here divergences appear in the product Jacobian, either when its elements diverge, or its columns all converge to large multiples of the eigenvector for the largest eigenvalue (orientation collapse), resulting in a zero determinant. In this case GSR corresponds to a set of operations on the product Jacobian; removing large scalar multipliers of the matrix and performing row reduction with pivoting. In the first interpretation, λ_1 is determined from the exponential growth rate of the first vector and $\lambda_1 + \lambda_2$ from the growth rate of the area defined by both vectors. In the second interpretation the Lyapunov exponents are determined from the eigenvalues of the long time product Jacobian. For the Hénon map the spectrum of Lyapunov exponents is approximately $(0.4, -1.6)$ bits per iteration for the parameter values of eqn (13.12).

13.5 Estimating the Lyapunov spectrum for experimental data

The spectral calculation of the last section appears useless for the problem of determining Lyapunov spectra from experimental data as it requires the equations of motion defining the system. As mentioned in section 13.2, in continuous systems governed by a discrete 1-D map, the extraction of discrete map data allows the estimation of the largest exponent by some conceptually simple, if numerically unstable algorithms [18]. There have also been attempts to estimate the dominant (smallest magnitude) negative exponent from experimental data by measuring the mean rate of decay of induced perturbations.

Substantial progress has recently been made in the general problem of spectral estimation from experimental data [19]. Utilising the alternative definition of the Lyapunov spectrum, all of the non-negative Lyapunov exponents may be estimated for any system in which samples of a single dynamical observable are available. This algorithm has been used successfully on model systems with known spectra, and early calculations on experimental data obtained from chemical and hydrodynamic strange attractors are promising. The algorithm is now briefly described.

The well known technique of phase-space reconstruction with delay coordinates [14,17] makes it possible to obtain an attractor whose Lyapunov spectrum is identical to that of the original attractor, from discrete-time samples of almost any dynamical observable. Given the time series $x(t_i)$, the new attractor is defined by the trajectory $(x(t_i), x(t_i+T), x(t_i+2T),\ldots, x(t_i+(n-1)T))$. For sufficiently large n, and almost all time delays T, an embedding of the original attractor is obtained. In what follows we assume that an attractor has been successfully reconstructed in this manner.

The long-term exponential growth rate of a j-volume element in an attractor is governed by the sum of the first j Lyapunov exponents, provided that the volume of the element is small enough that the linear approximation applies. In computing spectra from sets of differential equations it was convenient to use the linearised system, the tangent space of the centre of the sphere, to satisfy this constraint. We saw, however, that the evolution of volume elements and GSR might be performed in *phase* space, a space that is accessible with experimental data. GSR is not exactly applicable to experimental data, as data points will not be found at the precise locations of the replacement vector set. Thus, with experimental data it is possible to define initially small volume elements of any desired dimension, follow their evolution for short times, and even replace them when necessary, but it seems that their orientation must be lost upon replacement.

In ref. [19] we show that the errors involved in using *almost* orientation-preserving replacement elements decay exponentially fast in time, and , if certain requirements are met, do not accumulate from one replacement to the next. For example, the relative error in estimating λ_1 is approximately $\theta_M^2/\lambda_1 t_r$ after many replacement and evolution steps, where θ_M is the maximum error in a single replacement and t_r is the time between replacements (the expression is only valid for rather large t_r). If it is possible to perform replacements infrequently, Lyapunov exponents may be accurately estimated from experimental data. The decay of orientation errors, that is, the approach of errant volume elements to the appropriate phase-space orientations, is guaranteed in a chaotic system, and the more

chaotic the system, the faster the decay. In the Lorenz attractor, a 20°
orientation error in the replacement of each length element will result in a
10% error in λ_1, providing that replacements need not be performed
more frequently than once per orbit.

We now discuss the particular case of estimating the sum of the first two
Lyapunov exponents, which should illustrate the procedure for any number
of exponents. Given an attractor reconstructed in n-dimensions, an area
element is defined by three nearby points; the first delay coordinate point
and its two nearest neighbours in n-space. If the three points were
separated by at least one mean orbital period in the original time series,
we may consider the points to have started from distinct initial conditions.
The evolution of this element is monitored by simply looking ahead in the
time series to find the future location of each of its defining points. When
the triple begins to grow too large, or becomes so skewed that we expect
to make a large error in computing its area (a problem that arises if
external noise is present), we record the initial and final area of the
current element, and then look for a replacement element. The search for
replacements is somewhat involved, as it requires minimising both the size
of the two replacement vectors, and the angular separation between the
normals to the original and replacement elements. The initial area of the
jth element is denoted by $A(t_j)$ and the element is replaced at time t_{j+1}
when its area is $A'(t_{j+1})$ (see Fig. 13.5b). After m replacements spanning
a long time t we estimate

$$(13.19) \qquad (\lambda_1 + \lambda_2)_m = \frac{1}{t} \sum_{j=1}^{m} \log_2 \left[\frac{A'(t_{j+1})}{A(t_j)} \right]$$

This is identical to eqn (13.15) except that we are working in a reconstructed phase space with replacement elements that are always finite and
only approximately orientation-preserving. In the limit of an infinite
amount of noise-free data spanning an infinite number of orbits, this
quantity is exactly the sum of λ_1 and λ_2. In practice, experimental data
are noisy and span a finite number of orbits, so the accuracy of the
estimate depends on the quality and quantity of experimental data. In Fig.
13.5a the algorithm for estimating λ_1 is presented schematically.

Despite the similarity of our algorithm to the calculation for model
systems, it may not be used in general to determine negative exponents
from experimental data. In systems such as the nearly planar Lorenz
attractor, there is little or no *resolvable* fractal structure: thus we cannot
define volume elements whose dimension is larger than that of the planar
surface. Even in a system with resolvable fractal structure, such as the
1.26-dimensional Hénon attractor, areas decay much faster than lengths
grow, so that area elements must be replaced with great frequency. Since
post-transient attractor data are not effective for sampling contracting
phase-space directions, it seems reasonable that the decay of induced

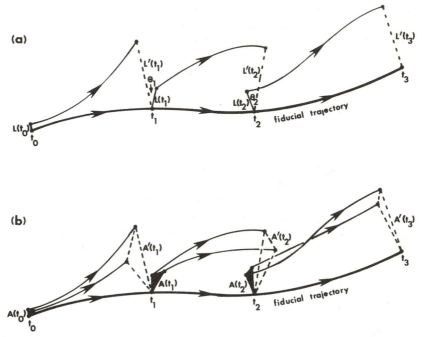

Fig. 13.5 A schematic representation of the procedure used to estimate Lyapunov exponents from experimental data. (a) The largest Lyapunov exponent λ_1 is computed from the growth of length elements. When the vector between the two data points has become large, a new point is chosen near the fiducial trajectory, minimising the replacement length L and the orientation error θ. (b) A similar procedure is followed to calculate $\lambda_1 + \lambda_2$ from the growth of area elements. When an area becomes too large or too skewed, two new points are chosen near the fiducial trajectory, minimising the replacement area A and the difference between the phase-space orientation of the original and replacement areas.

perturbations might provide a means of estimating negative exponents. This approach has several problems to contend with, of which the most important may be verifying that perturbations change only the state of the system (the current values of phase space variables) and not the system itself.

Our algorithm has been tested on many systems including the Hénon, Lorenz, and hyperchaos attractors (the last is a 3.005-dimensional attractor in a four-dimensional phase space with two positive Lyapunov exponents). The defining equations of motion were used solely to generate an observable sample of a one-dimensional coordinate projection. For these systems, the non-negative exponents were determined to within a few per cent of their known values. In ref. [3] we presented results for λ_1 for experimental data obtained from the Taylor-Couette hydrodynamic system as a function of the relative Reynolds number R/R_c. A transition to chaos from motion on a two-torus at $R/R_c \approx 12.2$ had already been suggested by other dynamical diagnostics (phase portraits, Poincaré sections, power spectra, fractional dimension) but was first usefully quantified by the calculation of $\lambda_1 (R/R_c)$.

A fundamental problem with the computation of Lyapunov exponents from experimental data is that the exponents are not rigorously defined in the presence of external noise. (See, for example, ref. [10].) Our calculations are based on the assumption that behaviour on (relatively noise-free) intermediate length scales is close to that on the smallest experimentally accessible (noise-dominated) length scales, which is assumed to be close to that on infinitesimal length scales in the (identically noise-free) underlying system. We have found that low-pass filtering of experimental data before exponent estimation has often reduced the effects of moderate amounts of external noise.

We conclude with the results of ref. [19] concerning data requirements. In each of the systems studied, 6 or 7 bits of data resolution sufficed for accurate exponent estimation. The number of data points required to estimate all of the non-negative exponents in a system of fractional dimension d is on the order of 10^d to 30^d, spanning between 10^{d-1} and 100^{d-1} orbits, where the value within these ranges depends on the complexity of the underlying 1–D map (if such a map exists). Exponent estimation appears to be prohibitively expensive for attractors of dimension greater than 3 or 4. The problem is the same one that arises in all of the currently popular techniques for estimating fractional dimension. To 'diagnose' a high-dimensional attractor, the attractor must be defined (filled out) with a number of points that depends *exponentially* on its dimension [4]. This problem appears insurmountable, but as the theory of dynamical diagnostics is incomplete, we are hesitant to proclaim such calculations impossible.

Acknowledgements

The author wishes to thank Jack B. Swift and Harry L. Swinney of the University of Texas Nonlinear Dynamics Laboratory for their contributions to this work. The efforts of Bob Hopkins and Paul Johnson in the preparation of the chapter are gratefully acknowledged. Financial support was provided by the Department of Energy, Office of Basic Energy Sciences contract DE-AS05-84ER13147.

References

[1] Abraham, N.B., Gollub, J.P. and Swinney, H.L. Testing nonlinear dynamics. *Physica 11-D*, 252–64 (1984), has an extensive list of references.
[2] Bennetin, G., Galgani, L. and Strelcyn, J-M. Lyapunov characteristic exponents for smooth dynamical systems and for Hamiltonian systems; a method for computing all of them. *Meccanica* **15**, 9–20 (1980).
[3] Brandstater, A., Swift, J., Swinney, H.L., Wolf, A., Farmer, J.D., Jen, E. and Crutchfield, J.P. Low-dimensional chaos in a hydrodynamic system. *Phys. Rev. Lett.* **51**, 1442–5 (1983).

[4] Greenside, H.S., Wolf, A., Swift, J. and Pignaturo, T. Impracticality of a box-counting algorithm for calculating the dimensionality of strange attractors. *Phys. Rev. A* **25**, 3453–6 (1982).

[5] Haken, H. At least one Lyapunov exponent vanishes if the trajectory of an attractor does not contain a fixed point. *Phys. Lett.* **94A**, 71–2 (1983).

[6] Hénon, M. A two-dimensional mapping with a strange attractor. *Comm. Math. Phys.* **50**, 69–77 (1976).

[7] Kaplan, J.L. and Yorke, J.A. Chaotic behaviour of multidimensional difference equations. In *Lecture Notes in Mathematics 730,* eds H.O. Peitgen and H.O. Walther. Springer, Berlin (1979).

[8] Lorenz, E.N. Deterministic nonperiodic flow. *J. Atmos. Sci.* **20**, 130–41 (1983).

[9] Lyapunov, A.M. Problème général de la stabilité du mouvement. *Ann. Math. Study* **17** (1947).

[10] Matsumoto, K. and Tsuda, I. Noise-induced order. *J. Stat. Phys.* **31**, 87 (1983).

[11] May, R. M. Simple mathematical models with very complicated dynamics. *Nature, Lond.* **261**, 459–67 (1976).

[12] Moore, J.T. *Elementary Linear and Matrix Algebra: the Viewpoint of Geometry.* McGraw-Hill, New York (1972).

[13] Oseledec, V.I. A multiplicative ergodic theorem. Lyapunov characteristic numbers for dynamical systems. *Trans. Moscow Math. Soc.* **19**, 197 (1968).

[14] Packard, N.H., Crutchfield, J.P., Farmer, J.D. and Shaw, R.S. Geometry from a time series. *Phys. Rev. Lett.* **45**, 712–16 (1980).

[15] Shaw, R. Modeling chaotic systems. In *Chaos and Order in Nature,* ed. H. Haken. Springer, Berlin (1981).

[16] Shimada, I. and Nagashima, T. A numerical approach to ergodic problem of dissipative dynamical systems. *Progr. Theor. Phys.* **61**, 1605–16 (1979).

[17] Takens, F. Detecting strange attractors in turbulence. In *Lecture Notes in Mathematics 898,* eds D.A. Rand and L-S. Young. Springer, Berlin (1981).

[18] Wolf, A. and Swift, J. Progress in computing Lyapunov exponents from experimental data. In *Statistical Physics and Chaos in Fusion Plasmas,* eds C.W. Horton, Jr. and L.E. Reichl, pp. 111–25 Wiley, New York (1984).

[19] Wolf, A., Swift, J.B., Swinney, H.L. and Vastano, J.A., Determining Lyapunov exponents from a time series. *Physica 16–D*, 285–317 (1985).

14

Estimating the fractal dimensions and entropies of strange attractors

P. Grassberger

Physics Department, University of Wuppertal,
Gauss-Str. 20, D-5600 Wuppertal 1,
Federal Republic of Germany

14.1   Information flow and dimension

Physical systems typically involve a huge number of degrees of freedom ($\gtrsim 10^{20}$, say). Since it is impossible to treat all of them explicitly, one performs some kind of 'coarse graining'. After this is done, one deals explicitly with few variables only, but in general these variables evolve nondeterministically with time. This was widely considered to be the only source of randomness in nature until the 'chaos revolution' of recent years spread the concept of deterministic chaos.

Deterministic chaos is also related to coarse graining, but of an essentially different kind. It results from the fact that we not only cannot deal with too many variables, but we also cannot deal with infinitely precise numbers. If we accept that space–time is continuous, this means that we must cut off the digital expansions for all coordinates somewhere.

In the textbook examples of classical mechanics, this cutoff has no effect. But there exist formally deterministic systems—and even very simple ones, with only few variables—where trajectories emerging from nearby initial conditions diverge exponentially. Due to this 'sensitive dependence on initial conditions', any ignorance about seemingly insignificant (and thus cut-off) digits spreads with time towards the significant digits, leading to an essentially unpredictable and 'chaotic' behaviour.

In this chapter, I shall concentrate on dissipative chaotic systems. The sets of points in state space towards which nearly all trajectories are converging are called strange attractors in that case.

Let me stress again that chaotic motion is fundamentally different from ordered motion. The unpredictability cannot be avoided by just making more precise measurements of the initial conditions. Assume that we want to measure the initial conditions very precisely, say with some error $\pm\varepsilon$

in each variable of state space. Then we have first to build a suitable measuring device. But this same device can be used later to measure the state again with the same precision, whereas the equations of motion are unable to predict it with the same accuracy.

If the distance between nearby trajectories increases exponentially—and this is expected by scale invariance, as long as this distance is much smaller than all typical length scales in the problem—the lack of information is independent of ε and proportional to the elapsed time.

Consider now a piece of trajectory between times t_1 and t_2. Assume that t_1 is sufficiently large that transients have already died out, and that the system is on its attractor with some time-invariant probability distribution. The information $\hat{S}_\varepsilon$ needed to specify this trajectory to an error $\pm \varepsilon$ during the whole interval $[t_1, t_2]$ consists then of two parts:

(a) the information S_ε needed to specify it at time t_1;
(b) the information needed to fix up the ignorance leaking in at a constant
     rate due to the ignorance about originally insignificant digits.

Thus, we expect

(14.1) $$\hat{S}_\varepsilon [t_1, t_2] \approx S_\varepsilon + (t_2 - t_1) \cdot K$$

for $\varepsilon \to 0$ and $t_2 - t_1 \to \infty$.

The constant K is called the Kolmogorov–Sinai or 'metric' entropy [6, 32, 51]. It is essentially the information flow rate in the limit of nearly error-free measurements [10, 49].

For predictable systems, one would have $K = 0$. For the other extreme of Brownian motion (or, rather, its mathematical idealisation as a Wiener process) one has $K = \infty$: even perfect knowledge of the state at some instant would not be sufficient to predict it in the near future.

Let us now look at the dependence of S_ε on the error ε. If we want to specify a point x on a fixed interval with accuracy $\pm \varepsilon$, the needed information (i.e. the number of significant bits) behaves for $\varepsilon \to 0$ like $\log_2 (1/\varepsilon)$. In the following, we shall always use natural logs instead of $\log_2$, implying that we measure information in 'nats' and not in 'bits'. For a point in D-dimensional space, the information is D times as big. Thus, we expect that

(14.2) $$S_\varepsilon \sim D \cdot \log (1/\varepsilon) \text{ for } \varepsilon \to 0$$

where D is called the information dimension of the attractor [4, 11–13].

We have to be somewhat more precise. The estimate $\log(1/\varepsilon)$ for the information stored in a point on an interval is true only if we know *a priori* that the point is indeed on the interval, and nothing more. Analogously, eqn (14.2) assumes *a priori* that the state is on the known attractor (i.e. any transients have died out), and is distributed according to some invariant 'measure' (= distribution), which is also assumed to be known from previous observations.

The typical case, assumed throughout the following, is that the system is ergodic and mixing. In that case, nearly all initial conditions within some suitable basin of attraction lead to the same invariant distribution, called the 'natural measure' [7]. But in all chaotic systems there also exist other invariant distributions, reached from initial conditions of measure zero. Quantities like D and K actually refer to one particular distribution. If nothing else is said, it will be the natural measure.

The most striking feature of strange attractors is that D is in general noninteger. Since D is closely related to the Hausdorff dimension (see section 14.3), the attractor is a 'fractal' in the sense of Mandelbrot [40]. The observation of noninteger D in Couette–Taylor flow [8] (see Fig. 14.1) and in other hydrodynamic systems [39] is indeed a beautiful proof that

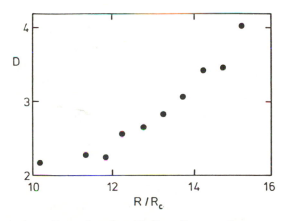

Fig: 14.1 Information dimension for Taylor–Couette flow, as a function of Reynolds number. (Adapted from [8].)

strange attractors occur in real physical systems. Studying D and ways to measure it will thus be a main concern of this chapter.

As we have already stressed, the motion on a strange attractor is unpredictable due to the divergence of nearby trajectories. This divergence is measured by the Lyapunov exponents [13, 41]. In order to define them, consider an infinitesimally small sphere in state space around some point $\mathbf{x}$ with radius ε. After a time $t \gg 0$, this sphere is transformed into an ellipsoid with semi-axes

$$(14.3) \qquad\qquad \varepsilon_i(t) \sim \varepsilon e^{\lambda_i t}$$

($i = 1, 2, \ldots f$; f = number of degrees of freedom) for nearly all $\mathbf{x}$. In the following, we shall always assume that the λ_i are ordered by magnitude, $\lambda_1 \geq \lambda_2 \geq \ldots \geq \lambda_f$. Since the system is dissipative by assumption, their sum is < 0. But since it is also chaotic, at least $\lambda_1 > 0$. The distance between two arbitrary nearby points will then increase as:

$$| \Delta x(t) | \sim e^{\lambda_i t}.$$

At nearly every point in the basin of attraction, the semi-axes λ_i define the axes of a local coordinate system, up to signs. Those directions with $\lambda_i > 0$ are called unstable, and those with $\lambda_i < 0$ are called stable. In addition, one can have 'central' directions with $\lambda_i = 0$. In particular, the direction of the velocity vector in continuous-time systems has $\lambda_i = 0$.

In the next section, we shall see that we can attribute a 'partial dimension' D_i to each of the stable, unstable, and central directions. Just as D measures how the information depends on an uncertainty common to all state variables, D_i measures its dependence on the uncertainty of the ith local coordinate only. Thus it is clear that

$$(14.4) \qquad\qquad D = \sum_i D_i$$

Stated differently, D_i is the density of information per bit of the ith coordinate. Since the Lyapunov exponent λ_i is just the speed of the information flow along that coordinate, the rate of information flowing from the 'insignificant digits' of x_i into the system is $D_i \times \lambda_i$. The Kolmogorov entropy, being the total rate of information flow, will then be given by [19, 24, 35, 43]

$$(14.5) \qquad\qquad K = \sum_i{}' D_i \cdot \lambda_i$$

where the sum extends over positive λ_i only.

The natural measure is characterised by maximal information density along the unstable directions: along these directions, each incoming bit is unpredictable, leading to $D_i = 1$. The resulting relation

$$K = \sum_i \lambda_i$$

has been proposed by Ruelle [47]. Measures with $D_i \neq 1$ along unstable directions arise, e.g. in the flow on repellers leading to transient chaos [18, 29, 45].

If the motion is invertible (which it always is for continuous time systems), information is leaving the system by convergence of points along the stable directions. The rate is again given by eqn (14.4) but this time the sum extends over all negative λ_i. Conservation of information (the average knowledge about the system stays constant with time) leads then to the simple relation

$$(14.6) \qquad\qquad \sum_i D_i \cdot \lambda_i = 0$$

A formula for D proposed by Kaplan and Yorke [14, 30] follows immediately if we assume that D is the maximum allowed by eqns (14.4) and (14.6), taking also into account that $0 \leq D_i \leq 1$. In section 14.2 we shall discuss the technical details of the above concepts and relations.

In addition to Kolmogorov entropy and information dimension, often-studied quantities are the topological entropy [1] and the fractal (or Hausdorff) dimension of the attractor. We shall see in section 14.3 how these (and similar quantities [16, 24]) are related to Renyi entropies [46] on the one hand, and to fluctuations in the local expansion rates (i.e. to deviations from $\varepsilon_i(t) \sim \exp(\lambda_i t)$, on the other.

Practical algorithms for evaluating these generalised dimensions and information flow rates will be discussed in section 14.4.

14.2 Formal developments

Technically, the information S_ε is defined via a partitioning of the attractor into (hyper-) cubes of size ε (see Fig. 14.2). Let us call p_i the

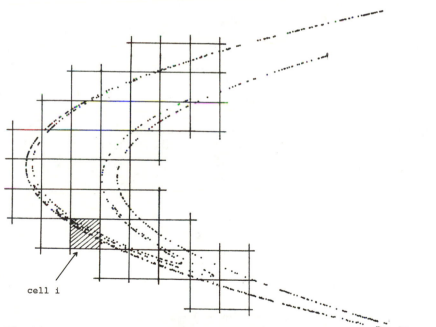

cell i

Fig: 14.2 Partitioning of space (Hénon map).

probability that an arbitrary point $\mathbf{x}(t)$ falls into cube i. That is, p_i is the 'mass' of cube i with respect to the natural measure μ:

$$(14.7) \qquad p_i = \int\limits_{\text{cube } i} d\mu(\mathbf{x})$$

Then, S_ε is defined as

(14.8)
$$S_\varepsilon = -\sum_i p_i \log p_i$$

Analogously, we can define $\hat{S}_\varepsilon$ $[t_1, t_2]$ by partitioning not space but space-time (see Fig. 14.3). Assume that the interval $[t_1, t_2]$ has been divided

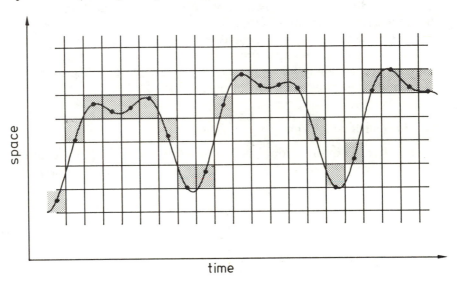

Fig: 14.3 Partitioning of space–time.

into n bins of length τ each. Let $p_{\{i_1...i_n\}}$ be the joint probability that $\mathbf{x}(t_1 + \tau) \in$ cube i, $\mathbf{x}(t_1 + 2\tau) \in$ cube $i_2 \dots \mathbf{x}(t_2) \in$ cube i_n. Then

(14.9)
$$S_\varepsilon [t_1, t_2] = -\sum_{\{i\}} p_{\{i_1...i_n\}} \log p_{\{i_1...i_n\}}$$

The information dimension is [4,13]

(14.10)
$$D = \lim_{\varepsilon \to 0} \frac{S_\varepsilon}{\log 1/\varepsilon}$$

and the metric entropy is [10,49]

(14.11)
$$K = \lim_{\varepsilon \to 0} \lim_{t_2 \to \infty} \frac{1}{t_2 - t_1} S_\varepsilon[t_1, t_2]$$

The equation defining S_ε can be interpreted as follows: S_ε is the average, weighted according to the natural measure, of the quantity

(14.12)
$$-\log p_i = -\log \int d\mu(\mathbf{x})$$

where the integral goes over a cube of size ε in a fixed mesh. But this should be the same, within a constant of order 1, as the average over a cube or ball of size ε not in a fixed mesh, but with its centre distributed randomly according to μ:

(14.13)
$$S_\varepsilon \approx - \int d\mu(\mathbf{x}) \cdot \log M_\varepsilon(\mathbf{x}) \equiv - <\log M_\varepsilon>$$

with

(14.14)
$$M_\varepsilon(\mathbf{x}) = \int_{|\mathbf{y}-\mathbf{x}|<\varepsilon} d\mu(\mathbf{y})$$

Equation (14.10) states then that the mass in a ball of size ε increases, in geometric average, like ε^D.

A completely analogous argument shows that

(14.15)
$$\hat{S}_\varepsilon[t_1, t_2] \approx - <\log \hat{M}_\varepsilon[t_1, t_2]>$$

where $\hat{M}_\varepsilon(\mathbf{x}, [t_1, t_2])$ is the mass of that domain around $\mathbf{x}(t_1)$, the trajectories emerging from which stay within a distance ε from $\mathbf{x}(t)$ for all $t_1 < t < t_2$ (see Fig. 14.4).

(14.16)
$$M_\varepsilon(\mathbf{x}, [t_1, t_2]) = \mu(B_\varepsilon(\mathbf{x}))$$

with

(14.17)
$$B_\varepsilon(\mathbf{x}) = \{\mathbf{y}(t_1) : |\mathbf{y}(t) - \mathbf{x}(t)| < \varepsilon \text{ for all } t \, [t_1, t_2]\}$$

As we noted in section 14.1, the stable and unstable directions define at (nearly) any point a foliation. We can thus generalise the above and consider, instead of balls, ellipsoids with axes ε_1, $\varepsilon_2 \dots$ along the thereby induced directions. The definitions of S_ε, $\hat{S}_\varepsilon$, M_ε, $\hat{M}_\varepsilon$ and B_ε are generalised in an obvious way to $S_{\varepsilon_1\varepsilon_2\dots}$, etc.

During time evolution, each ellipsoid will rotate (which is irrelevant for us), and the semi-axes will change as

(14.18)
$$\varepsilon_i \to \varepsilon_i \cdot e^{\Lambda_i(\mathbf{x},t)}$$

The exact behaviour of the dilatation factors Λ_i is also irrelevant for the moment, but their averages are by definition the Lyapunov exponents:

(14.19)
$$\lambda_i = \frac{1}{t} \int d\mu(\mathbf{x}) \, \Lambda_i(\mathbf{x}, t)$$

Let us now consider the domain $B_{\varepsilon_1\varepsilon_2} \dots (\mathbf{x})$. It will be an ellipsoid with semi-axes

(14.18a)
$$\approx \begin{cases} \varepsilon_i & \text{along the stable directions} \\ \varepsilon_i \, e^{-\Lambda_i(\mathbf{x},t)} & \text{along the unstable directions} \end{cases}$$

If $\hat{S}_{\varepsilon_1\varepsilon_2} \dots [t_1, t_2]$ is to increase linearly with t_2-t_1, with rate K, the mass of an ellipsoid should then scale like a power in each ε_i, at least for the unstable directions. But scale invariance anyhow suggests that it scales with each ε_i,

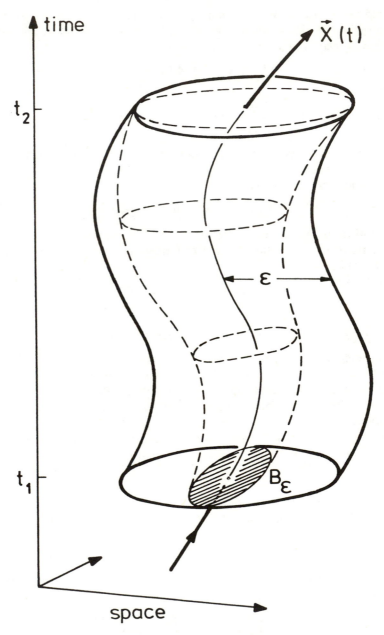

Fig: 14.4 The domain $B_\varepsilon(\mathbf{x})$ (shaded region) consists of those points, the trajectories emerging of which stay in an ε-sausage around $\mathbf{x}(t)$.

$$(14.20) \qquad \langle \log M_{\varepsilon_1 \varepsilon_2} \ldots \rangle \underset{\varepsilon_i \to 0}{\sim} \sum_i D_i \log \varepsilon_i$$

from which one obtains, using eqns (14.11) and (14.15), (14.19 – 20),

(14.21)
$$K = \sum_i{}' D_i \lambda_i$$

Here, the sum extends over all directions with positive ε_i only.

The constants D_i are the (information) dimensions along the ith (un-) stable manifold. They clearly satisfy

(14.22)
$$0 \le D_i \le 1$$

and

(14.23)
$$D = \sum_i D_i$$

Equation (14.20) shows that the attractor factorises essentially in a direct product of continua (corresponding to $D_i = 1$), discrete points ($D_i = 0$), and Cantor sets ($D_i \ne 0, \ne 1$), with orientation according to the (un-) stable directions. As an example, we show in Fig 14.5 the Hénon [28] attractor with $b = 0.3$ and $a = 1.4$, in which case $D_1 = 1$ and $D_2 \approx 0.25$ [24]. For a different example, see [37].

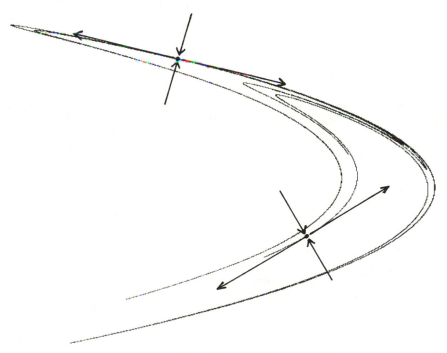

Fig: 14.5 Attractor of the Hénon map $(x,y) \rightarrow (1 + y - ax^2, bx)$ with $a = 1.4$, $b = 0.3$. The arrows indicate the stable and unstable directions.

As pointed out in section 14.1, we can apply the same argument to the time-reversed motion if it is also deterministic, and obtain eqn. (14.6). This equation, together with eqn (14.21), was first conjectured in ref. [24], and proven rigorously in ref. [35]. An upper bound on D compatible with that relation is obtained if $D_i = 1$ for all $i < j$, $D_i = 0$ for all $i > j$ and $0 \le D_j < 1$. Here, the integer j is uniquely determined by eqn (14.6). The bound is [34]

$$(14.24) \qquad D \le D_{\text{Lyap}} \equiv j + \frac{\sum\limits_{i=1}^{j} \lambda_j}{|\lambda_{j+1}|}$$

It was conjectured by Kaplan and Yorke [14, 30] that indeed $D = D_{\text{Lyap}}$ in all typical cases. If true, this represents by far the easiest way of computing D in analytically defined models. (In experimental situations, it is much less useful.) In two-dimensional maps, $D_1 = 1$ for the natural measure, and thus the conjecture is correct [56]. In higher dimensions, one knows several examples [2, 23,50] where $D < D_{\text{Lyap}}$, but all are 'untypical' (i.e. correspond to a set of measure zero in some control parameter). For non-natural measures, the Kaplan–Yorke conjecture does not apply in general, but eqns (14.6), (14.21) and (14.24) are still correct [18, 29, 35].

For a heuristic explanation of why a typical attractor should fill mostly the directions of least contraction, as required by the Kaplan–Yorke conjecture, consider a 3-dimensional map with $\lambda_1 > 0$ and $-\lambda_1 > \lambda_2 > \lambda_3$. The stable and unstable manifolds of a typical point X are shown schematically in Fig. 14.6. A trajectory coming close to X will follow closely the heavy line passing through the points Y_1, Y_2,... It is obvious from Fig 14.6 that in a typical case this line is tangent to the least stable direction (direction '2'). Thus, the attractor also will typically be extended in that direction rather than in direction '3'. The open question is simply whether 'typical' here means the same as in ref. [14], and whether this explanation still holds in higher dimensions.

14.3   Generalised entropies and dimensions

Shannon's definition of information is not the only one possible. It fulfils a number of important conditions (the Khinchin axioms [31, 46], but if one relaxes these conditions somewhat, a number of other information-like quantities can be defined.

The most important of these generalised entropies are the order-α Renyi informations, defined as [46]

$$(14.25) \qquad S_\varepsilon^{(\alpha)} = \frac{1}{1-\alpha} \log \sum_i p_i^\alpha$$

Here, α is any positive real number $\ne 1$, and the p_is are the same probabilities as in the preceding section. By de l'Hopital's rule one finds

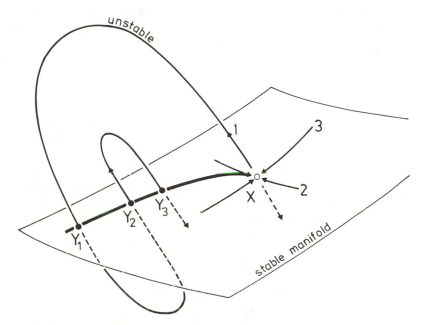

Fig: 14.6 Stable and unstable manifolds of a typical point X on an attractor of a 3-dimensional map. A trajectory coming close to X has to approach it along the heavy line, explaining why the attractor is extended along the less stable direction '2' instead of the more stable direction '3'.

(14.26)
$$S_\varepsilon = \lim_{\alpha \to 1} S_\varepsilon^{(\alpha)}$$

and the derivative with respect to α can be written as

(14.27)
$$\frac{\partial S_\varepsilon^{(\alpha)}}{\partial \alpha} = -(1-\alpha)^{-2} \sum_i z_i \log \frac{z_i}{p_i}$$

with $z_i = p_i^\alpha / \Sigma p_j^\alpha$. Since

$$\sum_i z_i = 1$$

we can interpret the right-hand sum in eqn (14.27) as a Kullback information gain [33,46], which is well known to be non-negative. Thus, the $S_\varepsilon^{(\alpha)}$ are indeed generalisations of the Shannon information, and they provide upper bounds (for $\alpha < 1$) and lower bounds (for $\alpha > 1$) for it.

The same construction can be made for the spatio-temporal informations $\hat{S}_\varepsilon [t_1, t_2]$, and order-$\alpha$ dimensions $D^{(\alpha)}$ and information flow rates $K^{(\alpha)}$ can be defined [16, 24] again by eqns (14.10) and (14.11). Due to the monotonicity of $S_\varepsilon^{(\alpha)}$, both $D^{(\alpha)}$ and $K^{(\alpha)}$ are also monotonically decreasing with α, and $D = \lim_{\alpha \to 1} D^{(\alpha)}$ and $K = \lim_{\alpha \to 1} K^{(\alpha)}$.

Of particular interest is the limit $\alpha \to 0$. In this limit, $S_\varepsilon^{(\alpha)}$ becomes the logarithm of the number of non-empty cubes; thus $D^{(0)}$ is just the fractal dimension [40] of the attractor, and $K^{(0)}$ is the topological entropy [1, 6, 9] of the flow *on* the attractor. (In most discussions of topological entropy, the flow in the whole basin of attraction is discussed. If this basin has fractal boundaries, the topological entropy receives a finite contribution from the flow near the boundaries.)

Finally, all $D^{(\alpha)}$ and $K^{(\alpha)}$ are invariant under a smooth coordinate change [16], i.e. they do not depend on the particular choice of coordinates. This is very important, as the choice of suitable coordinates is in general not unique (see section 14.4).

Like the Shannon information, the Renyi information can be approximately related to the mass of ε-balls. The difference is that now one has to use a different kind of averaging,

$$(14.28) \qquad S_\varepsilon^{(\alpha)} \approx \frac{1}{1-\alpha} \log <M_\varepsilon^{\alpha-1}>$$

instead of the geometrical average, and consequently

$$(14.29) \qquad <M_\varepsilon^{\alpha-1}> \underset{\varepsilon \to 0}{\sim} \varepsilon^{(\alpha-1)D(\alpha)}$$

and, similarly,

$$(14.30) \qquad <\hat{M}_\varepsilon[t_1,t_2]^{\alpha-1}> \underset{\substack{\varepsilon \to 0 \\ t_2-t_1 \to \infty}}{\sim} \varepsilon^{(\alpha-1)D(\alpha)} e^{(1-\alpha)(t_2-t_1)K(\alpha)}$$

Equations (14.29) and (14.30) have two important applications. First, consider the case $\alpha=2$. In this case, eqn (14.29) simplifies to

$$(14.31) \qquad \iint d\mu(\mathbf{x})\, d\mu(\mathbf{y})\, \Theta\,(\varepsilon - |\mathbf{x}-\mathbf{y}|) \sim \varepsilon^{D(2)}$$

The left-hand side is nothing but the integral over the two-point

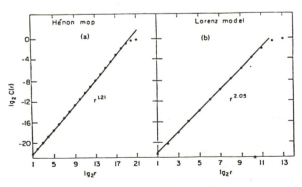

Fig. 14.7 Correlation integral, estimated from 15 000 iterations, for the Hénon map (left) and the Lorenz model. (From [23].)

correlation function. It measures the probability that two random points **x** and **y** (random, but distributed according to the measure μ) are within a distance ε. Because of that, $D^{(2)}$ was called a 'correlation exponent' in refs [22] and [23]. Its measurement proceeds by taking a long time series $[\mathbf{x}(t), \mathbf{x}(t+\tau), \mathbf{x}(t+2\tau), \ldots]$ and counting the number of pairs $[\mathbf{x}(t+n\tau), \mathbf{x}(t+k\tau)]$ with distance less than ε. This number should scale like $\varepsilon^{D^{(2)}}$. For the Hénon map and the Lorenz model [36], this scaling is shown in Fig. 14.7; for other systems it holds equally well [23].

The order-2 entropy $K^{(2)}$ can be obtained in exactly the same way by starting from eqn. (14.30) [24,53]. We just have to count the number of pairs $[\mathbf{x}(t+n\tau), \mathbf{x}(t+k\tau)]$ with the property that

[14.32]
$$|\mathbf{x}(t+n\tau) - \mathbf{x}(t+k\tau)| < \varepsilon$$
$$|\mathbf{x}(t+(n+1)\tau) - \mathbf{x}(t+(k+1)\tau)| < \varepsilon$$
$$\vdots$$
$$|\mathbf{x}(t+(n+d)\tau) - \mathbf{x}(t+(k+d)\tau)| < \varepsilon$$

This number should decrease $\sim \exp(-d\tau K^{(2)})$, tests of which can be found in [24].

Modifications of these algorithms, needed if the state space has very high dimension, and similar algorithms for estimating D and K directly, will be given in the next section.

For the second application of eqns (14.29) and (14.30), consider first their generalisations from ε-balls to ellipsoids. In complete analogy to eqn (14.20), and as suggested by scale invariance, we assume that

(14.33)
$$<M^{\alpha-1}_{\varepsilon_1 \varepsilon_2 \ldots}> \underset{\varepsilon_i \to 0}{\sim} \prod_i \varepsilon_i^{(\alpha-1)D_i(\alpha)}$$

Here again

(14.34)
$$0 \le D_i^{(\alpha)} \le 1$$
$$D^{(\alpha)} = \sum_i D_i^{(\alpha)}$$

and $D_i^{(\alpha)}$ can be considered as the order-α dimension of the measure along the ith direction. Direct numerical checks of eqn. (14.33) are not easy. The only nontrivial system for which it has been checked numerically [21] is the Mackey-Glass delay equation [38].

We want now to derive the analogue of the information flow equation (14.21) and of the information conservation law (14.6). But there is a subtle problem. As we have said, the Renyi informations do not satisfy one of the Khinchin axioms, namely postulate IV on page 548 of ref. [46]. This postulate says roughly that the information in a joint distribution is the sum of the informations corresponding to the single distributions. The Renyi informations satisfy it if and only if the single distributions are independent, i.e. if the joint distribution factorises. In our case, this means that we can expect something like conservation of Renyi entropies only if the flow is sufficiently mixing (which should be the case for natural measures).

For a mixing flow, the expansion coefficients $\Lambda_i(\mathbf{x}, t)$ should become for $t \to \infty$ like the sum of t independent random variables:

(14.35a)
$$<\Lambda_i(\mathbf{x},t)> \sim t \cdot \lambda_i$$

(14.35b)
$$<(\Lambda_i - \lambda_i)(\Lambda_k - \lambda_k)> \sim t \cdot Q_{ik}, \dots \text{ etc.}$$

In terms of the generating function, this means

(14.36)
$$<e^{\Sigma z_i \Lambda_i}> \underset{t \to \infty}{\sim} e^{tg(z_1, z_2, \dots)}$$

with

(14.37)
$$g(\mathbf{z}) = \sum_i z_i \lambda_i + \tfrac{1}{2} \sum_{i,k} z_i z_k Q_{ik} + \dots$$

Equation (14.35a) is just the definition of Lyapunov exponents, but eqn (14.35b) and the analogous equations for the higher moments are non-trivial. They have been tested in ref. [24], where very precise values of the covariance matrix $\mathbf{Q}$ were obtained for several attractors. For some models, even third-order correlation matrices could be computed, with eqn. (14.36) always satisfied.

The time invariance of the measure, the ansatz (14.33) for the mass of an ellipsoid, and eqn. (14.37) can now be taken together to obtain the very simple relation

(14.38)
$$g((1-\alpha)\mathbf{D}^{(\alpha)}) = 0$$

or, equivalently,

(14.39)
$$\sum_i D_i^{(\alpha)} \lambda_i = \frac{\alpha-1}{2} \sum_{i,k} D_i^{(\alpha)} D_k^{(\alpha)} Q_{ik} \pm \dots$$

This is obviously the generalisation of eqn (14.24). In order to obtain the analog of (14.21),

(14.40)
$$K^{(\alpha)} = \sum_i {}' D_i^{(\alpha)} \lambda_i - \frac{\alpha-1}{2} \sum_{i,k} {}' D_i^{(\alpha)} D_k^{(\alpha)} Q_{ik} \pm \dots$$

one has to start from eqn (14.30), generalised first to ellipsoids.

Both eqn (14.39) and eqn (14.40) have been tested numerically in several examples in refs [18], [24] and [29]. Although they were always fulfilled for natural measures on attractors, they were not obeyed for other invariant measures. Whether that means that these latter measures are nonmixing in the technical sense [6] is not clear.

14.4 Measuring information flow and dimension

The most straightforward way to measure K and D would be to use the definitions based on box counting. For systems with low-dimensional state space (i.e. with few variables), this is indeed feasible [12, 17, 18, 25], but for $\geq$ four variables the storage requirements become prohibitive, independently of the dimension of the attractor.

In that case, all practical methods are based on estimating the mass $M_\varepsilon(\mathbf{x})$ of typical ε-balls (for estimating D) respectively of typical domains $B_\varepsilon(\mathbf{x})$ (see eqn. (14.17) and Fig. 14.4) for estimating K.

In order to estimate $M_\varepsilon(\mathbf{x})$, one counts the number of points $\mathbf{y}_n$ in a time series $[\mathbf{y}_1 = \mathbf{y}(t), \ \mathbf{y}_2 = \mathbf{y}(t+\tau), \ \mathbf{y}_N = \mathbf{y}\ (t+(N-1)\tau)]$ which satisfy $|\mathbf{y}_n - \mathbf{x}| < \varepsilon$. For the distance, one can use here either the Euclidean norm or the norm given by the largest difference in any coordinate. In the latter case, one counts these $\mathbf{y}_n$ for which all $|\mathbf{y}_{n,k} - \mathbf{x}_k| < \varepsilon$ for all k. The point $\mathbf{x}$ should be arbitrarily placed on the attractor. In order to obtain the information dimension, taking one single point $\mathbf{x}$ is enough in principle [13,26], but then the time series has to be excessively long. Thus, in practice, one will average over several reference points $\mathbf{x}_1, \ldots \mathbf{x}_M$.

When determining D (and any $D^{(\alpha)}$ with $\alpha \neq 2$), one should take first the limit $N \to \infty$ (estimating thus $M_\varepsilon(\mathbf{x}_k)$ for fixed $\mathbf{x}_k$), and afterwards $M \to \infty$. The errors in $M_\varepsilon(\mathbf{x}_k)$ due to the finiteness of N cause systematic errors — because of the nonlinear averaging — for small ε, where the scaling laws (14.20) and (14.29) should be most precise. Thus, these scaling laws can only be tested down to that ε for which the ball around every $\mathbf{x}_k$ contains $\gg 1$ points.

This problem is absent only for $\alpha = 2$, in which case one averages the $M_\varepsilon(\mathbf{x}_k)$ linearly, and in which case the role of $\mathbf{y}_n$ and $\mathbf{x}_k$ is completely symmetric (see eqn. (14.31)). In that case, scaling can be tested further down to values of ε such that the *sum* of all balls around all $\mathbf{x}_k$s contains $\gg 1$ points. For that reason, the correlation exponent $D^{(2)}$ is the easiest generalised dimension to estimate, even if it is not the most interesting. Attractors with dimension $D^{(2)} \approx 7$ have been successfully analysed in this way, from time series of $\sim 10^4$ points [23]. Efficient algorithms (ref.[23], appendix) are essential in keeping computer time low.

In the case $\alpha = 2$ it is natural to take for $[\mathbf{x}_n]$ and $[\mathbf{y}_n]$ the same time series. This is particularly useful in cases where a long time series is not easy to obtain (as in most observations of natural phenomena), and one therefore wants to make maximal use of observed data. The numerical effort then increases quadratically with the number N of observations. For $\alpha \neq 2$, it seems optimal to take $N \gg M \gg 1$. If one took $N = M$, the scaling region would be rather small (some $\mathbf{x}_n$ would be in such lowly populated regions that only very large balls would contain many $\mathbf{y}_k$s).

At the other extreme it has been proposed [26,27] to use only one reference point **x**, since then the numerical effort is only proportional to N, and the information obtained from different reference points is not completely independent anyhow. But for the interesting small-distance limit, this information is nearly independent, and the reduction in computer time by taking only one reference point is usually marginal as compared with the effort involved in obtaining the time series.

All the above remarks also apply to the computation of generalised entropies. In this case one has to count the number of sequences $\{y_n, y_{n+1}, \ldots y_{n+d}\}$ in the time series for which

(14.41)
$$\begin{aligned}
|y_n - x_k| &< \varepsilon \\
|y_{n+1} - x_{k+1}| &< \varepsilon \\
&\vdots \\
|y_{n+d} - x_{k+d}| &< \varepsilon
\end{aligned}$$

Here, we have assumed that the x_k are also obtained from a time series (eventually $x_k = y_k$) with the same delay τ between successive measurements. The mean number of such sequences should decrease (for finite but small ε and $d \to \infty$) like $\exp(-d\tau K^{(\alpha)})$, where α depends on the way of averaging. The cases $\alpha=1$ (geometric averaging) and $\alpha=2$ have been studied in refs. [54] and [24], respectively.

A variant of the above methods of measuring $D^{(\alpha)}$ and $K^{(\alpha)}$ consists of the following. One measures the radii $R_j(x_k)$ of the smallest tubes around $\{x_k, \ldots x_{k+d}\}$ containing exactly j sequences $\{y_n \ldots y_{n+d}\}$. Their logarithmic average (averaged over all x_k) should behave like [3, 8, 26, 27]

(14.42)
$$\langle \log R_j \rangle \sim \frac{1}{D} \left(\log \frac{j}{N} + d\tau K \right).$$

The averages of powers of R_j should scale like [3,20]

(14.43)
$$\langle R_j^q \rangle \sim \left(\frac{j}{N} e^{d\tau K^{(\alpha)}} \right)^{q/D^{(\alpha)}}$$

with α given by

(14.44)
$$q = (1 - \alpha) D^{(\alpha)}$$

For this method, the range of distances over which one can test scaling is between the two ranges discussed above. For small distances (corresponding to $j \approx 1$), one again has systematic deviations from eqns (14.42) and (14.43), but this time the deviations can be computed exactly [20]. The main drawback of the method is that one has to order the points y_n according to their distances from x_k. This has to be done for each k, which enhances computation time and storage requirements considerably. Nevertheless, it seems that this method is quite efficient for computing information dimension [8] and other generalised dimensions [55] (notice, however, that in the latter paper some of the errors seem grossly underestimated [20]).

In the above discussion, we have assumed that we have determined all dynamical variables (i.e. all components of **x**) in the time series. In practical cases, this is rather unrealistic. The way out of this dilemma was pointed out by Takens [52] and Packard *et al.* [42].

First, one notices that a D-dimensional attractor should be representable faithfully in any $\mathbb{R}^d$ with dimension $d > D$. Actually this is not quite true, since information dimension is a purely measure-theoretic concept, and the smallest $\mathbb{R}^d$ in which a D-dimensional attractor is embeddable might have much higher dimension. But, in practice, such pathological cases do not seem to occur.

Secondly, a $\mathbb{R}^d$ useful for that purpose (called mock state space in the following) is spanned by d successive measurements of a (generic) time sequence of any single observable. Since $D_i^{(q)}$ and $K^{(q)}$ are invariants, they will be independent of the choice of observable and of the time sequence, except in singular cases.

Here some comments are in order about optimal choices for d and τ. In principle, even if one is only interested in the dimension and not in K, one should test several values of d, including values which are definitely much larger than D. One reason was discussed in ref. [23]: often a D-dimensional object just looks smoother if embedded in higher dimensional space then when represented in some D-dimensional space. A simple example is the surface of a sphere with constant mass density. When projected on to a plane, the density becomes singular at $r = R$. But too big values of d have the drawback that statistics get very bad (there simply are no pieces of the time series which run parallel for a very long time). Optimal compromises have thus to be found in each case.

The value of the time delay τ should not be chosen too big. One would otherwise run into the same problems of low statistics. Also, if the system is inherently noisy, different coordinates in the mock state space would no longer be deterministically related, resulting in a spuriously large value of D. On the other hand, a too small value of τ would mean that all coordinates in the mock state space are roughly equal, leading to a cigar-shaped and quasi-one-dimensional attractor. Experience has shown that optimal results are obtained if τ is somewhat smaller than the typical turnover time. (An exception is the estimate of partial dimensions in ref. [21], where $\tau \approx 1/6$ of the turnover time was optimal.)

Successful estimates of different $D^{(\alpha)}$s for various systems, both real and simulated ones, have been performed in this way in refs [8], [23], [24], [39] and [55]. In general, it was found there that the embedding dimension d had to be considerably bigger than $D^{(\alpha)}$, otherwise the latter was systematically underestimated.

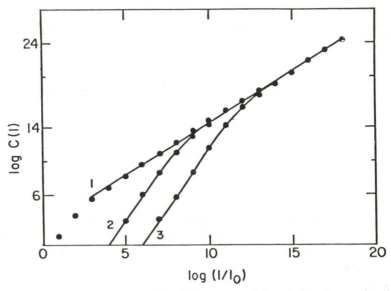

Fig. 14.8 Correlation integral of the Hénon map with added random noise. Noise levels are $|\delta x_n| = 0$ (curve 1), $|\delta x_n| \leqslant 0.5 \times 10^{-3}$ (curve 2), and $|\delta x_n| \leqslant 5 \times 10^{-2}$ (curve 3). (From [5].)

As a last remark, let us consider the influence of low-level noise on an otherwise deterministic attractor. Assume that the noise is white, with amplitude of order δ. Unless the system amplifies this noise excessively, the noise will not be felt on a length scale $\gg \delta$. On length scales $\ll \delta$, on the other hand, any deterministic structure is washed out completely, and the system is not confined to the D-dimensional attractor. Thus, the effective dimension (i.e. the slope in a doubly logarithmic plot of the correlation integral versus ε) will be D for $\varepsilon \gg \delta$, but will be the full dimension of state space for $\varepsilon \ll \delta$. This was indeed found [5] for the Hénon and Mackey-Glass equations (see Fig 14.8).

14.5   Conclusions

We have discussed the information flow leading to the unpredictability of chaotic motion. We have seen that it is a fundamental problem, rendering obsolete Laplace's 'superior intelligence' which can predict the fate of the universe from its initial conditions. Indeed, Lorenz's famous paper [36] arose simply from our inability to predict even the weather of our Earth for more than a few days.

This information flow rate was found to be a sum over all unstable directions of the product of the divergence rates for nearby trajectories (the Lyapunov exponents) times the partial dimensions of the attractor along these directions. The information-balance equation led then to a connection between dimension and Lyapunov exponents (a weakened form of the Kaplan–Yorke conjecture), which—unlike the Kaplan–Yorke conjecture— is true in *all* cases. Practical algorithms for computing dimensions and information flow rates were discussed.

In most of the discussions, we used only that the motion is stationary in the mean, i.e. that the considered distribution is invariant under time translations. It was not necessary to consider the fact that it is attractive. Accordingly, we obtain the same results also for repellers or other invariant distributions leading sometimes to metastable chaos.

A remark about Hamiltonian systems is in order here. All the above arguments apply to them as well, but, according to the conventional folklore, most of them are trivial there. This folklore claims that single chaotic orbits fill regions of finite volume on the energy shell. (In this case, all $D_i^{(\alpha)}$ are equal to 1 for a chaotic orbit, and eqns (14.6) and (14.39) are equivalent to Liouville's theorem.) For systems with a three-dimensional energy shell, this is indeed true. For higher-dimensional systems, the *sum* of all chaotic orbits fill a finite volume, and one assumes that all chaotic regions are connected by Arnold diffusion. But the latter need not be true [15], and numerical simulations [44] indeed suggest that typical chaotic trajectories fill fractal regions, with fractal dimension much less than the dimension of the energy shell. If this proves to be correct, the above considerations should apply nontrivially also to Hamiltonian systems.

Throughout the present chapter, we have assumed that some coarse graining is done, and have looked at the behaviour when this coarse graining is made finer and finer. Indeed, without coarse graining, it seems impossible to define information or entropy. Without considering the limit of infinitely fine coarse graining, it seems hard to distinguish rigorously between deterministic chaos and stochastic processes. There are also, of course, other indications of deterministic chaos, like Feigenbaum sequences, or flows which resemble Lorenz attractors. But the most clear-cut sign of determinism is that effective Kolmogorov entropy flow rates, and effective dimensions stay finite in the limit of fine coarse graining.

Whether existing or future observations are precise enough to see this behaviour is an open question which definitely deserves further study.

References

[1] Adler, R., Konheim, A. and McAndrew M. Topological entropy. *Trans. Amer. Math. Soc.* **114**, 309 (1965).
[2] Alexander, J.C. and Yorke, J.A. Fat baker's transformation. Univ. Maryland preprint (1982).
[3] Badii, R. and Politi, A. Hausdorff dimension and uniformity factor of strange attractors. *Phys. Rev. Lett.* **52**, 1661 (1984).
[4] Balatoni, J. and Renyi, A. On the notion of entropy. *Publ. Math. Inst. Hung. Acad. Sci.* **1**, 9 (1956) (selected papers of A. Renyi, vol. 1, p.558; P.Turan, ed., Budapest, 1976).

[5]   Ben-Mizrachi, A., Procaccia, I. and Grassberger, P. The characterization of experimental (noisy) strange attractions. *Phys. Rev.* **A29**, 975 (1984).
[6]   Billingsley, P. *Ergodic Theory and Information.* Wiley, New York (1965).
[7]   Bowen, R. and Ruelle, D. The ergodic theory of axiom A flows. *Inventiones Math.* **29**, 181 (1975).
[8]   Brandstaeter, A., Swift, J., Swinney, H.L., Wolf, A., Farmer, J.D., Jen, E. and Crutchfield, J.P. Low-dimensional chaos in a system with Avogadro's number of degrees of freedom. *Phys. Rev. Lett.* **51**, 1442 (1983).
[9]   Crutchfield, J.P. and Packard, N.H. Symbolic dynamics of one-dimensional maps: entropy, finite precision, and noise. *Int. J. Theor. Phys.* **21**, 433 (1982).
[10] Farmer, D. Order within chaos. Thesis, University of California, Santa Cruz (1981).
[11] Farmer, D. Information dimension and the probabilistic structure of chaos. *Z. Naturforsch.* **37a**, 1304 (1982).
[12] Farmer, D. Chaotic attractors of an infinite dimensional system.*Physica 4-D*, 366 (1982).
[13] Farmer, D., Ott, E. and Yorke, J. A. The dimension of chaotic attractors. *Physica 7-D* 153 (1983).
[14] Frederickson, P., Kaplan, J.L. and Yorke, J.A. The Lyapunov dimension of strange attractors. *J. Diff. Eqns* **49**, 185 (1983).
[15] Galgani, L. private discussion.
[16] Grassberger, P. Generalized dimensions of strange attractors. *Phys. Lett.* **97A**, 227 (1983).
[17] Grassberger, P. On the fractal dimensions of the Hénon attractor. *Phys. Lett.* **97A**, 224 (1983).
[18] Grassberger, P. Information flow and maximum entropy measures for 1-D maps. *Physica 14-D*, 365 (1985).
[19] Grassberger, P. Information aspects of strange attractors. In *Proceedings of NATO Workshop on Chaos in Astrophysics.* Palm Coast, Florida, eds. J. Perdang *et al.* (1984).
[20] Grassberger, P. Generalizations of the Hausdorff dimensions of fractal measures. *Phys. Lett.* **107A**, 101 (1985).
[21] Grassberger, P. Unpublished.
[22] Grassberger, P. and Procaccia, I. On the characterization of strange attractors. *Phys. Rev. Lett.* **50**, 346 (1983).
[23] Grassberger, P. and Procaccia, I. Measuring the strangeness of strange attractors. *Physica 9-D* 189 (1983).
[24] Grassberger, P. and Procaccia, I. Dimensions and entropies of strange attractors from a fluctuating dynamics approach. *Physica 13-D* 34 (1984).
[25] Greenside, H.S., Wolf, A., Swift, J. and Pignataro, T. Impracticality of a box-counting algorithm for calculating the dimensionality of strange attractors. *Phys. Rev.* **25A**, 3453 (1982).
[26] Guckenheimer, J. Dimension estimates for attractors. *Contemp. Maths* **28**, 357–67 (1984).
[27] Guckenheimer, J. and Buzyna, G. Dimension measurements for geostrophic turbulence. *Phys. Rev. Lett.* **51**, 1438 (1983).
[28] Hénon, M. A two-dimensional mapping with a strange attractor. *Comm. Math. Phys.* **50**, 69 (1976)
[29] Kantz, H. and Grassberger, P. Repellers, semi-atttractors, and long-lived chaotic transients. *Physica 17-D*, 75 (1985).
[30] Kaplan, J.L. and Yorke, J.A. Chaotic behaviour of multidimensional difference equations. In *Lecture Notes on* Mathematics. **730**, p.204. Springer, Berlin (1978).

[31] Khinchin, A.J. *Mathematical Foundations of Information Theory*. Dover, New York (1957).

[32] Kolmogorov, A.N. Entropy per unit time as a metric invariant of automorphisms. *Dokl. Akad. Nauk. SSSR* **124**, 754 (1959). (English summary in *Math. Rev.* **21**, 2035.)

[33] Kullback, S. and Leibler, R. A. On information and sufficiency. *Ann. Math. Statist.* **22**, 79 (1951).

[34] Ledrappier, F. Some relations between dimension and Lyapunov exponent. *Comm. Math. Phys.* **81**, 229 (1981)

[35] Ledrappier, F. and Young, L.-S., The metric entropy of diffeomorphisms, parts I and II. Berkeley preprints (1984)

[36] Lorenz, E.N. Deterministic nonperiodic flow. *J. Atmos Sci.* **20**, 130 (1963).

[37] Lorenz, E.N. The local structure of a chaotic attractor in four dimensions. *Physica 13-D* 90 (1984).

[38] Mackey, M.C. and Glass, L. Oscillations and chaos in physiological control systems. *Science* **197**, 287 (1977).

[39] Malraison, B., Atten, P., Berge, P. and Dubois, M. Turbulence—dimension of strange attractors. *J. Physique-Lettres* **44**, L897 (1983).

[40] Mandelbrot, B.B., *The Fractal Geometry of Nature*. W.H. Freeman, San Francisco (1982).

[41] Oseledec, V.I. Multiplicative ergodic theorem; Lyapunov characteristic numbers for dynamical systems. *Trans. Mosc. Math. Soc.* **19**, 197 (1968).

[42] Packard, N.H., Crutchfield, J.P., Farmer, J.D. and Shaw, R.S. Geometry from a time series. *Phys. Rev. Lett.* **45**, 712 (1980).

[43] Pesin, Ya. B., Characteristic Lyapunov exponents and smooth ergodic theory. *Russ. Math. Surveys* **32**, 55 (1977).

[44] Pettini M. *et al.* In *Proceeding of NATO Workshop on Chaos in Astrophysics*, Palm Coast, Florida, eds J. Perdang *et al.* (1984).

[45] Pianigiani, G. and Yorke, J.A. Expanding maps on sets which are almost invariant: decay and chaos. *Trans. Amer. Math. Soc.* **252**, 351 (1979).

[46] Renyi, A. *Probability Theory*. Elsevier North-Holland, Amsterdam (1970).

[47] Ruelle, D. Sensitive dependence on initial conditions and turbulent behaviour. In *Bifurcation Theory and its Applications in Scientific Disciplines*, *Ann. N.Y. Acad. Sci.* **136**, 229 (1981).

[48] Russel, D.A., Hanson, J.D. and Ott E. Dimension of strange attractors. *Phys. Rev. Lett.* **45**, 1175 (1980).

[49] Shaw, R. Strange attractors, chaotic behaviour, and information flow. *Z. Naturforsch.* **36a**, 80 (1981).

[50] Shtern, V.N., Attractor dimension for the generalized baker's transformation. *Phys. Lett.* **99A**, 268 (1983).

[51] Sinai, Ya.G. On the concept of entropy of a dynamical system. *Dokl. Akad. Nauk SSSR* **124**, 768 (1959).

[52] Takens, F. Detecting strange attractors in turbulence. In *Lecture Notes in Mathematics* **898**, 366. Springer, New York (1981).

[53] Takens, F. Invariants related to dimension and entropy. In *Atas. do 13 Coloquio Brasileiro de Matematica* (1984).

[54] Termonia, Y. Kolmogorov entropy from a time series. *Phys. Rev.* **A29**, 1612 (1984).

[55] Termonia, Y. and Alexandrowicz, Z. Fractal dimension of strange attractors from radius versus size of arbitrary clusters. *Phys. Rev. Lett.* **51**, 1265 (1983).

[56] Young, L.-S. Dimension, entropy, and Lyapunov exponents. *Ergodic Th. Dyn. Syst.* **2**, 109 (1982).

Part VI

Epilogue

15
How chaotic is the universe?

O. E. Rössler

Institut für Physikalische und Theoretische Chemie,
Universität Tubingen,
D-7400 Tubingen, Federal Republic of Germany

The title of the chapter is better than anything can possibly follow it. Originally I planned to write something nice and grandiose about Anaxagoras' invention of chaos as an explanation of the universe, and his ideas about transfinite iteration and the subtlety of the single 'immiscible' substance, the mind. Diesel automobile engines, the geyser Old Faithful, X-ray bursters in the sky, and autonomous nerve equations (including a 3-variable FitzHugh equation whose chaotic analogue computer solutions were shown to me by its inventor in late 1976) were then to follow suit—to illustrate the ubiquity of trajectorial mixing in simple differential systems populating the cosmos.

Yet, even though it would be tempting to (re-)consider these topics (cf.[13]) in detail, and perhaps to add a disclaimer about the validity of discrete computational models as an exhaustive description of nature (cf.[3]), something less pretentious will be done in the following. 'A return to the mothers' of concrete three-dimensional visualisation is to be proposed once more.

Look at a gas at equilibrium — a chaotic 'gas' (an artificial word that means 'chaos') of equal billiard balls. And feel the exhilaration of riding on such a ball like a Baron Munchhausen (or inside it — it makes for a perfect bumping cart). Or even better: lean against the perfect walls of the container of the gas (in a safe little niche) and watch and listen.

It is like watching snowflakes fall. It takes a little while to get in tune and see the laws behind the whirling: that I am moving upwards with the ground, at constant speed, for example. If the moving spheres are big enough and slow enough (and you are small enough in your niche to feel awed), you may suddenly 'see' — even if this should turn out false on later analysis — that every ball owns a territory: one Nth of the volume of the whole container is assigned to it. And you 'realise' that each ball is busy carrying out a duty: to contribute to the general pounding on the walls with the same vigour as all the others, on average. (Even if its mass were widely different from those of the others, the mean force contributed would be the same.) When all the particle masses are equal as assumed, this job is even audible (although, strictly speaking, there are no sound waves permitted): there is a fixed mean rate of collisions with the wall per unit area; it 'snows' at a constant rate.

In other words, there is not only a unit volume (and hence unit length) present, and a unit pressure (and hence unit energy), but also a unit time (per squared unit length). All of this is, of course, well known since the time of Waterston [20], the first billiards fan in physics.

Still, it is possible to step back a little more after this daydream, and remember that what we were looking at — these huge, floating spheres gliding by slowly and colliding silently and gracefully, all the while maintaining a fixed density in space-time — was chaos. More precisely it was hyperchaos with $3N-1$ positive Lyapunov characteristic exponents (that is, directions of repetitive stretching and folding-over in state-space [14]) present. This process of an unfathomable complexity (see [18] for some of the mathematical details for up to five balls) produces a perfect mixture in the sense of Anaxagoras (to mention his 'Fragment Number 12' again [14]).

Even if only a very artificial set of initial conditions had been chosen with infinite accuracy — so that all the balls were running on orthogonal tracks as it were in-between the collisions, and the whole system was therefore equivalent to a discrete system (a cellular automaton of the reversible type [6]) — one could already be sure that the motion performed by the system was in general of the maximum computational complexity possible and therefore beyond predictability in principle [6]. The above chaos contains these special solutions as unstable periodic ones (even if the period is infinite), embedded into the larger measure 'non-periodic' ones (cf. [8]).

Now the question to be looked at: suppose you were in the possession of a Hamiltonian system of the above type, implemented in a desk-top computer of that Laplacian type that is still not available on the market. Or, even better, suppose yourself implemented by a real-time machine of this kind. (This is not totally impossible — if one agrees to being gaseous, a 'gaseous vertebrate', say — since complicated dissipative structures containing potential observers are, in principle, realisable on the basis of the idealised Hamiltonian assumptions made; cf. [12].) And then ask yourself the question again: how chaotic is the universe?

This time it is a question not only of being, but also of appearing. If I were the only chaotic system in the universe, would not the *whole* universe appear chaotic to me? In other words, there are two senses in which the universe can be (or appear) chaotic. The less obvious second alternative is the one to be looked at in the following.

It could turn out, for example, that a universe that is chaotic itself *ceases* to be chaotic as soon as it is observed by an observer who is chaotic himself. This possibility would lose some of its implausibility if it could be shown that observer-internal chaos is the source of quantum mechanics.

In the following, a tiny little step in this latter direction will be attempted. In this attempt, only results already obtained above are to be used. There is, as we saw, both a characteristic unit energy and a characteristic unit time to every chaotic Hamiltonian system of the above type. In other words, there is a unit action to certain chaotic observers.

This is not too surprising a statement. It is implicit in the Gibbs entropy formula [7] and indeed follows from the Gibbs paradox. Take a volume of gas (of the above type) and place it beside another identical such volume that likewise is at equilibrium, but do not connect the two. Then you obtain an entropy for the composite system that is twice that of the original gas volume. (The equilibrium entropy per volume is the same if you look at two volumes instead of one. This is self-evident.) If you now remove the barrier between the two volumes so that each billiard ball can roam about twice its former space, the total entropy (and the entropy per volume) is *still* the same. This is what should be called the Gibbs paradox. For, if entropy has something to do with disorder and mixing, an increase should have occurred. Usually, the paradox is rather seen in the following fact. Suppose the two different gas volumes had actually contained two different kinds of billiard ball, but as much alike as you wish. In this case the new entropy per volume after the mixing is indeed much larger than before the removal of the barrier (by N times ln 2 times k larger). The phenomenon thus has something to do with the indistinguishability of particle types.

In the present context, this is of no concern since we had assumed just one type of billiard ball to be present. Under this assumption, it is easy to continue, however. Suppose we had been given just one of the two gas specimens above, but still wanted to insert a barrier into its middle. What would be the entropy per volume, in this case? You guess the trend. However much we continue to insert (or remove) barriers, the entropy per volume always is the same. In this way, we can even calculate the whole entropy. It is simply that of N volumes of gas of the present density, each containing just one particle. We thus only need to know the entropy of a single particle in that unit volume, V/N, and we are finished. Gibbs [7] thought that inserting phase-space volume might do — and indeed everything came out fine. Specifically, one obtains $S_G/k = 3N \cdot \ln\Phi_1$, where Φ_1 is the phase-space volume of a single degree of freedom per particle, $\Phi_1 = \bar{p} \cdot q \cdot 3.33$ where $\bar{p}$ is the mean momentum of each particle, q is the side length of a cube of volume V/N, and the constant factor has to do with the ratio of the surface to the volume of an N-sphere and is an algebraic expression that involves π and e (cf. [1] for a lucid exposition). The point is that Φ_1, the unit phase-space volume, is a unit action. This unit action is identical (except for a factor of order unity) to the one found in the above 'floating spheres' picture.

This dramatic coincidence between an empirically correct formula (it is valid at high temperatures in absolute terms if the last remaining constant — a unit action by which Φ_1 has to be divided in order to become a dimensionless quantity — is put equal to Planck's constant [16]) on the one hand, and a casually obtained statistical property of the same chaotic system (a fixed density in space-time) on the other, could be entirely accidental. Entropy may be a very atypical macroscopic quantity. A second objection: the equilibrium entropy S_G is not applicable to chaotic observers (who by definition must be open systems).

Interestingly, the author who first obtained the above-mentioned experimental result [16] and thereby (with Tetrode [19]) rediscovered Planck's constant in a formally ordinary differential equation ('ODE') (rather than partial differential equation — radiation) context, made an attempt to understand this finding in a statistical-mechanical (ODE) context [17]. Sackur, who does not quote Gibbs, thereby rediscovered the Gibbs unit cell (V/N) and stressed the theoretical significance of the mean cell passage time (Φ_1/kT) that goes with it. This picture is independent of openness or closedness conditions. Sackur, who died in 1914, was later criticised by Ehrenfest and Trkal [4] for his attempt to attribute reality to these cells. This leads back to the first objection above. Is entropy atypical in its observing the Sackur cell?

This is an open question. It could well be that macroscopic properties other than entropy are not subject to the Gibbs dividing principle. However, when it comes to homogeneous isothermal observers with just one particle class (considered here), it appears natural to guess otherwise. Formally, the problem even can be 'defined away' by requiring all 'macroscopic' observables to observe the Sackur cell. So far, no rigorous definition of macroscopicity seems to be in the literature (cf. [9]). *Something* of the microscopic reality has to be forgotten, as averaged-out (macroscopic) quantities are all that is left eventually as larger and larger particle numbers are admitted. To postulate specifically that it is the distinguishability of internal Sackur cells that is lost is admissible since this is much more fine-grained an assumption than is usually deemed necessary. Of course, this definition does not *solve* the problem. It only provides a hypothesis that is specific enough to be falsifiable.

Recently, an attempt was made to prove that, for homogeneous isothermal single-particle type observers, the Sackur cell constitutes a limit in principle to internal self-observation [15]. The argument was similar to one used by Popper [11] to show that physical observers can never completely observe themselves. It is presently unclear, nevertheless, whether indeed an argument on so general a level is needed to establish Gibbs' indistinguishability, as well as the above definition of macro-scopicity, in a classical framework. Maybe an intuitive argument that uses only the fact that the observer in question is chaotic can be found instead.

What would be the main implication if macroscopic observers (based on hyperchaos of the Sinai type) were indeed subject to the Gibbs cellularity? The consequence would be the same one which always follows when 'the state of information of an observer about his own state' is limited: 'indeterminacy' [10]. Von Neumann, who introduced this hypothetical explanation of uncertainty, immediately dismissed it because of its being incompatible with the quantum mechanical formalism. In the present more general (or more limited) context, this counter argument is of no concern.

Several implications beyond mere 'indeterminacy' follow if the above principle of free permutability of Sackur cells (every unit time interval) is used to specify macroscopic observers [15]. There exists a well-defined equivalence class of (from the point of view of the observer) equiprobable versions of himself. This new ensemble, in turn, permits definition of an ensemble of equiprobable measurement situations, to every single con-crete measurement situation. 'Irreducibly statistical' observations therefore exist. Their properties too can be specified in principle. It seems that the internal unit action of the observer can get imposed on the external world as an observational uncertainty of action.

All of this hinges on the reality of the Sackur cell, however. It is quite possible that this particular attempt to 'read an invariant into hyperchaos' did not hit the right target. The question of how an internally chaotic observer sees the universe has yet to yield definitive answers.

Two *general* possibilities open themselves up, which it may be worth while to discuss finally. The first is that indeed a finite 'something' will be found which causes observations to be lawfully indeterminate, but that this classical indeterminacy will pale before quantum mechanics, becoming completely absorbed into it eventually so that no trace remains. This would be in accordance with von Neumann's early (1932) view already mentioned. The other nontrivial possibility is that everything will lose shape so to speak. There would be quantum mechanics and its nonlocality [2], but there would also be other equally nontrivial and nonlocal phenomena. (Note that a classical observational uncertainty of action necessarily generates 'telescoping' effects concerning the duration of an observed event, so that locality in time at least would be violated.) The interplay could become very tangled. Therefore, a 'monistic' interpretation would have to be sought seriously once more in this case, but this time with quantum mechanics not in the dominating position.

Chaos as an object of study is interesting enough already. Why should one turn everything around and let chaos become an active participator in the very process of studying things? Is there indeed a need for an analogue to psychoanalysis in physics? I would say that if chaos became instrumental in bringing about a general shift of paradigm — from the usual detached, 'exophysical' way of looking at one's model worlds to an understanding, 'endophysical' one [5], this would only be another manifestation of the surprising vigour of this new concept.

Acknowledgements

I thank Michael Conrad, Joe Ford, Arnold Mandell and Karl Haubold for discussions.

References

[1]   Becker, R. *Theory of Heat*. Springer, Berlin (1967).
[2]   Bell, J.S. On the problem of hidden variables in quantum mechanics. *Rev. Mod. Phys.* **38**, 447–52 (1966).
[3]   Conrad, M. and Rössler, O.E. Example of a system which is computation universal but not effectively programmable. *Bull. Math. Biol.* **44**, 443–7 (1982).
[4]   Ehrenfest, P. and Trkal, V. Ableitung des Dissoziationsgleichgewichtes aus der Quantumtheorie und darauf beruhhende Berechnung der chemischen Konstanten. *Ann. d. Phys.* **65**, 609–28 (1921).
[5]   Finkelstein, D. (personal communication 1983) suggested the use of these terms instead of 'physics from without' and 'physics from within'.
[6]   Fredkin, E. Digital information mechanics. MIT preprint, March 1983.
[7]   Gibbs, J.W. *Fundamental principles of Statistical Mechanics*. New Haven (1902).
[8]   Li, T.Y. and Yorke, J.A. Period three implies chaos. *Am. Math. Monthly* **82**, 985–92 (1975).
[9]   Ludwig, G. Introduction to Theoretical Physics, vol. 4 (in German). Vieweg, Munich (1978).
[10] von Neumann, J. *Mathematical Foundations of Quantum Mechanics*. Princeton University Press, Princeton, NJ, see p. 439 (1955).
[11] Popper, K.R. Indeterminism in quantum physics and classical physics, Part I and II. *Brit. J. Phil. Sci.* **1**, 117–33,173–95, see p. 129 (1950).
[12] Prigogine, I. *Introduction to Thermodynamics of Irreversible Processes*. C.C. Thomas, Springfield (1955).
[13] Rössler, O.E. Different types of chaos in two simple differential equations. *Z. Naturforsch.* **31A**, 1664–70 (1976).
[14] Rössler, O.E. The chaotic hierarchy. *Z. Naturforsch.* **38A**, 788–802 (1983).
[15] Rössler, O.E. Indistinguishability implies quantization. Preprint, January 1985.
[16] Sackur, O. Die Anwendung der kinetischen Theorie der Gase auf chemische Probleme. *Ann. d. Phys.* **36**, 958–80 (1911).
[17] Sackur, O. Die universelle Bedeutung des sog. elementaren Wirkungs-quantum. *Ann. d. Phys.* **40**, 67–86 (1913).
[18] Sinai, J.G. Appendix to English translation of S. Krylov, *Works on the Foundations of Statistical Physics*. Princeton University Press, Princeton, NJ (1980).
[19] Tetrode, H. Die chemische Konstante der Gase und das elementare Wirkungs-quantum. *Ann. d. Phys.* **39**, 434–42 (1912).
[20] Waterston, J.J. Lecture read before the Royal Society, December 1845. Reprinted in *The Collected Papers of John James Waterston*, ed. Lord Rayleigh, p. 209. Edinburgh and London (1928).

Index